Fundamentals of Graphics Using MATLAB®

Fundamentals of Graphics Using MATLAB®

Ranjan Parekh

CRC Press is an imprint of the
Taylor & Francis Group, an **informa** business

CRC Press
Taylor & Francis Group
6000 Broken Sound Parkway NW, Suite 300
Boca Raton, FL 33487-2742

First issued in paperback 2021

Printed on acid-free paper

ISBN-13: 978-0-367-18482-7 (hbk)
ISBN-13: 978-1-03-208341-4 (pbk)

Visit the Taylor & Francis Web site at
http://www.taylorandfrancis.com

and the CRC Press Web site at
http://www.crcpress.com

Contents

Preface

THIS BOOK INTRODUCES FUNDAMENTAL CONCEPTS AND PRINCIPLES OF 2D AND 3D graphics and is written for under- and postgraduate students studying graphics- and/ or multimedia-related subjects. Most of the books on graphics use C programming environments to illustrate practical implementations. This book deviates from this common practice and illustrates the use of MATLAB® for the purpose. MATLAB by MathWorks, Inc. is a data analysis and visualization tool suitable for algorithmic development and simulation applications. One of the advantages of MATLAB is that it contains large libraries of in-built functions which can be utilized to reduce program-development time as compared to other contemporary programming environments. It is assumed that the student has basic knowledge of MATLAB, especially various matrix operations and plotting functions. The MATLAB codes have been provided as answers to specific examples, and the reader can simply copy and paste the codes to execute them. In general, the codes display answers to expected results like equation of curves, blending functions, and transformation matrices as well as plot the final results to provide a visual representation of the solution. The objectives of this book are, first, to demonstrate how MATLAB can be used to solve problems in graphics and, second, to help the student gain an in-depth knowledge about the subject matter through visual representations and practical examples.

This book is roughly divided into two parts: 2D graphics and 3D graphics, although in some places both of these concepts overlap mainly to highlight the differences between them or for using simpler concepts to prepare the reader for more complex ones.

The first part of this book mainly deals with concepts and problems related to 2D graphics, and spans over five chapters: (1) Interpolating Splines, (2) Blending Functions and Hybrid Splines, (3) Approximating Splines, (4) 2D Transformations, and (5) Spline Properties.

Chapter 1 provides an introduction regarding the various types of interpolating splines and their representations using polynomials. The theoretical concepts about how spline equations are derived and the matrix algebra involved are discussed in detail followed by numerical examples and MATLAB codes to illustrate the processes. Most of the examples are followed by graphical plots to enable the reader visualize how the equations get translated into corresponding curves given their start points, end points, and other related parameters. The chapter also highlights the differences in these procedures for both standard or spatial form and parametric form of the spline equations using linear, quadratic, and cubic variants.

Chapter 2 introduces the concept of blending functions and how these functions are used to derive equations for hybrid splines which pass through only a subset of their control points or where conditions other than control points are used for deriving their equations. Specifically, the chapter deals with the Hermite spline, Cardinal spline, Catmull–Rom spline, and Bezier spline. For Bezier splines, both the quadratic and cubic variants are discussed along with Bernstein polynomials used to formulate their blending functions. As in other chapters, the theoretical concepts are followed by numerical examples, MATLAB codes and graphical plots for visualization. The chapter ends with a discussion about how one spline type can be converted to another.

Chapter 3 discusses how polynomial equations are derived for approximating splines that do not pass through any of their control points and how their blending functions are computed. Specifically, the chapter provides detailed discussions about the Cox de Boor algorithm and how it can used to derive equations for linear, quadratic, and cubic B-splines. Essentially, B-splines consist of multiple curve segments with continuity at join points. Values of the parametric variable at the join points are stored in a vector called the knot vector. If the knot values are equally spaced, then the resulting spline is called uniform B-spline; otherwise, it is referred to as non-uniform. B-splines are called open-uniform when knot vector values are repeated. The chapter provides representations of the knot vector and illustrates how the spacing in the vector generates the above-mentioned variants. As before, the theoretical concepts are followed by numerical examples, MATLAB codes, and graphical plots for visualization.

Chapter 4 formally introduces a 2D coordinate system and then lays the foundations of a homogeneous coordinate system using which all the transformations can be represented in a uniform manner. Two-dimensional transformations are used to change the location, orientation, and shapes of splines in 2D plane. These transformations are translation, rotation, scaling, reflection, and shear applied individually or in combination of two or more; hence, they are known as composite transformations. Given known coordinates of a point, each of these transformations is represented by a matrix which when multiplied to the original coordinates produces a new set of transformed coordinates. The transformation matrices are first derived, and then their applications are illustrated using examples, MATLAB codes, and graphical plots. Both affine and perspective transformation types are discussed. The chapter ends with a discussion on viewing transformations used for mapping a window to a viewport, and coordinate system transformation used for mapping between multiple coordinate systems.

Chapter 5 enumerates some of the common properties of splines and how these can be calculated from spline equations. First, it discusses the critical points namely minimum and maximum of spline curves. Additionally for splines of degree 3 or more, the point of inflection (POI) is of interest. Next, it discusses how the tangent and normal to a spline curve can be calculated. The tangent to a curve is the derivative of the curve equation, while the normal is the line perpendicular to the tangent. The third property is calculation of length of a spline curve between any two given points, both for spatial and parametric equations. The fourth property is to calculate the area under a curve, which is bounded by a primary axis and two horizontal or vertical lines. An extension to this is calculation of area bounded by two curves. The fifth property is calculation of centroid of an area, the point of the center of gravity for plates of uniform density. The chapter ends with a discussion on interpolation and curve fitting for data points and a list of some commonly used built-in MATLAB functions for plotting 2D graphs and plots.

The second part of this book focuses on concepts and problems related to 3D graphics and spans over the remaining four chapters namely (6) Vectors, (7) 3D Transformations, (8) Surfaces, and (9) Projections.

Chapter 6 introduces the concept of vectors and their mathematical representations in 2D and 3D spaces. Vectors involve both magnitude and direction. They are represented in terms of orthogonal reference components of unit magnitudes along the primary axes together with a set of scaling factors. The chapter discusses how vectors can be added and multiplied together. Vector products can either be scalar, called a dot product, or vector, called a cross product. Using these concepts, the chapter then provides details of how vector equations of lines and planes can be derived. Next, the chapter discusses how vectors can be aligned to specific directions and finally how vector equations can be represented using homogeneous coordinates. The chapter ends with a section on how the tangent vector and the normal vector can be calculated for a curve. As before, the theoretical concepts are followed by numerical examples, MATLAB codes, and graphical plots for visualization.

Chapter 7 demonstrates how 3D transformations can be treated as extensions of 2D transformations. These are used to change the location, orientation, and shapes of splines in 3D space. These transformations are translation, rotation, scaling, reflection, shear applied individually or in combination of two or more, known as composite transformation. This chapter formally introduces a 3D coordinate system and then uses homogeneous coordinates to derive transformation matrices for the above operations. Their applications are then illustrated using examples, MATLAB codes and graphical plots for visualization. The latter part of the chapter deals with vector alignment in 3D space and uses these concepts to derive rotation matrices in 3D space around vectors and arbitrary lines.

Chapter 8 takes a look at how surfaces can be created and represented using parametric and implicit equations, and how the nature of the surface depends on the parameters of the equations. Depending on creation process, surfaces can be categorized as extruded and surfaces of revolution, both of which are discussed with examples and graphical plots. The chapter then takes a look at how tangent planes of surfaces can be computed and provides methods for computing area and volume of surfaces. The latter part of the chapter deals with surface appearances namely how textures can be mapped on surfaces and how illumination models can be used to determine brightness intensities at a point on the surface. The chapter ends with a discussion on some commonly used built-in MATLAB functions for plotting 3D graphs.

Chapter 9 studies various types of projections and derives matrices for each. Projection is used to map a higher-dimensional object to a lower-dimensional view. Projection can be of two types: parallel and perspective. In parallel projection, projection lines are parallel to each other, while in perspective projection, projection lines appear to converge to a reference point. Parallel projection can again be of two types: orthographic and oblique. In parallel orthographic projection, the projection lines are perpendicular to the view plane, while in parallel oblique projection, the projection lines can be oriented at any arbitrary angle to the view plane. Usually for 3D projection, parallel orthographic projection can also be sub-divided into two types: multi-view and axonometric. In multi-view projection, the projection occurs on the primary planes i.e. *XY*-, *YZ*-, or *XZ*-planes, while in axonometric projection, the projection occurs on any arbitrary plane. The chapter illustrates each type of projection using examples, MATLAB codes, and graphical plots for visualization.

Each chapter is followed by a summarized list of salient points discussed in the chapter. A set of review questions and a list of practice problems are provided at the end of each chapter for self-evaluation. This book contains more than 90 solved numerical examples with their corresponding MATLAB codes and an additional 90 problems given for practice. Readers are encouraged to execute the codes given in the examples and also write their own codes to solve the practice problems. Most of the MATLAB codes given in this book will require MATLAB version 2015 or later to execute properly. Some of the functions mentioned have been specifically introduced from version 2016 and these have been mentioned at the appropriate places. The usage of about 70 different MATLAB functions related to graphics and plotting have been demonstrated in this book and a list of these functions with a short description is provided at Appendix I. Readers are asked to use MATLAB help utilities to get further information on these. The MATLAB codes are written in a verbose manner for a better understanding of the readers who are new to the subject matter. Some of the codes could have been written in a more compact manner but that might have reduced their comprehensibility. Around 170 figures have been included in this book to help the readers get proper visualization cues of the problems especially for 3D environments. Answers to the practice problems are provided in Appendix II.

All readers are encouraged to provide feedback about the content matter of the book as well as any omissions or typing errors. The author can be contacted at ranjan_parekh@yahoo.com.

Ranjan Parekh
Jadavpur University
Calcutta 700032, India
2019

MATLAB® is a registered trademark of The MathWorks, Inc. For product information, please contact:

The MathWorks, Inc.
3 Apple Hill Drive
Natick, MA 01760-2098 USA
Tel: 508-647-7000
Fax: 508-647-7001
E-mail: info@mathworks.com
Web: www.mathworks.com

Author

Dr. Ranjan Parekh, PhD (Engineering), is Professor at the School of Education Technology, Jadavpur University, Calcutta, India, and is involved with teaching subjects related to Graphics and Multimedia at the post graduate level. His research interests include multimedia information processing, pattern recognition, and computer vision. He is also the author of *Principles of Multimedia* (McGraw Hill, 2012; http://www.mhhe.com/parekh/multimedia2).

CHAPTER 1

Interpolating Splines

1.1 INTRODUCTION

Splines are irregular curve segments with known mathematical properties. Splines are frequently encountered in vector graphics when graphic objects are required to have a defined shape in 2D planes (Figure 1.1a) or 3D space (Figure 1.1b) or moved along a specified path (Figure 1.1c). Based on the coordinates of some of the points on the curves, or slopes of lines along the curves, the graphics system needs to calculate a mathematical representation of the curve before storing them onto a disk. This representation usually takes the form of "vectors" or a series of values stored in matrices. The values are calculated using an orthogonal 2D coordinate system consisting of the origin, *X*-axis, and *Y*-axis. These coordinate axes are often called the primary or principal axes.

The term "spline" has been derived from the ship building industry where it is used to refer to wooden planks bent between wooden posts for building the curved hull of ships (O'Rourke, 2003). The location of the fixed posts controlled the shape of the plank. In graphics, we use specific points along the spline curve to control the shape of the spline and hence they are aptly referred to as "control points," shortened as CPs. Depending on the relationship between the CPs and the actual curves, the splines can be broadly categorized into three types: (1) interpolating splines, where the spline actually goes through the CPs; (2) approximating splines, where the spline goes near the CPs but not actually through them; and (3) hybrid splines, where the spline goes through some of the CPs but not through all (Hearn and Baker, 1996) (see Figure 1.2).

Splines are mathematically modeled using polynomials. Polynomials are expressions constructed from variables and constants, and involve addition, subtraction, multiplication, and non-negative integer exponents. A polynomial can be 0 (zero) or a sum of non-zero terms. Each term consists of a constant, called coefficient, multiplied by a variable. The exponent of the variable is called its degree. The first example is a valid polynomial, but the second example is not, because the variable is associated with a division operation and also because of the fractional exponent. The general *n*th degree polynomial is shown in the third example.

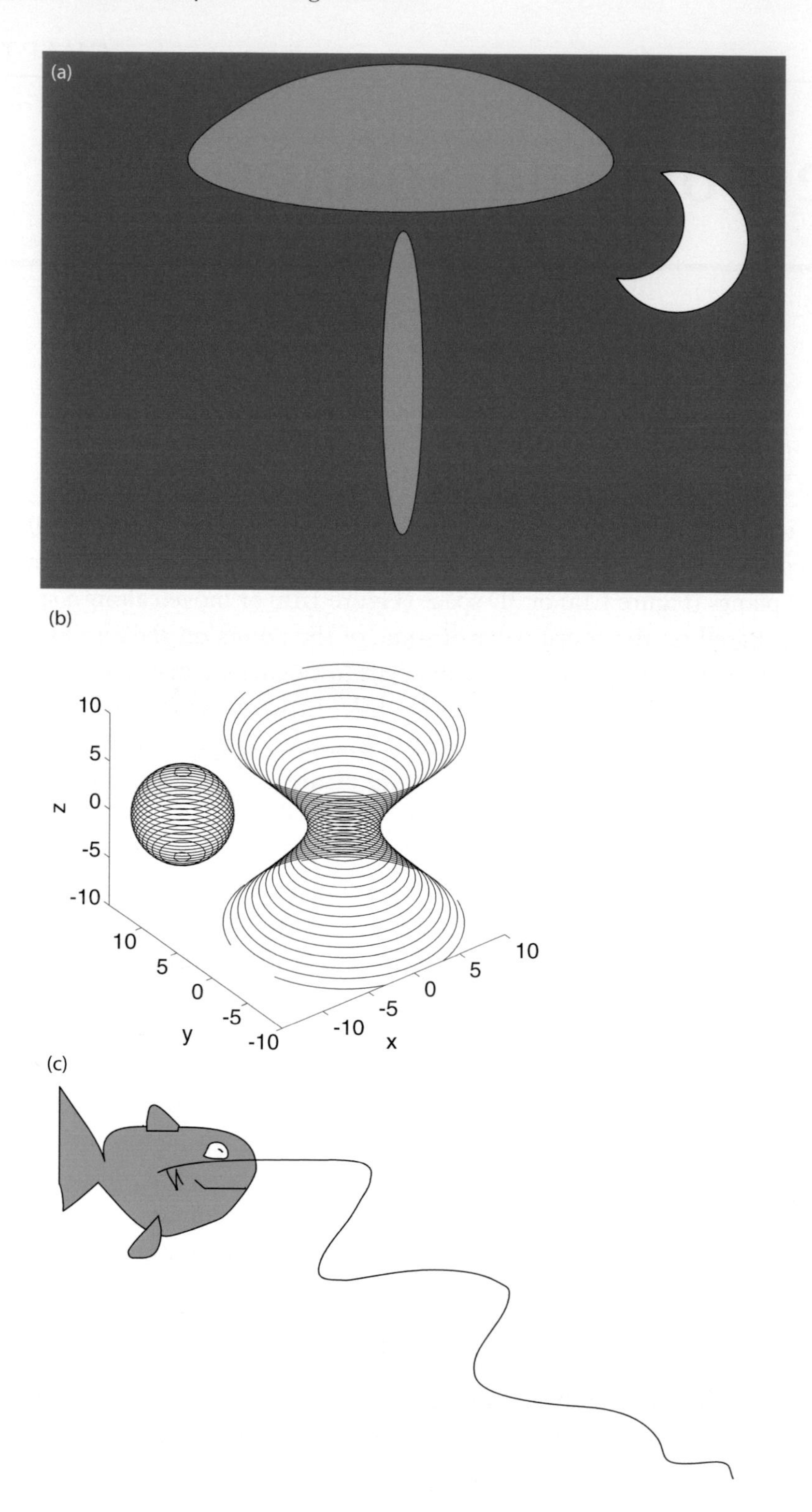

FIGURE 1.1 Use of splines in graphics for creating (a) 2D shapes (b) 3D surfaces (c) motion path trajectory

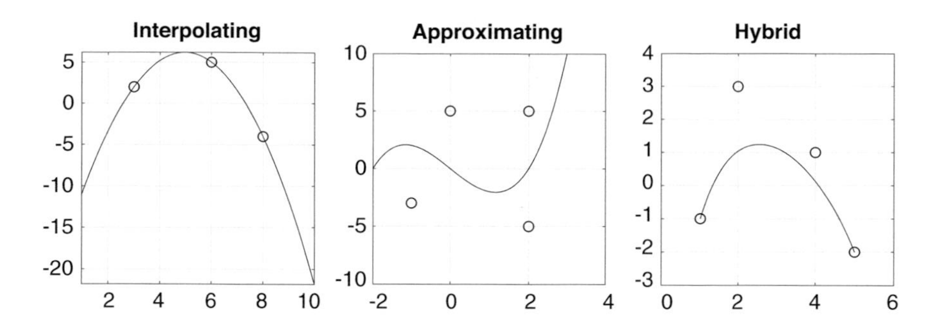

FIGURE 1.2 Types of splines.

$$1-2x+3x^2$$

$$1-\frac{2}{x}+3x^{2.5}$$

$$a_n x^n + a_{n-1} x^{n-1} + \cdots + a_1 x + a_0$$

A polynomial equation is written when one polynomial is set equal to another. It can either be in explicit form e.g. $y=f(x)$ when either side of the equation contains variables of explicit type, or it can be in implicit form e.g. $f(x,y)=0$ where multiple types of variables can be on the same side. Examples are shown below:

$$y=x^2$$

$$x^3+y^3-5xy=0$$

Polynomial equations can also be represented in parametric form where the variables are expressed as functions of another variable t e.g. $x=f(t), y=g(t)$. The advantage of parametric equations is that the variables x and y do not need to be constrained by a single equation and can be changed independently of each other, which offers more flexibility for representing complex curves. As a convention, the value of t is usually taken to lie between 0 and 1 unless otherwise specified. The value of $t=0$ corresponds to the start point and $t=1$ to the end point of the spline curve. Examples are shown below:

$$x=t, y=t^2$$

$$x=r\cdot\cos t, y=r\cdot\sin t$$

A polynomial equation is frequently represented using graphs, which are useful in visually depicting how one variable changes with another. The graph of a zero polynomial i.e. $f(x)=0$ is the X-axis. The graph of a zero degree polynomial represented by $f(x)=a$, where a is a constant, is a line parallel to the X-axis at a distance a from it. The graph of a degree 1 polynomial, represented by $f(x)=a+bx$, is a straight line with a slope b and intercept a. The graph

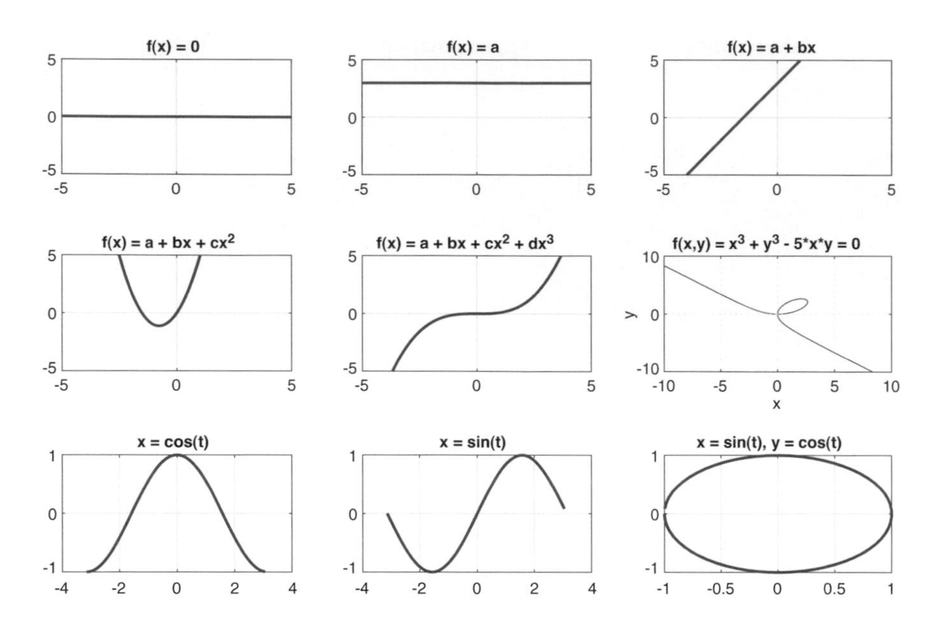

FIGURE 1.3 Graphs of polynomial equations.

of a degree 2 polynomial, represented by $f(x)=a+bx+cx^2$, is a parabolic curve and can be specified if at least three points are known on the curve. The graph of a degree 3 polynomial, represented by $f(x)=a+bx+cx^2+dx^3$, is a cubic curve and can be specified if at least four points are known on the curve. Instead of explicit equations, implicit equations $f(x,y)=0$ can also be plotted by varying the independent variable by fixed intervals and computing the corresponding values of the dependent variables. Plots of parametric equations consist of three different graphs: the first is the t vs. x graph generated from the function $x=f(t)$, the second is the t vs. y graph generated from the function $y=g(t)$, while the third graph is x vs. y generated by plotting the x and y values from the same values of t obtained from the previous plots. Thus, even if we always do not generate an equation between x and y by eliminating t, we can always plot a graph of x vs. y. Figure 1.3 shows graphs of various polynomial equations.

In the following sections, we take a look at few types of interpolating splines and how their equations are derived.

1.2 LINEAR SPLINE (STANDARD FORMS)

A linear spline is a straight line represented by a first-degree polynomial and can be generated given two points are known along it. Standard form of a linear spline implies the spline equation is computed in the spatial domain i.e. the x-y plane. Let the given points be $P_1(x_1, y_1)$ and $P_2(x_2, y_2)$. Choose a starting linear equation that is written in matrix form

$$y=a+bx=\begin{bmatrix} 1 & x \end{bmatrix}\begin{bmatrix} a \\ b \end{bmatrix} \tag{1.1}$$

Substitute the given points in the starting equation to generate two equations. Two equations are sufficient to solve for the two unknown coefficients a and b

$$y_1 = a + bx_1$$
$$y_2 = a + bx_2 \tag{1.2}$$

The two equations are written in matrix form $Y = C \cdot A$, where C is the constraint matrix and A is the coefficient matrix:

$$\begin{bmatrix} y_1 \\ y_2 \end{bmatrix} = \begin{bmatrix} 1 & x_1 \\ 1 & x_2 \end{bmatrix} \begin{bmatrix} a \\ b \end{bmatrix} \tag{1.3}$$

The equations are solved to find the values of the unknown coefficients. Thus, we have $A = \text{inv}(C) \cdot Y$

$$\begin{bmatrix} a \\ b \end{bmatrix} = \begin{bmatrix} 1 & x_1 \\ 1 & x_2 \end{bmatrix}^{-1} \begin{bmatrix} y_1 \\ y_2 \end{bmatrix} \tag{1.4}$$

The values of the coefficients are substituted in the starting equation to arrive at the equation of the spline.

$$y = \begin{bmatrix} 1 & x \end{bmatrix} \begin{bmatrix} 1 & x_1 \\ 1 & x_2 \end{bmatrix}^{-1} \begin{bmatrix} y_1 \\ y_2 \end{bmatrix} \tag{1.5}$$

Example 1.1

Find the equation of a line through points $P_1(3, 2)$ and $P_2(8, -4)$.

Choose a starting equation

$$y = a + bx = \begin{bmatrix} 1 & x \end{bmatrix} \begin{bmatrix} a \\ b \end{bmatrix}$$

Substitute given points in the equation

$$2 = a + b(3)$$
$$-4 = a + b(8)$$

Write in matrix form $Y = C \cdot A$

$$\begin{bmatrix} 2 \\ -4 \end{bmatrix} = \begin{bmatrix} 1 & 3 \\ 1 & 8 \end{bmatrix} \begin{bmatrix} a \\ b \end{bmatrix}$$

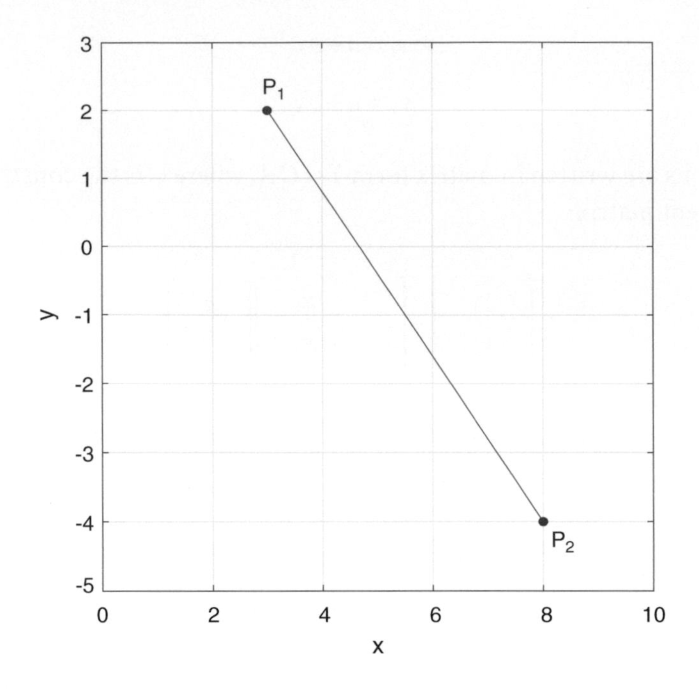

FIGURE 1.4 Plot for Example 1.1.

Solve the matrix equation $A = C^{-1} \cdot Y$

$$\begin{bmatrix} a \\ b \end{bmatrix} = \begin{bmatrix} 1.6 & -0.6 \\ -0.2 & 0.2 \end{bmatrix} \begin{bmatrix} 2 \\ -4 \end{bmatrix} = \begin{bmatrix} 5.6 \\ -1.2 \end{bmatrix}$$

Substitute the coefficient values in the starting equation

$$y = 5.6 - 1.2x$$

Verification: In most graphics problems, it is usually possible to verify the results obtained by substituting the given data in it. Putting $x = 3$, we get $y = 2$, and putting $x = 8$, we get $y = -4$. Hence, the line does indeed pass through the given points (Figure 1.4).

MATLAB® Code 1.1

```
clear all; clc;
syms x;
x1 = 3; y1 = 2;
x2 = 8; y2 = -4;
X = [x1 x2]; Y = [y1 y2];
C = [1 x1; 1 x2];
A = inv(C)*Y';
a = A(1);
```

```
b = A(2);
fprintf('Required equation : \n');
y = a + b*x;
y = vpa(y)

%plotting
xx = linspace(x1, x2);
yy = subs(y, x, xx);
plot(xx, yy, 'b-'); hold on;
scatter(X, Y, 20, 'r', 'filled');
xlabel('x'); ylabel('y');
grid; axis square;
axis([0 10 -5 3]);
d = 0.5;
text(x1+d, y1, 'P_1');
text(x2+d, y2, 'P_2');
hold off;
```

NOTE

`%`: signifies a comment line
`axis`: controls appearance of the axes of the plot, specifies ordered range of values to display
`clc`: clears workspace of previous text
`clear`: clears memory of all stored variables
`fprintf`: prints out strings and values using formatting options
`grid`: turns on display of grid lines in a plot
`hold`: holds the current graph state so that subsequent commands can add to the same graph
`inv`: computes inverse of a matrix
`linspace`: creates 100 linearly spaced values between the two end-points specified
`plot`: creates a graphical plot from a set of values
`scatter`: type of plot where the data is represented by colored circles
`subs`: substitutes symbolic variable with a matrix of values for generating a plot
`syms`: declares the arguments following as symbolic variables
`text`: inserts textual strings at specified locations in the graph
`title`: displays a title on top of the graph
`vpa`: displays symbolic values as variable precision floating point values
`xlabel, ylabel`: puts text labels along the corresponding primary axes

An alternate form of the standard line equation can be formed where, instead of two given points, only one point and the slope of the line are given. This aspect is discussed below.

Let the given points be $P_1(x_1, y_1)$ and s be the slope of the line. Choose a starting linear equation that is written in matrix form as before in Equation (1.1).

$$y = a + bx = \begin{bmatrix} 1 & x \end{bmatrix} \begin{bmatrix} a \\ b \end{bmatrix}$$

Calculate derivative of the starting equation

$$y' = \frac{dy}{dx} = b \tag{1.6}$$

Substitute the given values in the starting equation to generate two equations

$$\begin{aligned} y_1 &= a + bx_1 \\ s &= b \end{aligned} \tag{1.7}$$

The two equations are written in matrix form $Y = C{\cdot}A$ as before

$$\begin{bmatrix} y_1 \\ s \end{bmatrix} = \begin{bmatrix} 1 & x_1 \\ 0 & 1 \end{bmatrix} \begin{bmatrix} a \\ b \end{bmatrix} \tag{1.8}$$

The equations are solved to find the values of the coefficients: $A = C^{-1}{\cdot}Y$

$$\begin{bmatrix} a \\ b \end{bmatrix} = \begin{bmatrix} 1 & x_1 \\ 0 & 1 \end{bmatrix}^{-1} \begin{bmatrix} y_1 \\ s \end{bmatrix} \tag{1.9}$$

The values of the coefficients are substituted in the starting equation to arrive at the equation of the spline.

$$y = \begin{bmatrix} 1 & x \end{bmatrix} \begin{bmatrix} 1 & x_1 \\ 0 & 1 \end{bmatrix}^{-1} \begin{bmatrix} y_1 \\ s \end{bmatrix} \tag{1.10}$$

Example 1.2

Find the equation of a line through the point P(–1, 1) and having slope 2.

Choose a starting equation

$$y = a + bx = \begin{bmatrix} 1 & x \end{bmatrix} \begin{bmatrix} a \\ b \end{bmatrix}$$

Calculate derivative of the starting equation

$$y' = \frac{dy}{dx} = b$$

Substitute given values in the equation

$$1 = a + b(-1)$$

$$2 = b$$

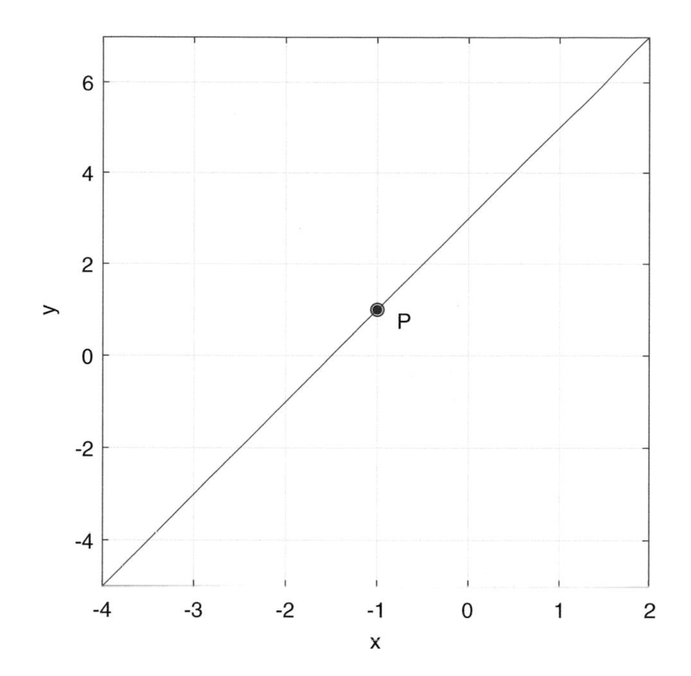

FIGURE 1.5 Plot for Example 1.2.

Write in matrix form $Y = C{\cdot}A$

$$\begin{bmatrix} 1 \\ 2 \end{bmatrix} = \begin{bmatrix} 1 & -1 \\ 0 & 1 \end{bmatrix} \begin{bmatrix} a \\ b \end{bmatrix}$$

Solve the matrix equation $A = C^{-1}{\cdot}Y$

$$\begin{bmatrix} a \\ b \end{bmatrix} = \begin{bmatrix} 1 & 1 \\ 0 & 1 \end{bmatrix} \begin{bmatrix} 1 \\ 2 \end{bmatrix} = \begin{bmatrix} 3 \\ 2 \end{bmatrix}$$

Substitute the coefficient values in the starting equation

$$y = 2x + 3$$

Verification: Putting $x = -1$ in the above equation, we get $y = 1$. Also, slope $\frac{dy}{dx} = 2$ (Figure 1.5).

MATLAB Code 1.2

```
clear all; clc;
syms x;
x1 = -1; y1 = 1; s = 2;
Y = [y1 s];
C = [1 x1; 0 1];
```

```
A = inv(C)*Y';
a = A(1);
b = A(2);
fprintf('Required equation:\n');
y = a + b*x;
y = vpa(y)

%plotting
xx = linspace(x1-3, x1+3);
yy = subs(y, x, xx);
plot(xx, yy, 'b-', x1, y1, 'bo'); hold on;
scatter(x1, y1, 20, 'r', 'filled');
xlabel('x'); ylabel('y');
grid; axis square; axis tight;
text(x1+0.5, y1, 'P');
hold off;
```

1.3 LINEAR SPLINE (PARAMETRIC FORM)

A linear spline can also be represented by parametric equations. Let the given points through which the spline passes be P_0 and P_1. As per the convention mentioned before $P_0 \equiv P(0)$ i.e. the first point corresponds to the point where $t = 0$. Similarly, $P_1 \equiv P(1)$ i.e. the second point corresponds to the point where $t = 1$. Note that in some cases, the first point may not always correspond to $t = 0$ or the last point may not always correspond to $t = 1$. We will discuss these issues subsequently.

Choose a starting linear parametric equation written in matrix form

$$P(t) = a + bt = \begin{bmatrix} 1 & t \end{bmatrix} \begin{bmatrix} a \\ b \end{bmatrix} \tag{1.11}$$

Substitute given points in the starting equation by choosing $t = 0$ at start and $t = 1$ at end.

$$\begin{aligned} P_0 &= a + b(0) \\ P_1 &= a + b(1) \end{aligned} \tag{1.12}$$

Write equations in matrix form $G = C{\cdot}A$, where G is called the geometry matrix

$$\begin{bmatrix} P_0 \\ P_1 \end{bmatrix} = \begin{bmatrix} 1 & 0 \\ 1 & 1 \end{bmatrix} \begin{bmatrix} a \\ b \end{bmatrix} \tag{1.13}$$

Solve the equation for A i.e. $A = C^{-1}{\cdot}G = B{\cdot}G$, where B is called the basis matrix

$$\begin{bmatrix} a \\ b \end{bmatrix} = \begin{bmatrix} 1 & 0 \\ 1 & 1 \end{bmatrix}^{-1} \begin{bmatrix} P_0 \\ P_1 \end{bmatrix} \tag{1.14}$$

Substitute the coefficient values in the starting equation

$$P(t)=\begin{bmatrix} 1 & t \end{bmatrix}\begin{bmatrix} 1 & 0 \\ 1 & 1 \end{bmatrix}^{-1}\begin{bmatrix} P_0 \\ P_1 \end{bmatrix} \quad (1.15)$$

NOTE

In reality, the parametric equations should be written separately for x and y i.e. $x(t)=a_x+b_x \cdot t$ and $y(t)=a_y+b_y \cdot t$. However, we use a compact notation here by substituting $P(t)=[x(t), y(t)], a=\left[a_x, a_y\right], b=\left[b_x, b_y\right]$. After solving for a and b, we separate out the individual components and substitute them in the respective equations for x and y.

Example 1.3

Find the equation of a line through points $P_0(3, 2)$ and $P_1(8, -4)$ in parametric form.

Choose a starting equation.

$$P(t)=a+bt=\begin{bmatrix} 1 & t \end{bmatrix}\begin{bmatrix} a \\ b \end{bmatrix}$$

Write equations in matrix form $G = C \cdot A$, where G is called the geometry matrix

$$\begin{bmatrix} 3 & 2 \\ 8 & -4 \end{bmatrix}=\begin{bmatrix} 1 & 0 \\ 1 & 1 \end{bmatrix}\begin{bmatrix} a \\ b \end{bmatrix}$$

Solve the equation for A i.e. $A = C^{-1} \cdot G = B \cdot G$

$$\begin{bmatrix} a \\ b \end{bmatrix}=\begin{bmatrix} 1 & 0 \\ 1 & 1 \end{bmatrix}^{-1}\begin{bmatrix} 3 & 2 \\ 8 & -4 \end{bmatrix}=\begin{bmatrix} 3 & 2 \\ 5 & -6 \end{bmatrix}$$

Substitute the coefficient values in the starting equation

$$P(t)=\begin{bmatrix} 1 & t \end{bmatrix}\begin{bmatrix} 3 & 2 \\ 5 & -6 \end{bmatrix}$$

The required parametric equations are obtained by separating out the x and y components

$$x=3+5t$$

$$y=2-6t$$

Verification: $x(0)=3, x(1)=8, y(0)=2, y(1)=-4$ (Figure 1.6).

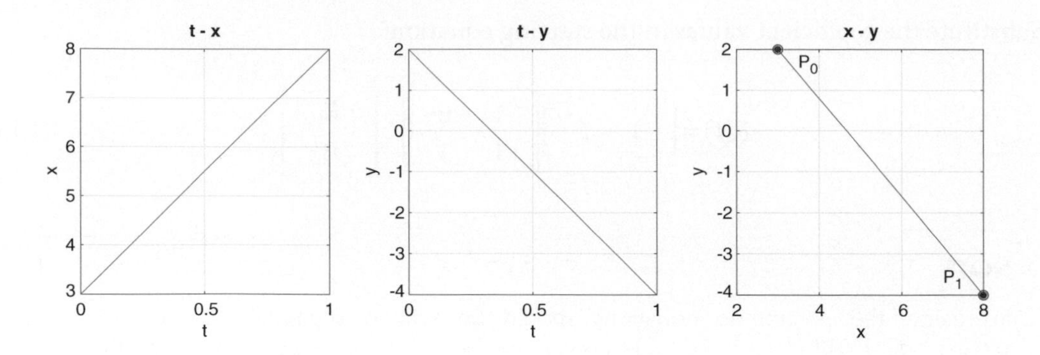

FIGURE 1.6 Plots for Example 1.3.

MATLAB Code 1.3

```
clear all; clc;
syms t;
x1 = 3; y1 = 2;
x2 = 8; y2 = -4;
X = [x1 x2]; Y = [y1 y2];
G = [X ; Y];
C = [1 0; 1 1];
A = inv(C)*G';
ax = A(1,1); ay = A(1,2);
bx = A(2,1); by = A(2,2);

fprintf('Required equations : \n');
x = ax + bx*t; x = vpa(x)
y = ay + by*t; y = vpa(y)

%plotting
tt = linspace(0,1);
xx = subs(x, t, tt);
yy = subs(y, t, tt);

subplot(131), plot(tt, xx); grid; axis square;
xlabel('t'); ylabel('x'); title('t - x');
subplot(132), plot(tt, yy); grid; axis square;
xlabel('t'); ylabel('y'); title('t - y');
subplot(133), plot(xx, yy, 'b-', X, Y, 'bo');
grid; axis square; hold on;
scatter(X, Y, 20, 'r', 'filled');
xlabel('x'); ylabel('y'); title('x - y');
text(x1+1, y1-0.5, 'P_0');
text(x2-1, y2+0.5, 'P_1');
hold off;
```

NOTE

`subplot`: displays multiple plots within a single figure window

1.4 QUADRATIC SPLINE (STANDARD FORM)

A quadratic spline is a parabolic curve represented by a second-degree polynomial equation and can be generated if at least three points are known along the curve. Standard form of a quadratic spline implies the spline equation is computed in the spatial domain i.e. the x-y plane. Let the given points be $P_1(x_1, y_1)$, $P_2(x_2, y_2)$, and $P_3(x_3, y_3)$. Choose a starting quadratic equation, which is written in matrix form

$$y = a + bx + cx^2 = \begin{bmatrix} 1 & x & x^2 \end{bmatrix} \begin{bmatrix} a \\ b \\ c \end{bmatrix} \tag{1.16}$$

Substitute the given points in the starting equation to generate three equations. They are sufficient to solve for the three unknown coefficients a, b, and c.

$$\begin{aligned} y_1 &= a + bx_1 + cx_1^{\,2} \\ y_2 &= a + bx_2 + cx_2^{\,2} \\ y_3 &= a + bx_3 + cx_3^{\,2} \end{aligned} \tag{1.17}$$

The three equations are written in matrix form $Y = C{\cdot}A$ as before

$$\begin{bmatrix} y_1 \\ y_2 \\ y_3 \end{bmatrix} = \begin{bmatrix} 1 & x_1 & x_1^{\,2} \\ 1 & x_2 & x_2^{\,2} \\ 1 & x_3 & x_3^{\,2} \end{bmatrix} \begin{bmatrix} a \\ b \\ c \end{bmatrix} \tag{1.18}$$

The equations are solved to find the values of the coefficients: $A = C^{-1}{\cdot}Y$

$$\begin{bmatrix} a \\ b \\ c \end{bmatrix} = \begin{bmatrix} 1 & x_1 & x_1^{\,2} \\ 1 & x_2 & x_2^{\,2} \\ 1 & x_3 & x_3^{\,2} \end{bmatrix}^{-1} \begin{bmatrix} y_1 \\ y_2 \\ y_3 \end{bmatrix} \tag{1.19}$$

The values of the coefficients are substituted in the starting equation to arrive at the equation of the spline

$$y = \begin{bmatrix} 1 & x & x^2 \end{bmatrix} \begin{bmatrix} 1 & x_1 & x_1^{\,2} \\ 1 & x_2 & x_2^{\,2} \\ 1 & x_3 & x_3^{\,2} \end{bmatrix}^{-1} \begin{bmatrix} y_1 \\ y_2 \\ y_3 \end{bmatrix} \tag{1.20}$$

Example 1.4

Find the equation of a quadratic spline through points $P_1(3, 2)$, $P_2(6, 5)$, and $P_3(8, -4)$.

Choose starting equation

$$y = a + bx + cx^2 = \begin{bmatrix} 1 & x & x^2 \end{bmatrix} \begin{bmatrix} a \\ b \\ c \end{bmatrix}$$

Write equations in matrix form $Y = C \cdot A$

$$\begin{bmatrix} 2 \\ -4 \\ 5 \end{bmatrix} = \begin{bmatrix} 1 & 3 & 9 \\ 1 & 8 & 64 \\ 1 & 6 & 36 \end{bmatrix} \begin{bmatrix} a \\ b \\ c \end{bmatrix}$$

Solve for A: $A = C^{-1} \cdot Y$

$$\begin{bmatrix} a \\ b \\ c \end{bmatrix} = \begin{bmatrix} -20.8 \\ 10.9 \\ -1.1 \end{bmatrix}$$

Substitute in the starting equation

$$y = -20.8 + 10.9x - 1.1x^2$$

Verification: $y(3) = 2$, $y(6) = 5$, $y(8) = -4$ (Figure 1.7).

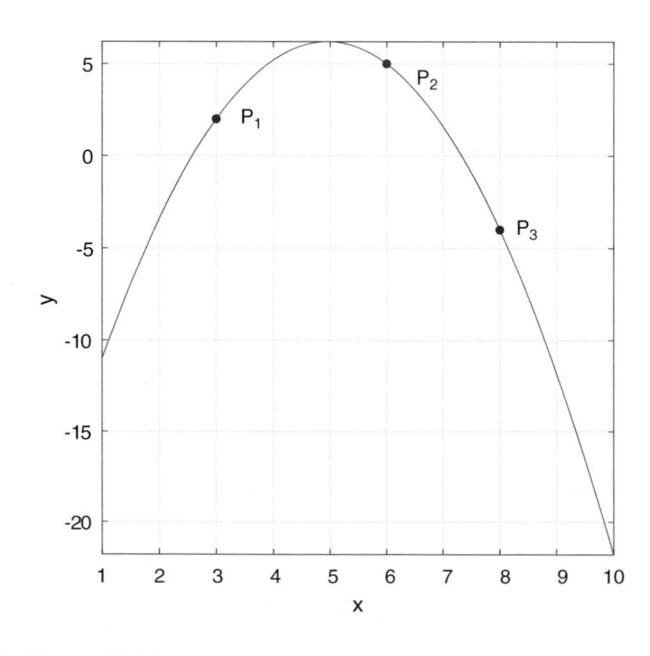

FIGURE 1.7 Plot for Example 1.4.

MATLAB Code 1.4

```
clear all; clc;
syms x;
x1 = 3; y1 = 2;
x2 = 6; y2 = 5;
x3 = 8; y3 = -4;
X = [x1, x2, x3];
Y = [y1, y2, y3];
C = [1, x1, x1^2; 1, x2, x2^2; 1, x3, x3^2];
A = inv(C)*Y';
a = A(1); b = A(2); c = A(3);
fprintf('Required equation : \n');
y = a + b*x + c*x^2; y = vpa(y)

%plotting
d = 0.5;
xx = linspace(x1-2, x3+2);
yy = subs(y, x, xx);
plot(xx,yy, 'b-');
hold on; grid;
scatter(X, Y, 20, 'r', 'filled');
xlabel('x'); ylabel('y');
axis square; axis tight;
text(x1+d, y1, 'P_1');
text(x2+d, y2, 'P_2');
text(x3+d, y3, 'P_3');
hold off;
```

1.5 QUADRATIC SPLINE (PARAMETRIC FORM)

A quadratic spline can also be represented by parametric equations. Let the given points be P_0, P_1, and P_2. Here, P_0 is the starting point i.e. $P_0 \equiv P(0)$ and P_2 is the end point i.e. $P_2 \equiv P(1)$. To determine the curve uniquely an additional piece of information is required regarding the value of parameter t at the middle point P_1. Let the value of t at P_1 be k, where $0 \le k \le 1$ i.e. $P_1 \equiv P(k)$. Different values of k referred to as the sub-division ratio, will give rise to curves with the same start and end points but having different shapes.

Choose a starting parametric quadratic equation written in matrix form

$$P(t) = a + bt + ct^2 = \begin{bmatrix} 1 & t & t^2 \end{bmatrix} \begin{bmatrix} a \\ b \\ c \end{bmatrix} \tag{1.21}$$

Substitute the given points in the starting equation by choosing $t = 0$ at start, $t = k$ at the middle, and $t = 1$ at end.

$$P_0 = a + b(0) + c(0)^2$$
$$P_1 = a + b(k) + c(k)^2 \qquad (1.22)$$
$$P_2 = a + b(1) + c(1)^2$$

Write equations in matrix form $G = C \cdot A$

$$\begin{bmatrix} P_0 \\ P_1 \\ P_2 \end{bmatrix} = \begin{bmatrix} 1 & 0 & 0 \\ 1 & k & k^2 \\ 1 & 1 & 1 \end{bmatrix} \begin{bmatrix} a \\ b \\ c \end{bmatrix} \qquad (1.23)$$

Solve the equation for A i.e. $A = C^{-1} \cdot G = B \cdot G$

$$\begin{bmatrix} a \\ b \\ c \end{bmatrix} = \begin{bmatrix} 1 & 0 & 0 \\ 1 & k & k^2 \\ 1 & 1 & 1 \end{bmatrix}^{-1} \begin{bmatrix} P_0 \\ P_1 \\ P_2 \end{bmatrix} \qquad (1.24)$$

Substitute the coefficient values in the starting equation

$$P(t) = \begin{bmatrix} 1 & t & t^2 \end{bmatrix} \begin{bmatrix} 1 & 0 & 0 \\ 1 & k & k^2 \\ 1 & 1 & 1 \end{bmatrix}^{-1} \begin{bmatrix} P_0 \\ P_1 \\ P_2 \end{bmatrix} \qquad (1.25)$$

Example 1.5

Find the equation of a quadratic spline through points $P_0(3, 2)$, $P_1(8, -4)$, and $P_2(6, 5)$ in parametric form with sub-division ratio $k = 0.8$.

Choose starting equation

$$P(t) = a + bt + ct^2 = \begin{bmatrix} 1 & t & t^2 \end{bmatrix} \begin{bmatrix} a \\ b \\ c \end{bmatrix}$$

Write equations in matrix form $G = C \cdot A$

$$\begin{bmatrix} 3 & 2 \\ 8 & -4 \\ 6 & 5 \end{bmatrix} = \begin{bmatrix} 1 & 0 & 0 \\ 1 & 0.8 & 0.64 \\ 1 & 1 & 1 \end{bmatrix} \begin{bmatrix} a \\ b \\ c \end{bmatrix}$$

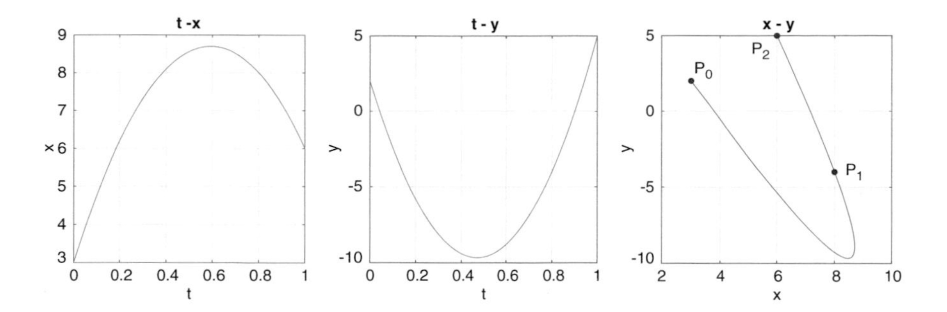

FIGURE 1.8 Plots for Example 1.5.

Solve the equation for A i.e. $A = C^{-1} \cdot G = B \cdot G$

$$\begin{bmatrix} a \\ b \\ c \end{bmatrix} = \begin{bmatrix} 3 & 2 \\ 19.25 & -49.5 \\ -16.25 & 52.5 \end{bmatrix}$$

Substitute the coefficient values in the starting equation

$$x = 3 + 19.25t - 16.25t^2$$

$$y = 2 - 49.5t + 52.5t^2$$

Verification: $x(0) = 3$, $y(0) = 2$, $x(0.8) = 8$, $y(0.8) = -0.4$, $x(1) = 6$, $y(1) = 5$ (Figure 1.8).

MATLAB Code 1.5

```
clear all; clc;
syms t;
x0 = 3; y0 = 2;
x1 = 8; y1 = -4;
x2 = 6; y2 = 5;
k = 0.8;
G = [x0, y0 ; x1, y1 ; x2, y2];
X = [x0 ; x1 ; x2]; Y = [y0 ; y1 ; y2];
C = [1, 0, 0; 1, k, k^2; 1, 1, 1];
A = inv(C)*G;
ax = A(1,1); ay = A(1,2);
bx = A(2,1); by = A(2,2);
cx = A(3,1); cy = A(3,2);
fprintf('Required equations: \n');
x = ax + bx*t + cx*t^2 ; x = vpa(x)
y = ay + by*t + cy*t^2 ; y = vpa(y)

%plotting
tt = linspace(0,1);
xx = subs(x, t, tt);
```

```
yy = subs(y, t, tt);
subplot(131), plot(tt, xx);
xlabel('t'); ylabel('x'); title('t - x');
grid; axis square;
subplot(132), plot(tt, yy);
xlabel('t'); ylabel('y'); title('t - y');
grid; axis square;
subplot(133), plot(xx, yy, 'b-');
hold on; grid; axis square;
scatter(X, Y, 20, 'r', 'filled');
xlabel('x'); ylabel('y'); title('x - y');
d = 0.5;
text(x0+d, y0, 'P_0');
text(x1+d, y1, 'P_1');
text(x2-1, y2-1, 'P_2');
hold off;
```

1.6 CUBIC SPLINE (STANDARD FORM)

A cubic spline is represented by a third-degree polynomial and can be generated if at least four points along the curve are known. Standard form of a cubic spline implies the spline equation is computed in the spatial domain i.e. the x-y plane. Let the given points be $P_1(x_1, y_1)$, $P_2(x_2, y_2)$, $P_3(x_3, y_3)$, and $P_4(x_4, y_4)$. Choose a starting cubic equation, which is written in matrix form

$$y = a + bx + cx^2 + dx^3 = \begin{bmatrix} 1 & x & x^2 & x^3 \end{bmatrix} \begin{bmatrix} a \\ b \\ c \\ d \end{bmatrix} \tag{1.26}$$

Substitute the given points in the starting equation to generate four equations. Four equations are sufficient to solve the four unknown coefficients a, b, c, and d

$$\begin{aligned} y_1 &= a + bx_1 + cx_1^{\,2} + dx_1^{\,3} \\ y_2 &= a + bx_2 + cx_2^{\,2} + dx_2^{\,3} \\ y_3 &= a + bx_3 + cx_3^{\,2} + dx_3^{\,3} \\ y_4 &= a + bx_4 + cx_4^{\,2} + dx_4^{\,3} \end{aligned} \tag{1.27}$$

The four equations are written in matrix form $Y = C{\cdot}A$

$$\begin{bmatrix} y_1 \\ y_2 \\ y_3 \\ y_4 \end{bmatrix} = \begin{bmatrix} 1 & x_1 & x_1^{\,2} & x_1^{\,3} \\ 1 & x_2 & x_2^{\,2} & x_2^{\,3} \\ 1 & x_3 & x_3^{\,2} & x_3^{\,3} \\ 1 & x_4 & x_4^{\,2} & x_4^{\,3} \end{bmatrix} \begin{bmatrix} a \\ b \\ c \\ d \end{bmatrix} \tag{1.28}$$

The equations are solved to find the values of the coefficients: $A = C^{-1} \cdot Y$

$$\begin{bmatrix} a \\ b \\ c \\ d \end{bmatrix} = \begin{bmatrix} 1 & x_1 & x_1^2 & x_1^3 \\ 1 & x_2 & x_2^2 & x_2^3 \\ 1 & x_3 & x_3^2 & x_3^3 \\ 1 & x_4 & x_4^2 & x_4^3 \end{bmatrix}^{-1} \begin{bmatrix} y_1 \\ y_2 \\ y_3 \\ y_4 \end{bmatrix} \tag{1.29}$$

The values of the coefficients are substituted in the starting equation to arrive at the equation of the spline.

$$y = \begin{bmatrix} 1 & x & x^2 & x^3 \end{bmatrix} \begin{bmatrix} 1 & x_1 & x_1^2 & x_1^3 \\ 1 & x_2 & x_2^2 & x_2^3 \\ 1 & x_3 & x_3^2 & x_3^3 \\ 1 & x_4 & x_4^2 & x_4^3 \end{bmatrix}^{-1} \begin{bmatrix} y_1 \\ y_2 \\ y_3 \\ y_4 \end{bmatrix} \tag{1.30}$$

Example 1.6

Find the equation of a cubic spline through points $P_1(-1, 2)$, $P_2(0, 0)$, $P_3(1, -2)$, and $P_4(2, 0)$.

Choose starting equation

$$y = a + bx + cx^2 + dx^3 = \begin{bmatrix} 1 & x & x^2 & x^3 \end{bmatrix} \begin{bmatrix} a \\ b \\ c \\ d \end{bmatrix}$$

Write equations in matrix form $Y = C \cdot A$

$$\begin{bmatrix} 2 \\ 0 \\ -2 \\ 0 \end{bmatrix} = \begin{bmatrix} 1 & -1 & 1 & -1 \\ 1 & 0 & 0 & 0 \\ 1 & 1 & 1 & 1 \\ 1 & 2 & 4 & 8 \end{bmatrix} \begin{bmatrix} a \\ b \\ c \\ d \end{bmatrix}$$

Solve for A: $A = C^{-1} \cdot Y$

$$\begin{bmatrix} a \\ b \\ c \\ d \end{bmatrix} = \begin{bmatrix} 0 \\ -2.67 \\ 0 \\ 0.67 \end{bmatrix}$$

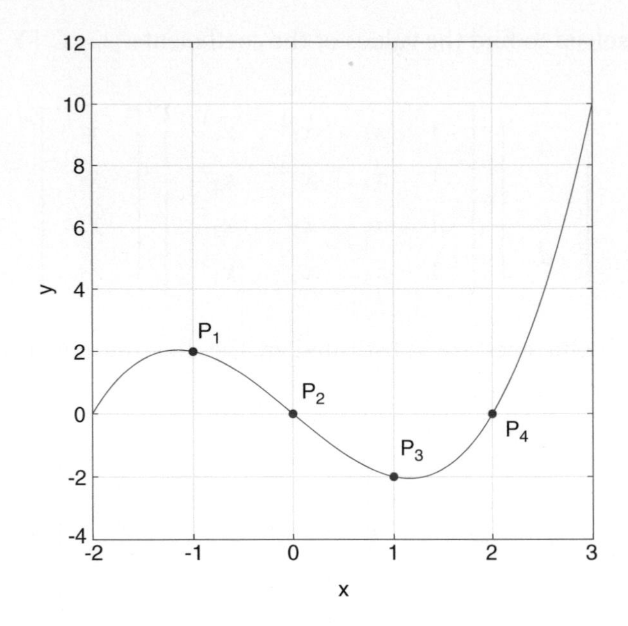

FIGURE 1.9 Plot for Example 1.6.

Substitute in the starting equation

$$y = -2.67x + 0.67x^3$$

Verification: $y(-1) = 2$, $y(0) = 0$, $y(1) = -2$, $y(2) = 0$ (Figure 1.9).

MATLAB Code 1.6

```
clear all; clc;
syms x;
x1 = -1; y1 = 2;
x2 = 0; y2 = 0;
x3 = 1; y3 = -2;
x4 = 2; y4 = 0;
X = [x1 ; x2 ; x3 ; x4];
Y = [y1 ; y2 ; y3 ; y4];
C = [1, x1, x1^2, x1^3; 1, x2, x2^2, x2^3; 1, x3, x3^2, x3^3; 1, x4, x4^2, x4^3];
A = inv(C)*Y;
a = A(1); b = A(2); c = A(3); d = A(4);
fprintf('Required equation  : \n');
y = a + b*x + c*x^2 + d*x^3; y = vpa(y, 3)

%plotting
X = [x1, x2, x3, x4]; m = min(X); n = max(X);
xx = linspace(m - 1, n + 1);
yy = subs(y, x, xx);
plot(xx,yy, 'b');
hold on; grid;
scatter(X, Y, 20, 'r', 'filled');
xlabel('x'); ylabel('y'); axis square;
e = 1;
text(x1, y1+e, 'P_1');
text(x2, y2+e, 'P_2');
text(x3, y3+e, 'P_3');
text(x4, y4+2*e, 'P_4');
hold off;
```

1.7 CUBIC SPLINE (PARAMETRIC FORM)

A cubic spline can also be represented by parametric equations. Let the given points be P_0, P_1, P_2, and P_3. Here, P_0 is the starting point i.e. $P_0 \equiv P(0)$ and P_3 is the end point of the curve i.e. $P_3 \equiv P(1)$. To determine the curve uniquely two additional pieces of information are required regarding the value of parameter t at the middle points P_1 and P_2. Let these values of t be m and n, where $0 \leq m,\ n \leq 1$ i.e. $P_1 \equiv P(m)$ and $P_2 \equiv P(n)$. Different values of m and n referred to as the sub-division ratios, will give rise to curves with the same start and end points but having different shapes.

Choose starting equation written in matrix form

$$P(t) = a + bt + ct^2 + dt^3 = \begin{bmatrix} 1 & t & t^2 & t^3 \end{bmatrix} \begin{bmatrix} a \\ b \\ c \\ d \end{bmatrix} \tag{1.31}$$

Substitute the given points in the starting equation by choosing $t = 0$ at start, $t = m, n$ at the middle points, and $t = 1$ at end.

$$\begin{aligned} P_0 &= a + b(0) + c(0)^2 \\ P_1 &= a + b(m) + c(m)^2 \\ P_2 &= a + b(n) + c(n)^2 \\ P_3 &= a + b(1) + c(1)^2 \end{aligned} \tag{1.32}$$

Write equations in matrix form $G = C{\cdot}A$

$$\begin{bmatrix} P_0 \\ P_1 \\ P_2 \\ P_3 \end{bmatrix} = \begin{bmatrix} 1 & 0 & 0 & 0 \\ 1 & m & m^2 & m^3 \\ 1 & n & n^2 & n^3 \\ 1 & 1 & 1 & 1 \end{bmatrix} \begin{bmatrix} a \\ b \\ c \\ d \end{bmatrix} \tag{1.33}$$

Solve the equation for A: $A = C^{-1}{\cdot}G = B{\cdot}G$

$$\begin{bmatrix} a \\ b \\ c \\ d \end{bmatrix} = \begin{bmatrix} 1 & 0 & 0 & 0 \\ 1 & m & m^2 & m^3 \\ 1 & n & n^2 & n^3 \\ 1 & 1 & 1 & 1 \end{bmatrix}^{-1} \begin{bmatrix} P_0 \\ P_1 \\ P_2 \\ P_3 \end{bmatrix} \tag{1.34}$$

Substitute the coefficient values in the starting equation

$$P(t)=\begin{bmatrix} 1 & t & t^2 & t^3 \end{bmatrix}\begin{bmatrix} 1 & 0 & 0 & 0 \\ 1 & m & m^2 & m^3 \\ 1 & n & n^2 & n^3 \\ 1 & 1 & 1 & 1 \end{bmatrix}^{-1}\begin{bmatrix} P_0 \\ P_1 \\ P_2 \\ P_3 \end{bmatrix} \tag{1.35}$$

Example 1.7

Find the equation of a cubic spline through points (–1, 2), (0, 0), (1, –2), and (2, 0) in parametric form with sub-division ratios m = 0.1 and n = 0.9.

Choose starting equation written in matrix form

$$P(t)=a+bt+ct^2+dt^3=\begin{bmatrix} 1 & t & t^2 & t^3 \end{bmatrix}\begin{bmatrix} a \\ b \\ c \\ d \end{bmatrix}$$

Write equations in matrix form $G = C{\cdot}A$

$$\begin{bmatrix} -1 & 2 \\ 0 & 0 \\ 1 & -2 \\ 2 & 0 \end{bmatrix}=\begin{bmatrix} 1 & 0 & 0 & 0 \\ 1 & 0.1 & 0.01 & 0.001 \\ 1 & 0.9 & 0.81 & 0.729 \\ 1 & 1 & 1 & 1 \end{bmatrix}\begin{bmatrix} a \\ b \\ c \\ d \end{bmatrix}$$

Solve the equation for A i.e. $A = C^{-1}{\cdot}G = B{\cdot}G$

$$\begin{bmatrix} a \\ b \\ c \\ d \end{bmatrix}=\begin{bmatrix} -1 & 2 \\ 12.7222 & -21.4444 \\ -29.1667 & 13.8889 \\ 19.4444 & 5.5556 \end{bmatrix}$$

Substitute the coefficient values in the starting equation

$$x=-1+12.7222t-29.1667t^2+19.4444t^3$$

$$y=2-21.4444t+13.8889t^2+5.5556t^3$$

Verification: $x(0) = -1$, $x(0.1) = 0$, $x(0.9) = 1$, $x(1) = 2$, $y(0) = 2$, $y(0.1) = 0$, $y(0.9) = -2$, $y(1)=0$

The actual values might differ slightly due to round-off errors (Figure 1.10).

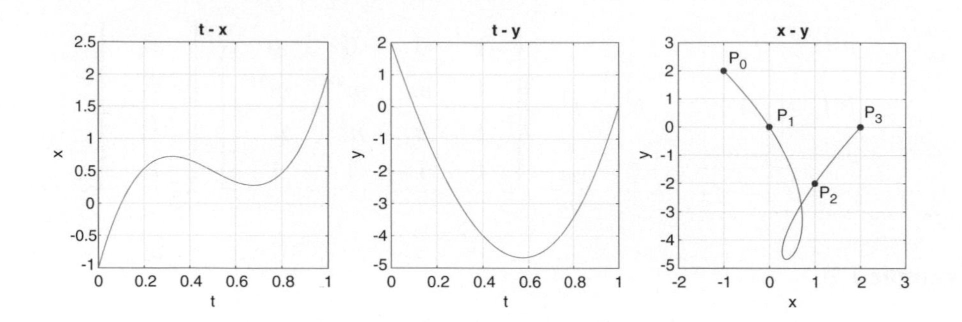

FIGURE 1.10 Plots for Example 1.7.

MATLAB Code 1.7

```
clear all; clc;
syms t;
x0 = -1; y0 = 2;
x1 = 0; y1 = 0;
x2 = 1; y2 = -2;
x3 = 2; y3 = 0;
m = 0.1; n = 0.9;
P = [x0 y0 ; x1 y1 ; x2 y2 ; x3 y3];
X = [x0 ; x1 ; x2 ; x3]; Y = [y0 ; y1 ; y2 ; y3];
C = [1, 0, 0, 0; 1, m, m^2, m^3; 1, n, n^2, n^3; 1, 1, 1, 1];
A = inv(C)*P;
ax = A(1,1); ay = A(1,2); bx = A(2,1); by = A(2,2);
cx = A(3,1); cy = A(3,2); dx = A(4,1); dy = A(4,2);
fprintf('Required equations : \n');
x = ax + bx*t + cx*t^2 + dx*t^3; x = vpa(x, 3)
y = ay + by*t + cy*t^2 + dy*t^3; y = vpa(y, 3)

%plotting
tt = linspace(0,1);
xx = subs(x, t, tt);
yy = subs(y, t, tt);
subplot(131), plot(tt,xx); grid;
xlabel('t'); ylabel('x'); title('t - x'); axis square;
subplot(132), plot(tt,yy); grid;
xlabel('t'); ylabel('y'); title('t - y'); axis square;
subplot(133), plot(xx,yy,'b-'); grid;
xlabel('x'); ylabel('y'); title('x - y'); axis square;
hold on;
scatter(X, Y, 20, 'r', 'filled');
axis([-2 3 -5 3]);
e = 0.5;
text(x0+e, y0, 'P_0');
text(x1+e, y1, 'P_1');
```

```
text(x2+e, y2, 'P_2');
text(x3+e, y3, 'P_3');
hold off;
```

1.8 PIECEWISE SPLINES (STANDARD FORM)

Complex curves cannot be appropriately modeled using cubic splines. They are typically S-shaped curves while complex curves may contain a number of twists and turns. One option is to model the curves using higher order splines; however, they need higher degree equations to be solved, which increases the computational overhead and time delay of the system. Moreover, higher degree splines are too sensitive to slight changes in CPs, which is typically not desirable since we generally want slight changes of the splines to be made by small adjustments of their CPs and do not favor drastic changes in shape. Such curves are best modeled by using multiple cubic splines joined end to end. These are known as piecewise splines.

Consider four given points P_1, P_2, P_3, and P_4 and it is required to find equations of piecewise splines through them. Essentially, this means that instead of a single cubic spline passing through the four points it is required to find three separate splines passing through each pair of points as shown in Figure 1.11.

Let the coordinates of the given points be $P_1(x_1, y_1)$, $P_2(x_2, y_2)$, $P_3(x_3, y_3)$, and $P_4(x_4, y_4)$. Let the three cubic curve segments be designated as A, B, and C between points P_1 and P_2, P_2 and P_3, and P_3 and P_4, respectively. As before, let starting cubic equations be of the form $y = a + bx + cx^2 + dx^3$. Since now there are three curve segments, there needs to be three different sets of coefficients as follows:

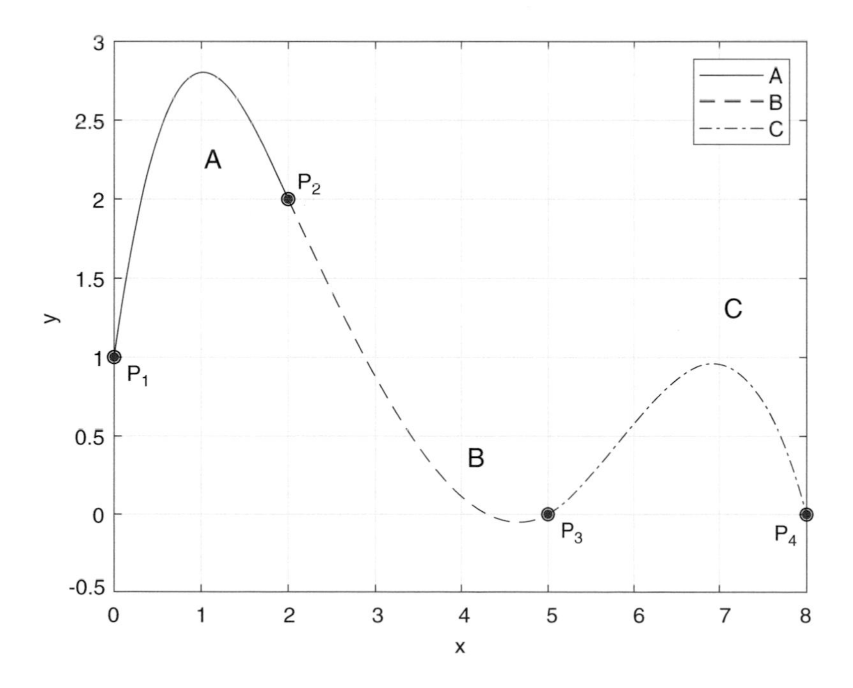

FIGURE 1.11 Piecewise splines.

$$\begin{aligned} &A\text{: } y = a_1 + b_1 x + c_1 x^2 + d_1 x^3 \\ &B\text{: } y = a_2 + b_2 x + c_2 x^2 + d_2 x^3 \\ &C\text{: } y = a_3 + b_3 x + c_3 x^2 + d_3 x^3 \end{aligned} \tag{1.36}$$

So altogether there are 12 different unknowns and at least 12 different equations are needed to solve them.

In order to formulate these 12 equations, various constraints are used to ensure that three separate spline segments join together to form a single smooth curve. The first constraint is known as C^0 continuity condition, which states that in order to form a smooth curve the three splines should physically meet at their joining points (Hearn and Baker, 1996). In other words, spline A should pass through points P_1 and P_2, spline B should pass through points P_2 and P_3, and spline C should pass through points P_3 and P_4. Substituting the point coordinates in the respective starting equations the following six equations are obtained. If $S(P_k)$ denotes segment S passing through point P_k, we can write:

$$\begin{aligned} &A(P_1)\text{: } y_1 = a_1 + b_1 x_1 + c_1 x_1^2 + d_1 x_1^3 \\ &A(P_2)\text{: } y_2 = a_1 + b_1 x_2 + c_1 x_2^2 + d_1 x_2^3 \\ &B(P_2)\text{: } y_2 = a_2 + b_2 x_2 + c_2 x_2^2 + d_2 x_2^3 \\ &B(P_3)\text{: } y_3 = a_2 + b_2 x_3 + c_2 x_3^2 + d_2 x_3^3 \\ &C(P_3)\text{: } y_3 = a_3 + b_3 x_3 + c_3 x_3^2 + d_3 x_3^3 \\ &C(P_4)\text{: } y_4 = a_3 + b_3 x_4 + c_3 x_4^2 + d_3 x_4^3 \end{aligned} \tag{1.37}$$

The second constraint to be obeyed is known as C^1 continuity condition, which states that to form a smooth curve the slopes of the individual spline segments should be equal at their meeting points (Hearn and Baker, 1996). Taking the derivative of the spline equations the following are obtained:

$$\begin{aligned} &A'\text{: } y' = b_1 + 2c_1 \cdot x + 3d_1 \cdot x^2 \\ &B'\text{: } y' = b_2 + 2c_2 \cdot x + 3d_2 \cdot x^2 \\ &C'\text{: } y' = b_3 + 2c_3 \cdot x + 3d_3 \cdot x^2 \end{aligned} \tag{1.38}$$

In this case: slope of A at P_2 = slope of B at P_2. If $S'(P_k)$ denotes slope of segment S at point P_k we have:

$$A'(P_2) = B'(P_2)\text{: } b_1 + 2c_1 \cdot x_2 + 3d_1 \cdot x_2^2 = b_2 + 2c_2 \cdot x_2 + 3d_2 \cdot x_2^2$$

Rearranging:

$$0 = -b_1 - 2c_1 \cdot x_2 - 3d_1 \cdot x_2^2 + b_2 + 2c_2 \cdot x_2 + 3d_2 \cdot x_2^2 \tag{1.39}$$

Also, slope of B at P_3 = slope of C at P_3

$$B'(P_3)=C'(P_3):\ b_2+2c_2\cdot x_3+3d_2\cdot {x_3}^2=b_3+2c_3\cdot x_3+3d_3\cdot {x_3}^2$$

Rearranging:

$$0=-b_2-2c_2\cdot x_3-3d_2\cdot {x_3}^2+b_3+2c_3\cdot x_3+3d_3\cdot {x_3}^2 \tag{1.40}$$

The third constraint to be obeyed is known as C^2 continuity condition, which states that to form a smooth curve the curvatures of the individual spline segments should be equal at their meeting points (Hearn and Baker, 1996). Taking the double derivative of the spline equations the following are obtained:

$$\begin{aligned} A''&: y''=2c_1+6d_1\cdot x \\ B''&: y''=2c_2+6d_2\cdot x \\ C''&: y''=2c_3+6d_3\cdot x \end{aligned} \tag{1.41}$$

In this case, curvature of A at P_2 = curvature of B at P_2. If $S''(P_k)$ denotes curvature of segment S at point P_k, we have:

$$A''(P_2)=B''(P_2):\ 2c_1+6d_1\cdot x_2-2c_2+6d_2\cdot x_2$$

Rearranging:

$$0=-2c_1-6d_1\cdot x_2+2c_2+6d_2\cdot x_2 \tag{1.42}$$

Also, curvature of B at P_3 = curvature of C at P_3

$$B''(P_3)=C''(P_3):\ 2c_2+6d_2\cdot x_3=2c_3+6d_3\cdot x_3$$

Rearranging:

$$0=-2c_2-6d_2\cdot x_3+2c_3+6d_3\cdot x_3 \tag{1.43}$$

The last constraint to be taken into consideration pertains to end-point conditions. The starting slope of spline A and the ending slope of spline C should also be known in order to specify the splines unambiguously. Let the start and end-point slopes be s_1 and s_2, respectively. Thus in this case, $s_1 = A'(P_1)$ and $s_2 = C'(P_4)$. Inserting the slope values in the derivative equations, the following are obtained:

$$\begin{aligned} s_1&=b_1+2c_1\cdot x_1+3d_1\cdot {x_1}^2 \\ s_2&=b_3+2c_3\cdot x_4+3d_3\cdot {x_4}^2 \end{aligned} \tag{1.44}$$

To find a solution to this system, all the 12 equations are plugged into the matrix form $Y = C \cdot A$:

$$\begin{bmatrix} y_1 \\ y_2 \\ y_2 \\ y_3 \\ y_3 \\ y_4 \\ 0 \\ 0 \\ 0 \\ 0 \\ s_1 \\ s_2 \end{bmatrix} = \begin{bmatrix} 1 & x_1 & x_1^2 & x_1^3 & 0 & 0 & 0 & 0 & 0 & 0 & 0 & 0 \\ 1 & x_2 & x_2^2 & x_2^3 & 0 & 0 & 0 & 0 & 0 & 0 & 0 & 0 \\ 0 & 0 & 0 & 0 & 1 & x_2 & x_2^2 & x_2^3 & 0 & 0 & 0 & 0 \\ 0 & 0 & 0 & 0 & 1 & x_3 & x_3^2 & x_3^3 & 0 & 0 & 0 & 0 \\ 0 & 0 & 0 & 0 & 0 & 0 & 0 & 0 & 1 & x_3 & x_3^2 & x_3^3 \\ 0 & 0 & 0 & 0 & 0 & 0 & 0 & 0 & 1 & x_4 & x_4^2 & x_4^3 \\ 0 & -1 & -2x_2 & -3x_2^2 & 0 & 1 & 2x_2 & 3x_2^2 & 0 & 0 & 0 & 0 \\ 0 & 0 & 0 & 0 & 0 & -1 & -2x_3 & -3x_3^2 & 0 & 1 & 2x_3 & 3x_3^2 \\ 0 & 0 & -2 & -6x_2 & 0 & 0 & 2 & 6x_2 & 0 & 0 & 0 & 0 \\ 0 & 0 & 0 & 0 & 0 & 0 & -2 & -6x_3 & 0 & 0 & 2 & 6x_3 \\ 0 & 1 & 2x_1 & 3x_1^2 & 0 & 0 & 0 & 0 & 0 & 0 & 0 & 0 \\ 0 & 0 & 0 & 0 & 0 & 0 & 0 & 0 & 0 & 1 & 2x_4 & 3x_4^2 \end{bmatrix} \begin{bmatrix} a_1 \\ b_1 \\ c_1 \\ d_1 \\ a_2 \\ b_2 \\ c_2 \\ d_2 \\ a_3 \\ b_3 \\ c_3 \\ d_3 \end{bmatrix} \tag{1.45}$$

The solution of this is: $A = \text{inv}(C) \cdot Y$

Example 1.8

Find piecewise cubic equation of a curve passing through $P_1(0, 1)$, $P_2(2, 2)$, $P_3(5, 0)$, and $P_4(8, 0)$. Slopes at first and last points are 4 and –2, respectively

Plugging the given values into the solution matrix

$$\begin{bmatrix} 1 \\ 2 \\ 2 \\ 0 \\ 0 \\ 0 \\ 0 \\ 0 \\ 0 \\ 0 \\ 4 \\ -2 \end{bmatrix} = \begin{bmatrix} 1 & 0 & 0 & 0 & 0 & 0 & 0 & 0 & 0 & 0 & 0 & 0 \\ 1 & 2 & 4 & 8 & 0 & 0 & 0 & 0 & 0 & 0 & 0 & 0 \\ 0 & 0 & 0 & 0 & 1 & 2 & 4 & 8 & 0 & 0 & 0 & 0 \\ 0 & 0 & 0 & 0 & 1 & 5 & 25 & 125 & 0 & 0 & 0 & 0 \\ 0 & 0 & 0 & 0 & 0 & 0 & 0 & 0 & 1 & 5 & 25 & 125 \\ 0 & 0 & 0 & 0 & 0 & 0 & 0 & 0 & 1 & 8 & 64 & 512 \\ 0 & -1 & -4 & -12 & 0 & 1 & 4 & 12 & 0 & 0 & 0 & 0 \\ 0 & 0 & 0 & 0 & 0 & -1 & -10 & -75 & 0 & 1 & 10 & 75 \\ 0 & 0 & -2 & -12 & 0 & 0 & 2 & 12 & 0 & 0 & 0 & 0 \\ 0 & 0 & 0 & 0 & 0 & 0 & -2 & -30 & 0 & 0 & 2 & 30 \\ 0 & 1 & 0 & 0 & 0 & 0 & 0 & 0 & 0 & 0 & 0 & 0 \\ 0 & 0 & 0 & 0 & 0 & 0 & 0 & 0 & 0 & 1 & 16 & 192 \end{bmatrix} \begin{bmatrix} a_1 \\ b_1 \\ c_1 \\ d_1 \\ a_2 \\ b_2 \\ c_2 \\ d_2 \\ a_3 \\ b_3 \\ c_3 \\ d_3 \end{bmatrix}$$

Solving the coefficients and substituting into the starting equations, the required solution is

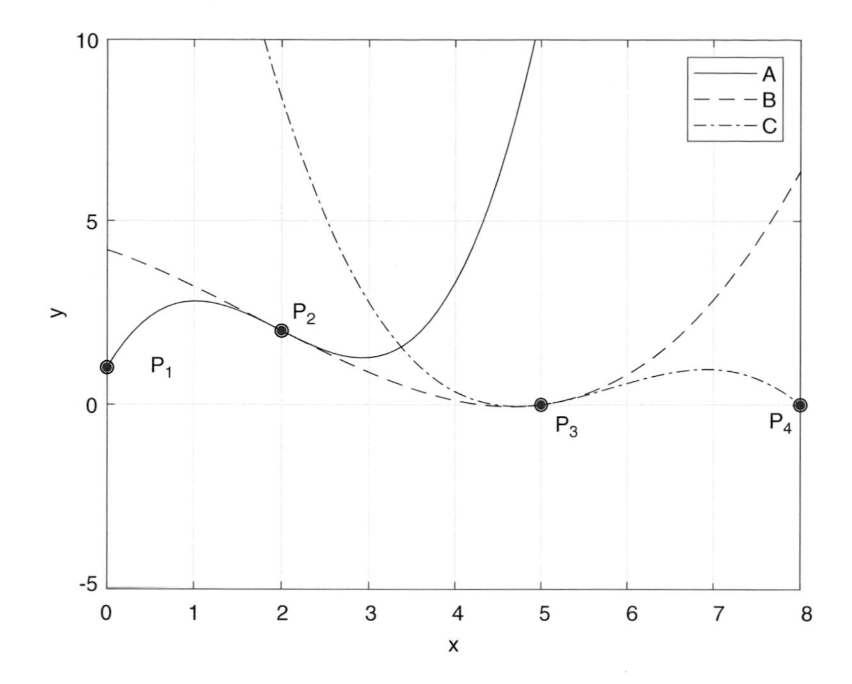

FIGURE 1.12 Plot for Example 1.8.

$$A: y = (1) + (4)\cdot x + (-2.64)\cdot x^2 + (0.45)\cdot x^3$$

$$B: y = (4.20) + (-0.8)\cdot x + (-0.24)\cdot x^2 + (0.05)\cdot x^3$$

$$C: y = (33.68) + (-18.49)\cdot x + (3.29)\cdot x^2 + (-0.19)\cdot x^3$$

Verification: A: $y(0) = 1$, B: $y(2) = 2$, C: $y(5) = 0$ (Figure 1.12).

MATLAB Code 1.8

```
clear all; format compact; clc;
syms x;
x1 = 0; y1 = 1;
x2 - 2; y2 = 2;
x3 = 5; y3 = 0;
x4 = 8; y4 = 0;
s1 = 4; s2 = -2;
X1 = [x1, x2, x3, x4];
Y1 = [y1, y2, y3, y4];
Y = [y1; y2; y2; y3; y3; y4; 0; 0; 0; 0; s1; s2];
C = [1, x1, x1^2, x1^3, 0, 0, 0, 0, 0, 0, 0, 0;
     1, x2, x2^2, x2^3, 0, 0, 0, 0, 0, 0, 0, 0;
     0, 0, 0, 0, 1, x2, x2^2, x2^3, 0, 0, 0, 0;
     0, 0, 0, 0, 1, x3, x3^2, x3^3, 0, 0, 0, 0;
     0, 0, 0, 0, 0, 0, 0, 0, 1, x3, x3^2, x3^3;
     0, 0, 0, 0, 0, 0, 0, 0, 1, x4, x4^2, x4^3;
```

```
    0, -1, -2*x2, -3*x2^2, 0, 1, 2*x2, 3*x2^2, 0, 0, 0, 0;
    0, 0, 0, 0, 0, -1, -2*x3, -3*x3^2, 0, 1, 2*x3, 3*x3^2;
    0, 0, -2, -6*x2, 0, 0, 2, 6*x2, 0, 0, 0, 0;
    0, 0, 0, 0, 0, 0, -2, -6*x3, 0, 0, 2, 6*x3;
    0, 1, 2*x1, 3*x1^2, 0, 0, 0, 0, 0, 0, 0, 0;
    0, 0, 0, 0, 0, 0, 0, 0, 0, 1, 2*x4, 3*x4^2];

A = inv(C)*Y ;
a1=A(1); b1=A(2); c1=A(3); d1=A(4);
a2=A(5); b2=A(6); c2=A(7); d2=A(8);
a3=A(9); b3=A(10); c3=A(11); d3=A(12);
fprintf('Equations for segments : \n');
yA = a1 + b1*x + c1*x^2 + d1*x^3; yA = vpa(yA, 3)
yB = a2 + b2*x + c2*x^2 + d2*x^3; yB = vpa(yB, 3)
yC = a3 + b3*x + c3*x^2 + d3*x^3; yC = vpa(yC, 3)

%plotting
xx = 0:0.1:9;
yp1 = subs(yA, x, xx);
yp2 = subs(yB, x, xx);
yp3 = subs(yC, x, xx);
plot(xx, yp1, 'k-', xx, yp2, 'k--', xx, yp3, 'k-.');
axis([0, 8, -5, 10]); grid on; hold on;
plot(X1, Y1, 'ko');
scatter(X1, Y1, 20, 'r', 'filled');
text(0.5,1,'P_1');
text(2.5,2,'P_2');
text(5.5,-0.5,'P_3');
text(7.5,0,'P_4');
legend('A', 'B', 'C'); xlabel('x'); ylabel('y');
hold off;
%verification
vrf1 = eval(subs(yA, x, x1))  %should return y1
vrf2 = eval(subs(yB, x, x2))  %should return y2
vrf3 = eval(subs(yC, x, x3))  %should return y3
```

NOTE

From Figure 1.12, it can be observed that *A*, *B*, and *C* are three different splines having different shapes, which is expected since they have different equations. However, the continuity conditions have constrained them in such a way that they have formed a single smooth curve only within the interval P_1 to P_4. Beyond their respective intervals they have diverged out into different trajectories.

`eval`: evaluates an expression
`legend`: designates different color or line types in a graph using textual strings

1.9 PIECEWISE SPLINES (PARAMETRIC FORM)

To end this chapter and to explain a very important concept of domain conversion a more complicated form of piecewise spline is discussed, which involves parametric equations. Since now each spline segment needs to be represented by two equations (x vs. t and y vs. t) the number of unknowns is effectively doubled to 24. However, the essential idea of forming the equations by using constraint conditions remains the same and can be visualized as an extended version of the ideas explained in the last section. To bring down the complexity of the situation a simplified assumption is used here: that the x vs. t relations are linear instead of cubic. This will reduce the total number of unknowns so that the situation can be more readily comprehended. However, the main reason for discussing this section is to make the reader aware that when given conditions are specified in the spatial domain and the required equations need to obtained in the parametric domain (or vice versa) then the values cannot simply be substituted in the constraint equations (as done in the previous case) but needs conversion from one domain to another (as will be illustrated below).

As before, let the coordinates of the given points be $P_1(x_1, y_1)$, $P_2(x_2, y_2)$, $P_3(x_3, y_3)$, and $P_4(x_4, y_4)$. Let the three cubic curve segments be designated as A, B, and C. This time we assume that the $x-t$ relations are linear so the starting equations are of the form: $x = m + nt$.

The equations for the separate splines therefore become:

$$\begin{aligned} A&: x = m_1 + n_1 t \\ B&: x = m_2 + n_2 t \\ C&: x = m_3 + n_3 t \end{aligned} \tag{1.46}$$

For spline A, at the starting point $t = 0$ and $x = x_1$ and at the ending point $t = 1$ and $x = x_2$. This provides the solution:

$$\begin{aligned} m_1 &= x_1 \\ n_1 &= x_2 - x_1 \end{aligned} \tag{1.47}$$

For spline B, at the starting point $t = 0$ and $x = x_2$ and at the ending point $t = 1$ and $x = x_3$. This provides the solution:

$$\begin{aligned} m_2 &= x_2 \\ n_2 &= x_3 - x_2 \end{aligned} \tag{1.48}$$

For spline C, at the starting point $t = 0$ and $x = x_3$ and at the ending point $t = 1$ and $x = x_4$. This provides the solution:

$$\begin{aligned} m_3 &= x_3 \\ n_3 &= x_3 - x_4 \end{aligned} \tag{1.49}$$

That was quite easily done and so the attention is now turned toward a more harder problem of dealing with the cubic y vs. t relations. Here, as expected, the starting relation is of the form: $y = a + bt + ct^2 + dt^3$

Equations for the separate splines become:

$$\begin{aligned} &A: y = a_1 + b_1 t + c_1 t^2 + d_1 t^3 \\ &B: y = a_2 + b_2 t + c_2 t^2 + d_2 t^3 \\ &C: y = a_3 + b_3 t + c_3 t^2 + d_3 t^3 \end{aligned} \tag{1.50}$$

The first derivatives of the equations are given in the following:

$$\begin{aligned} &A': y' = b_1 + 2c_1 \cdot t + 3d_1 \cdot t^2 \\ &B': y' = b_2 + 2c_2 \cdot t + 3d_2 \cdot t^2 \\ &C': y' = b_3 + 2c_3 \cdot t + 3d_3 \cdot t^2 \end{aligned} \tag{1.51}$$

The second derivatives of the equations are also as follows:

$$\begin{aligned} &A'': y = 2c_1 + 6d_1 \cdot t \\ &B'': y = 2c_2 + 6d_2 \cdot t \\ &C'': y = 2c_3 + 6d_3 \cdot t \end{aligned} \tag{1.52}$$

For the first constraint pertaining to C^0 continuity condition to be applied, it should be noted that for spline A at the starting point $t = 0$ and $y = y_1$ and at the ending point $t = 1$ and $y = y_2$. This provides the solution:

$$\begin{aligned} y_1 &= a_1 \\ y_2 - y_1 &= b_1 + c_1 + d_1 \end{aligned} \tag{1.53}$$

For spline B at the starting point $t = 0$ and $y = y_2$ and at the ending point $t = 1$ and $y = y_3$. This provides the solution:

$$\begin{aligned} y_2 &= a_2 \\ y_3 - y_2 &= b_2 + c_2 + d_2 \end{aligned} \tag{1.54}$$

For spline C at the starting point $t = 0$ and $y = y_3$ and at the ending point $t = 1$ and $y = y_4$. This provides the solution:

$$\begin{aligned} y_3 &= a_3 \\ y_4 - y_3 &= b_3 + c_3 + d_3 \end{aligned} \tag{1.55}$$

The second constraint to be obeyed is the C^1 continuity condition, which states that to form a smooth curve the slopes of the individual spline segments should be equal at their meeting points. In this case slope of A at point P_2 should be equal to the slope of B at point P_2 i.e. A' (at $t = 1$) = B' (at $t = 0$). But the slopes are equal in spatial domain (physical slopes in space) and the derivative equations (as shown above) are calculated in the parametric domain (y vs. t) so they cannot simply be equated, rather some kind of conversion from one domain to another is first needed. For arriving at the conversion factor the chain rule of differentiation is utilized.

By chain rule of differentiation:

$$\frac{dy}{dx}=\left(\frac{dy}{dt}\right)\cdot\left(\frac{dt}{dx}\right)=y'\cdot\left(\frac{\Delta t}{\Delta x}\right)=y'\cdot\frac{1-0}{\Delta x}=y'\cdot\left(\frac{1}{\Delta x}\right)$$

$$\frac{d^2y}{dx^2}=\frac{d}{dx}\left(\frac{dy}{dx}\right)=\frac{d}{dt}\left(\frac{dy}{dx}\right)\cdot\left(\frac{dt}{dx}\right)=\frac{d}{dt}\left\{\left(\frac{dy}{dt}\right)\cdot\left(\frac{dt}{dx}\right)\right\}\cdot\left(\frac{dt}{dx}\right)=\frac{d}{dt}\left(\frac{dy}{dt}\right)\cdot\left(\frac{dt}{dx}\right)^2=y''\cdot\left(\frac{1}{\Delta x}\right)^2 \tag{1.56}$$

This specifies the required conversion factor: that the derivative in spatial domain is equal to the derivative in parametric domain multiplied by a scaling factor of $(1/\Delta x)$ for a specific curve segment. Similarly the double derivative in spatial domain is equal to the double derivative in parametric domain multiplied by the scaling factor $(1/\Delta x)^2$.

Plugging these multipliers into the C^1 constraint equations:

Slope of A at P_2 = slope of B at P_2

$A'(t-1) - B'(t=0)$ [in spatial domain]

$$\frac{b_1+2c_1+3d_1}{x_2-x_1}=\frac{b_2}{x_3-x_2}$$

Rearranging:

$$(x_3-x_2)\cdot b_1+2(x_3-x_2)\cdot c_1+3(x_3-x_2)\cdot d_1-(x_2-x_1)\cdot b_2=0 \tag{1.57}$$

Slope of B at P_3 = slope of C at P_3

$B'(t=1) = C'(t=0)$ [in spatial domain]

$$\frac{b_2+2c_2+3d_2}{x_3-x_2}=\frac{b_3}{x_4-x_3}$$

Rearranging:

$$(x_4-x_3)\cdot b_2+2(x_4-x_3)\cdot c_2+3(x_4-x_3)\cdot d_2-(x_3-x_2)\cdot b_3=0 \tag{1.58}$$

In a similar way using the domain conversion multipliers for the C^2 constraint equations:

Curvature of A at P_2 = curvature of B at P_2

$A''(t=1) = B''(t=0)$ [in spatial domain]

$$\frac{2c_1+6d_1}{(x_2-x_1)^2}=\frac{2c_2}{(x_3-x_2)^2}$$

Rearranging and dividing both sides by 2:

$$(x_3 - x_2)^2 \cdot c_1 + (x_3 - x_2)^2 \cdot 3d_1 - (x_2 - x_1)^2 \cdot c_2 = 0 \quad (1.59)$$

Curvature of B at P_3 = curvature of C at P_3

$B''(t = 1) = C''(t = 0)$ [in spatial domain]

$$\frac{2c_2 + 6d_2}{(x_3 - x_2)^2} = \frac{2c_3}{(x_4 - x_3)^2}$$

Rearranging and dividing both sides by 2:

$$(x_4 - x_3)^2 \cdot c_2 + (x_4 - x_3)^2 \cdot 3d_2 - (x_3 - x_2)^2 \cdot c_3 = 0 \quad (1.60)$$

The domain conversion factors are also required for the end-point conditions.

Let s_1 be the slope at start point of segment A and s_2 be the end-point slope of segment C. Then:

$$s_1 = A'(t = 0):\ s_1 = \frac{b_1}{(x_2 - x_1)}$$

$$b_1 = s_1 \cdot (x_2 - x_1) \quad (1.61)$$

$$s_2 = C'(t = 1):\ s_2 = \frac{b_3 + 2c_3 + 3d_3}{(x_4 - x_3)}$$

$$b_3 + 2c_3 + 3d_3 = s_2 \cdot (x_4 - x_3) \quad (1.62)$$

Write all nine equations in the form $G = C{\cdot}A$

$$b_3 + 2c_3 + 3d_3 = s_2 \cdot (x_4 - x_3) \quad (1.63)$$

$$\begin{bmatrix} y_2 - y_1 \\ y_3 - y_2 \\ y_4 - y_3 \\ 0 \\ 0 \\ 0 \\ 0 \\ s_1(x_2 - x_1) \\ s_2(x_4 - x_3) \end{bmatrix} = \begin{bmatrix} 1 & 1 & 1 & 0 & 0 & 0 & 0 & 0 & 0 \\ 0 & 0 & 0 & 1 & 1 & 1 & 0 & 0 & 0 \\ 0 & 0 & 0 & 0 & 0 & 0 & 1 & 1 & 1 \\ (x_3 - x_2) & 2(x_3 - x_2) & 3(x_3 - x_2) & -(x_2 - x_1) & 0 & 0 & 0 & 0 & 0 \\ 0 & 0 & 0 & (x_4 - x_3) & 2(x_4 - x_3) & 3(x_4 - x_3) & -(x_3 - x_2) & 0 & 0 \\ 0 & (x_3 - x_2)^2 & 3(x_3 - x_2)^2 & 0 & -(x_2 - x_1)^2 & 0 & 0 & 0 & 0 \\ 0 & 0 & 0 & 0 & (x_4 - x_3)^2 & 3(x_4 - x_3)^2 & 0 & -(x_3 - x_2)^2 & 0 \\ 1 & 0 & 0 & 0 & 0 & 0 & 0 & 0 & 0 \\ 0 & 0 & 0 & 0 & 0 & 0 & 1 & 2 & 3 \end{bmatrix} \begin{bmatrix} b_1 \\ c_1 \\ d_1 \\ b_2 \\ c_2 \\ d_2 \\ b_3 \\ c_3 \\ d_3 \end{bmatrix}$$

The solution of this is: $A = \text{inv}(C) \cdot G$

Example 1.9

Find piecewise parametric cubic equation of a curve passing through $P_1(0, 1)$, $P_2(2, 2)$, $P_3(5, 0)$, and $P_4(8, 0)$. Slopes at first and last points are 2 and 1, respectively. Assume linear relation between t and x.

Since $x - t$ relations are linear, let the starting equations are of the form: $x = m + nt$. Equations for the separate spines:

$$A: x = m_1 + n_1 t$$

$$B: x = m_2 + n_2 t$$

$$C: x = m_3 + n_3 t$$

Substituting given points into the above equations and solving for the unknown coefficients:

$$m_1 = x_1 = 0, n_1 = x_2 - x_1 = 2, m_2 = x_2 = 2, n_2 = x_3 - x_2 = 3, m_3 = x_3 = 5, n_3 = x_4 - x_3 = 3$$

The required $x - t$ relations are:

$$A: x = 2t$$

$$B: x = 2 + 3t$$

$$C: x = 5 + 3t$$

Since $y - t$ relations are cubic, let the starting equations are of the form: $y = a + b \cdot t + c \cdot t^2 + d \cdot t^3$.

Equations for the separate splines:

$$A: y = a_1 + b_1 t + c_1 t^2 + d_1 t^3$$

$$B: y = a_2 + b_2 t + c_2 t^2 + d_2 t^3$$

$$C: y = a_3 + b_3 t + c_3 t^2 + d_3 t^3$$

Plugging the given values into the constraint matrix:

$$\begin{bmatrix} 1 \\ -2 \\ 0 \\ 0 \\ 0 \\ 0 \\ 0 \\ 4 \\ 3 \end{bmatrix} = \begin{bmatrix} 1 & 1 & 1 & 0 & 0 & 0 & 0 & 0 & 0 \\ 0 & 0 & 0 & 1 & 1 & 1 & 0 & 0 & 0 \\ 0 & 0 & 0 & 0 & 0 & 0 & 1 & 1 & 1 \\ 3 & 6 & 9 & -2 & 0 & 0 & 0 & 0 & 0 \\ 0 & 0 & 0 & 3 & 6 & 9 & -3 & 0 & 0 \\ 0 & 9 & 27 & 0 & -4 & 0 & 0 & 0 & 0 \\ 0 & 0 & 0 & 0 & 9 & 27 & 0 & -9 & 0 \\ 1 & 0 & 0 & 0 & 0 & 0 & 0 & 0 & 0 \\ 0 & 0 & 0 & 0 & 0 & 9 & 1 & 2 & 3 \end{bmatrix} \begin{bmatrix} b_1 \\ c_1 \\ d_1 \\ b_2 \\ c_2 \\ d_2 \\ b_3 \\ c_3 \\ d_3 \end{bmatrix}$$

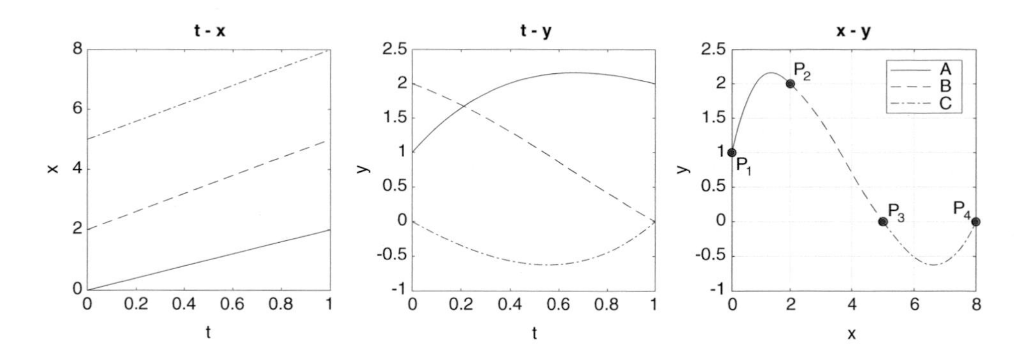

FIGURE 1.13 Plots for Example 1.9.

Solving and substituting into the starting equations:

$$A\colon y=(1)+(4)\cdot t+(-4.158)\cdot t^2+(1.158)\cdot t^3$$

$$B\colon y=(2)+(-1.263)\cdot t+(-1.539)\cdot t^2+(0.803)\cdot t^3$$

$$C\colon y=(-1.934)\cdot t+(0.868)\cdot t^2+(1.066)\cdot t^3$$

Verification: For A: $t=0$ produces $y=1$, $t=1$ produces $y=2$, for B: $t=0$ produces $y=2$, $t=1$ produces $y=0$, and for C: $t=0$ produces $y=0$, $t=1$ produces $y=0$ (Figure 1.13).

MATLAB Code 1.9

```
clear all; clc;
syms t;
x1 = 0; y1 = 1;
x2 = 2; y2 = 2;
x3 = 5; y3 = 0;
x4 = 8; y4 = 0;
s1 = 2; s2 = 1;

C=[ 1, 1, 1, 0, 0, 0, 0, 0, 0;
    0, 0, 0, 1, 1, 1, 0, 0, 0;
    0, 0, 0, 0, 0, 0, 1, 1, 1;
    (x3-x2), 2*(x3-x2), 3*(x3-x2), -(x2-x1), 0, 0, 0, 0, 0;
    0, 0, 0, (x4-x3), 2*(x4-x3), 3*(x4-x3), -(x3-x2), 0, 0;
    0, (x3-x2)^2, 3*(x3-x2)^2, 0, -(x2-x1)^2, 0, 0, 0, 0;
    0, 0, 0, 0, (x4-x3)^2, 3*(x4-x3)^2, 0, -(x3-x2)^2, 0;
    1, 0, 0, 0, 0, 0, 0, 0, 0;
    0, 0, 0, 0, 0, 0, 1, 2, 3
];

G = [y2-y1, y3-y2, y4-y3, 0, 0, 0, 0, s1*(x2-x1), s2*(x4-x3)];

A = inv(C)*G';
```

```
aA = y1; bA = A(1); cA = A(2); dA = A(3);
aB = y2; bB = A(4); cB = A(5); dB = A(6);
aC = y3; bC = A(7); cC = A(8); dC = A(9);

fprintf('Equations of segments :\n')

xA = x1 + (x2-x1)*t; xA = vpa(xA)
yA = aA + bA*t + cA*t^2 + dA*t^3; yA = vpa(yA, 3)
xB = x2 + (x3-x2)*t; xB = vpa(xB)
yB = aB + bB*t + cB*t^2 + dB*t^3; yB = vpa(yB, 3)
xC = x3 + (x4-x3)*t; xC = vpa(xC)
yC = aC + bC*t + cC*t^2 + dC*t^3; yC = vpa(yC, 3)

%plotting
tt = linspace(0,1);
xa = subs(xA, t, tt);
ya = subs(yA, t, tt);
xb = subs(xB, t, tt);
yb = subs(yB, t, tt);
xc = subs(xC, t, tt);
yc = subs(yC, t, tt);

subplot(131); plot(tt,xa, 'k-', tt, xb, 'k--', tt, xc, 'k-.');
xlabel('t'); ylabel('x'); title('t - x'); axis square;
subplot(132); plot(tt,ya, 'k-', tt, yb, 'k--', tt, yc, 'k-.');
xlabel('t'); ylabel('y'); title('t - y'); axis square;
subplot(133); X = [x1 x2 x3 x4]; Y = [y1 y2 y3 y4];
plot(xa,ya,'k-', xb,yb, 'k--',xc,yc,'k-.', X, Y, 'ko'); hold on;
scatter(X, Y, 20, 'r', 'filled'); grid;

text(0.5,1,'P_1');
text(2.5,2,'P_2');
text(5.5,0,'P_3');
text(7,0,'P_4');
legend('A', 'B', 'C');
xlabel('x'); ylabel('y'); title('x - y'); axis square;
hold off;
```

NOTE

The importance of the end-point conditions can be observed here. Compare Examples 1.8 and 1.9. Even though the CPs remain the same in both cases, the shape of the piecewise curve has changed, as shown in Figures 1.12 and 1.13, due to changes in end-point slopes only.

1.10 CHAPTER SUMMARY

The following points summarize the topics discussed in this chapter:

- Splines are irregular curve segments with known mathematical properties.
- The shape of the spline is determined by "CPs."
- Interpolating splines actually go through all of its CPs.
- Hybrid splines go through some of its CPs but not through others.
- Approximating splines in general does not go through any of its CPs.
- Splines are mathematically modeled using polynomials equations in the form $y = f(x)$.
- Polynomial equations can also be represented in parametric form $x = f(t), y = g(t)$.
- A linear spline is represented by a first-degree polynomial equation $y = a + bx$.
- A quadratic spline is represented by a second-degree polynomial equation $y = a + bx + cx^2$.
- A cubic spline is represented by a third-degree polynomial equation $y = a + bx + cx^2 + dx^3$.
- Spatial set of equations can be expressed in the matrix form $Y = C{\cdot}A$, whose solution is $A = \text{inv}(C){\cdot}Y$.
- Parametric set of equations can be expressed in the matrix form $G = C{\cdot}A$, whose solution is $A = \text{inv}(C){\cdot}G$.
- Complex curves are modeled by multiple cubic splines joined end to end, known as piecewise splines.
- Conversion between the spatial and parametric domains can be done using chain rules of differentiation.

1.11 REVIEW QUESTIONS

1. What is meant by "spline"? From where has the word originated?
2. What are the different types of splines possible?
3. What is meant by polynomial equation?
4. Differentiate between the standard form and parametric form of representation.
5. What is a constraint matrix, coefficient matrix, geometry matrix, and basis matrix?
6. Explain the difference between the notations P_0 and $P(0)$.
7. What is a linear spline and how is it represented using a polynomial?

8. What is a quadratic spline and how is it represented using a polynomial?
9. What is a cubic spline and how is it represented using a polynomial?
10. What are sub-division ratios used in parametric forms of quadratic and cubic splines?
11. Why is a parametric form of a spline represented using three different plots?
12. What are piecewise splines and why are they necessary?
13. What are the C^0, C^1, and C^2 continuity conditions?
14. What is meant by end-point condition?
15. How is it possible to convert values between spatial and parametric domains?

1.12 PRACTICE PROBLEMS

1. Find the equation of a linear spline passing through (–3, –3) and having slope –3.
2. Find equation of a quadratic curve passing through (0, 0), (π/2, 1), and (π, 0).
3. Find the equation of a quadratic curve through (0, 2), (–2, 0), and (2, –2) in parametric form with $k = 0.4$.
4. Derive parametric equations of a quadratic curve, which goes through three points (–2, 1), (–1, 2), and (2, –1) in such a way that the middle point divides the curve in the ratio 1:2.
5. Find the equation of a cubic interpolating spline through the points (3, 2), (8, –4), (6, 5), and (1, 0).
6. Find the equation of a cubic spline passing through four points (3, 2), (8, –4), (6, 5), and (1, 0), in parametric form, with sub-division ratios $m = 0.1$ and $n = 0.7$.
7. Find piecewise cubic equation of a curve passing through (–5, –2), (–1, –1), (5, 0), and (7, –2). Slopes at first and last points are 1 and 1, respectively.
8. Find the equation of a quadratic curve in the form $y = f(x)$, which passes through the three points $(k, -k)$, $(0, k)$, and $(-k, 0)$, where k is a constant.
9. For what value(s) of k will two curve segments having equations $y = k + 2k \cdot x - 3k \cdot x^2$ and $y = 3k - 2k \cdot x + k \cdot x^2$ satisfy the C^1 continuity condition at point $P(k, -k)$, where k is a constant?
10. Find piecewise parametric cubic equation of a curve passing through (1, –2), (2, –3), (3, –4), and (4, –5). Slopes at first and last points are 5 and –6, respectively. Assume x vs. t relations to be linear.

CHAPTER 2

Blending Functions and Hybrid Splines

2.1 INTRODUCTION

In the previous chapter, we have discussed how equations of interpolating splines can be derived by substituting the coordinate information into the starting equatisons. Apart from interpolating splines, there are other types of splines, which do not always pass through the CPs or all the CPs may not be known or conditions other than CPs need to be used for deriving their equations. For such splines, the techniques previously discussed would not be applicable and a new set of techniques need to be devised. The newer techniques are designed in such a way so that spline equations can be made independent of the coordinates of the CPs. Such techniques are applicable to hybrid splines, which only pass through a subset of the CPs, and approximating splines, which in general do not pass through any of its CPs. This has led to the concept of blending functions (BFs), which provides us with ways of determining where a spline is located in the neighborhood of a CP. In the following sections, the concept of BFs is explained and then applied to both interpolating and non-interpolating splines. The latter part of this chapter lays the foundation of hybrid splines. Hybrid splines are those which pass through few CPs but not through others or boundary conditions other than CPs are used to derive their equations. To fully define hybrid splines therefore, some additional constraints are often specified, for example slope of the curve at a point. Four types of hybrid splines are discussed, namely, Hermite spline, Cardinal spline, Catmull–Rom (C–R) spline, and Bezier spline. In each case, we discuss mostly the cubic curves in parametric form, as they are most often used in graphics and moreover, the ideas discussed can be extended to other degree of curves also.

2.2 BLENDING FUNCTIONS

The concept of "BFs" was proposed as a means for calculating equations of curves, which are non-interpolating i.e. they do not pass through some or all of the CPs (Hearn and Baker, 1996). For interpolating curves, the coordinates of the CPs are substituted in the starting equations to solve for the unknown coefficients. However, if the curve does not pass through the CPs then this method does not work and a new approach is required to determine the trajectory of a point on the curve with respect to the CPs.

Consider a spline curve whose CPs are at locations P_0, P_1, P_2, and P_3. In general, none of the CPs are actually located on the curve but would be around it somewhere in the neighborhood (see Figure 2.1). Let us assume that four masses L_0, L_1, L_2, and L_3 are located on the CPs P_0, P_1, P_2, and P_3, which exert gravitational pull on the spline. Then the center of mass is located at P given by

$$P = \frac{L_0 P_0 + L_1 P_1 + L_2 P_2 + L_3 P_3}{L_0 + L_1 + L_2 + L_3} \tag{2.1}$$

Now, consider that the masses are not constant but change their values as a function of the parametric variable $t (0 \le t \le 1)$ i.e. $L_i = f(t)$. Then, the center of mass P will also shift to different points. If all these points are joined together as t takes on various values from 0 to 1, then the locus of P would indicate the actual spline we are interested in. The functions $L_i = f(t)$ are known as BFs. It is obvious that the total number of BFs would be equal to the total number of masses which in turn would be equal to the total number of CPs. Thus for a cubic curve there are in general four BFs.

To reduce complexity of computations, an additional constraint is applied: $\Sigma L_i = 1$ i.e. $L_0 + L_1 + L_2 + L_3 = 1$. This reduces the denominator of the above equation to 1 and gives us

$$P = L_0 P_0 + L_1 P_1 + L_2 P_2 + L_3 P_3 \tag{2.2}$$

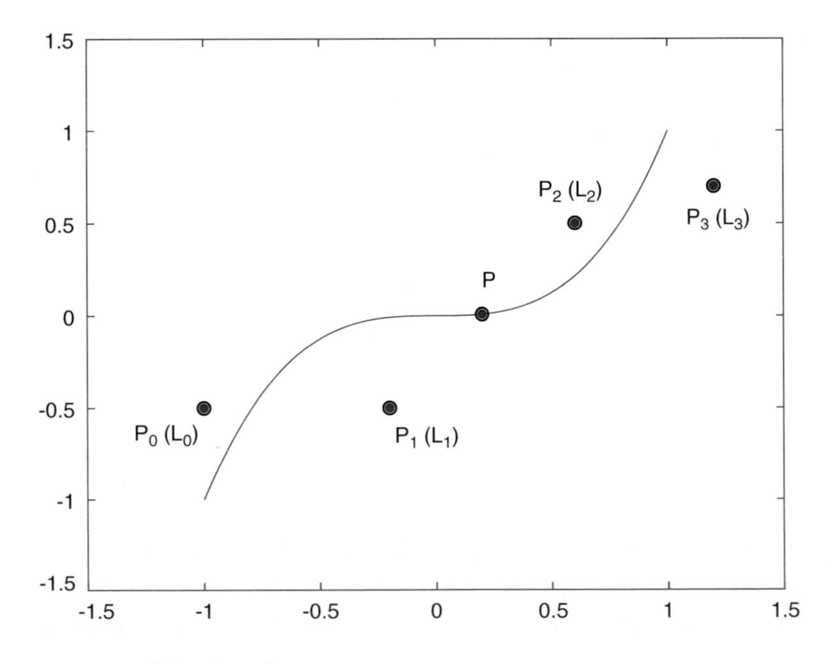

FIGURE 2.1 Concept of blending functions.

It might be obvious to reflect that larger the mass at a CP, more will be its gravitational pull on the spline and nearer will the spline be drawn towards that point. Writing the above in matrix form:

$$P = L_0P_0 + L_1P_1 + L_2P_2 + L_3P_3 = \begin{bmatrix} L_0 & L_1 & L_2 & L_3 \end{bmatrix} \begin{bmatrix} P_0 \\ P_1 \\ P_2 \\ P_3 \end{bmatrix} = L \cdot G \tag{2.3}$$

The matrix L provides us with a characterization of any point P on the spline independent of the CPs located in matrix G. Matrix L is referred to as the "BF matrix" while the individual elements L_0, L_1, L_2, and L_3 are the BFs (sometimes also referred to as "basis functions"). The matrix G is the geometry matrix as it defines the locations of the CPs, which determine the geometry of the curve.

From the previous chapter, we know that in general for a parametric cubic curve:

$$P(t) = a + bt + ct^2 + dt^3 = \begin{bmatrix} 1 & t & t^2 & t^3 \end{bmatrix} \begin{bmatrix} a \\ b \\ c \\ d \end{bmatrix} = T \cdot A = T \cdot (B \cdot G) = (T \cdot B) \cdot G = L \cdot G \tag{2.4}$$

where A is the coefficient matrix, B is the basis matrix, T is the parametric matrix, G is the geometry matrix, and L is the BF matrix. The above expression shows us that $L = T \cdot B$ i.e. the *BF* matrix L is nothing but the product of the parametric matrix T and the basis matrix B.

Below we summarize all the relevant relations, which will be required for solving numerical problems:

$$\begin{aligned} G &= C \cdot A \\ B &= \text{inv}(C) \\ A &= B \cdot G \\ P &= L \cdot G \\ L &= T \cdot B \\ P &= T \cdot A \\ P &= T \cdot B \cdot G \end{aligned} \tag{2.5}$$

Example 2.1

A cubic curve has the following BFs, where a, b, and c are constants: $L_0 = -4at$, $L_1 = at^3$, $L_2 = bt^2$, $L_3 = -c$. ***Find its basis matrix.***

From Equation (2.4), $L = T{\cdot}B$

For a cubic curve, $L = \begin{bmatrix} L_0 & L_1 & L_2 & L_3 \end{bmatrix} = \begin{bmatrix} 1 & t & t^2 & t^3 \end{bmatrix} \cdot B$

$$\text{Let } B = \begin{bmatrix} x_{11} & x_{12} & x_{13} & x_{14} \\ x_{21} & x_{22} & x_{23} & x_{24} \\ x_{31} & x_{32} & x_{33} & x_{34} \\ x_{41} & x_{42} & x_{43} & x_{44} \end{bmatrix}$$

From inspection, $x_{11} = 0$, $x_{21} = -4a$, $x_{31} = 0$, $x_{41} = 0$

Similarly, the other terms are obtained in a likewise manner and compiled together to form the basis matrix:

$$B = \begin{bmatrix} 0 & 0 & 0 & -c \\ -4a & 0 & 0 & 0 \\ 0 & 0 & b & 0 \\ 0 & a & 0 & 0 \end{bmatrix}$$

Verification: The correctness of the basis matrix could be verified by computing the product $T{\cdot}B$, which would give us the given L matrix back.

MATLAB® Code 2.1

```
clear; clc;
syms a b c t;
T = [1, t, t^2, t^3];
L01 = 0;    L02 = -4*a*t;    L03 = 0;        L04 = 0;
L11 = 0;    L12 = 0;         L13 = 0;        L14 = a*t^3;
L21 = 0;    L22 = 0;         L23 = b*t^2;    L24 = 0;
L31 = -c;   L32 = 0;         L33 = 0;        L34 = 0;

% Let B = [x11 x12 x13 x14 ; x21 x22 x23 x24 ; x31 x32 x33 x34 ; x41 x42 x43 x44]

x11 = L01 / T(1);
x21 = L02 / T(2);
x31 = L03 / T(3);
x41 = L04 / T(4);

x12 = L11 / T(1);
x22 = L12 / T(2);
x32 = L13 / T(3);
x42 = L14 / T(4);

x13 = L21 / T(1);
x23 = L22 / T(2);
x33 = L23 / T(3);
x43 = L24 / T(4);

x14 = L31 / T(1);
x24 = L32 / T(2);
x34 = L33 / T(3);
x44 = L34 / T(4);

B = [x11 x12 x13 x14 ; x21 x22 x23 x24 ; x31 x32 x33 x34 ; x41 x42 x43 x44]

% verification
fprintf('Verification : L = T*B\n');
T*B
```

2.3 BLENDING FUNCTIONS OF INTERPOLATING SPLINES

Even though the concept of BFs has been derived for non-interpolating curves, we can readily apply it to interpolating curves too. In fact that is what we are going to do now since we are already familiar with interpolating curves and this will help us assimilate the concept of BFs more easily.

For linear interpolating splines, we have seen in Section 1.3 that

$$B = \begin{bmatrix} 1 & 0 \\ 1 & 1 \end{bmatrix}^{-1} = \begin{bmatrix} 1 & 0 \\ -1 & 1 \end{bmatrix}$$

This gives us:

$$L(t) = T \cdot B = \begin{bmatrix} 1 & t \end{bmatrix} \begin{bmatrix} 1 & 0 \\ -1 & 1 \end{bmatrix} = \begin{bmatrix} 1-t, & t \end{bmatrix} \quad (2.6)$$

Separating out the components, individual BFs are:

$$\begin{aligned} L_0 &= 1 - t \\ L_1 &= t \end{aligned} \quad (2.7)$$

By varying t over all possible values from 0 to 1, we can plot the above BFs as shown in Figure 2.2.

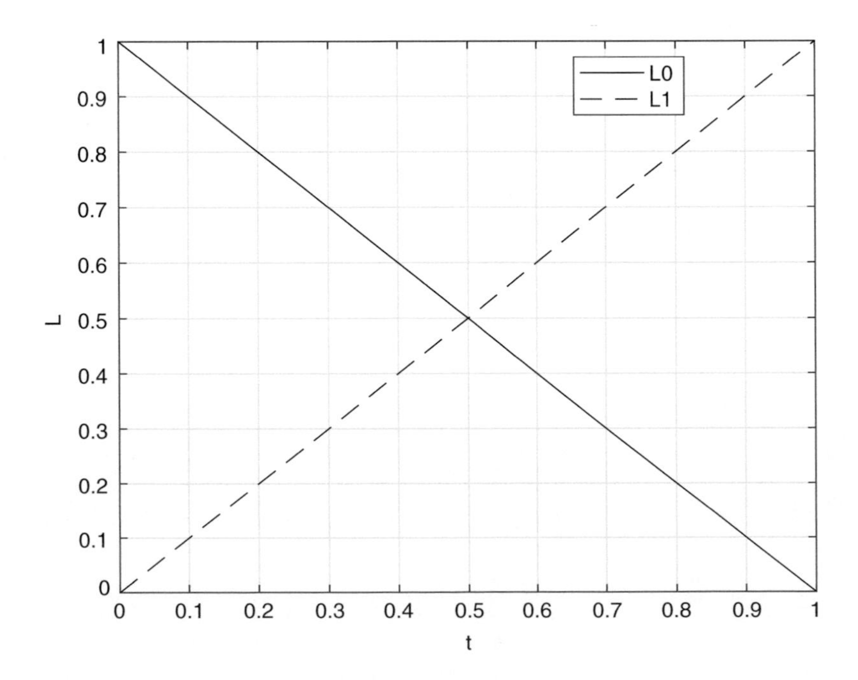

FIGURE 2.2 Blending functions of linear interpolating splines.

One observation which can immediately be made from the graph is that the constraint $\sum L_i = 1$ holds good i.e. for any value of t the sum of the values of L_0 and L_1 always equals 1. That is expected since $L_0 + L_1 == (1 - t) + t = 1$

The following important points can now be inferred from the above discussion:

- The BFs depict how the masses at the CPs change with varying values of t
- For each value of t, sum of the BF values is always equal to 1
- Although BFs are functions of t, at each CP t has a specific value and so all BFs become scalar constants

For quadratic interpolating splines, we have seen in Section 1.5 the following, where k is the sub-division ratio

$$B = \begin{bmatrix} 1 & 0 & 0 \\ 1 & k & k^2 \\ 1 & 1 & 1 \end{bmatrix}^{-1}$$

This gives us:

$$L(t) = T \cdot B = \begin{bmatrix} 1 & t & t^2 \end{bmatrix} \begin{bmatrix} 1 & 0 & 0 \\ 1 & k & k^2 \\ 1 & 1 & 1 \end{bmatrix}^{-1} \tag{2.8}$$

Example 2.2

Find the BFs of a quadratic spline having sub-division ratio k = 0.8.

From Equation (2.8)

$$L(t) = \begin{bmatrix} 1 & t & t^2 \end{bmatrix} \begin{bmatrix} 1 & 0 & 0 \\ 1 & 0.8 & 0.64 \\ 1 & 1 & 1 \end{bmatrix}^{-1} = \begin{bmatrix} 1 & t & t^2 \end{bmatrix} \begin{bmatrix} 1 & 0 & 0 \\ -2.25 & 6.25 & -4 \\ 1.25 & -6.25 & 5 \end{bmatrix}$$

Simplifying:

$$L(t) = \left[1 - 2.25t + 1.25t^2, 6.25t - 6.25t^2, -4t + 5t^2\right]$$

Separating out the component BFs (Figure 2.3)

$$L_0 = 1 - 2.25t + 1.25t^2$$

$$L_1 = 6.25t - 6.25t^2$$

$$L_2 = -4t + 5t^2$$

MATLAB Code 2.2

```
clear all; clc;
syms t; k = 0.8;
C = [1 0 0; 1 k k^2; 1 1 1];
B = inv(C);
L = [1 t t^2]*B;
fprintf('Blending functions are :\n');
disp(L(1)), disp(L(2)), disp(L(3));

%plotting

%Method-1
figure,
subplot(221), ezplot(L(1), [0,1]);
subplot(222), ezplot(L(2), [0,1]);
subplot(223), ezplot(L(3), [0,1]);

%Method-2
figure,
tt = 0:.01:1;
L0 = subs(L(1), t, tt);
L1 = subs(L(2), t, tt);
L2 = subs(L(3), t, tt);
plot(tt, L0, 'k-', tt, L1, 'k--', tt, L2, 'k-.');
grid;
xlabel('t');
ylabel('L');
legend('L0','L1','L2');
```

NOTE

Two methods for plotting the BFs are shown in the code above. In the first method shown in Figure 2.3a each BF is plotted individually. In the second method shown in Figure 2.3b, all the three BFs are plotted together. The second plot indicates how the BFs are related to each other. For example, it shows that only one BF is non-zero at each of the CPs i.e. at $t = 0$, at $t = 0.8$ and at $t = 1$.

`disp`: displays the symbolic expressions without additional line gaps.
`ezplot`: Plots symbolic variables directly without converting to matrix values
`figure`: Generates a new window to display figures.

For cubic interpolating splines, we have seen in Section 1.7 the following, where m and n are the sub-division ratios

$$B = \begin{bmatrix} 1 & 0 & 0 & 0 \\ 1 & m & m^2 & m^3 \\ 1 & n & n^2 & n^3 \\ 1 & 1 & 1 & 1 \end{bmatrix}^{-1}$$

This gives us:

$$L(t)=T\cdot B=\begin{bmatrix} 1 & t & t^2 \end{bmatrix}\begin{bmatrix} 1 & 0 & 0 & 0 \\ 1 & m & m^2 & m^3 \\ 1 & n & n^2 & n^3 \\ 1 & 1 & 1 & 1 \end{bmatrix}^{-1} \tag{2.9}$$

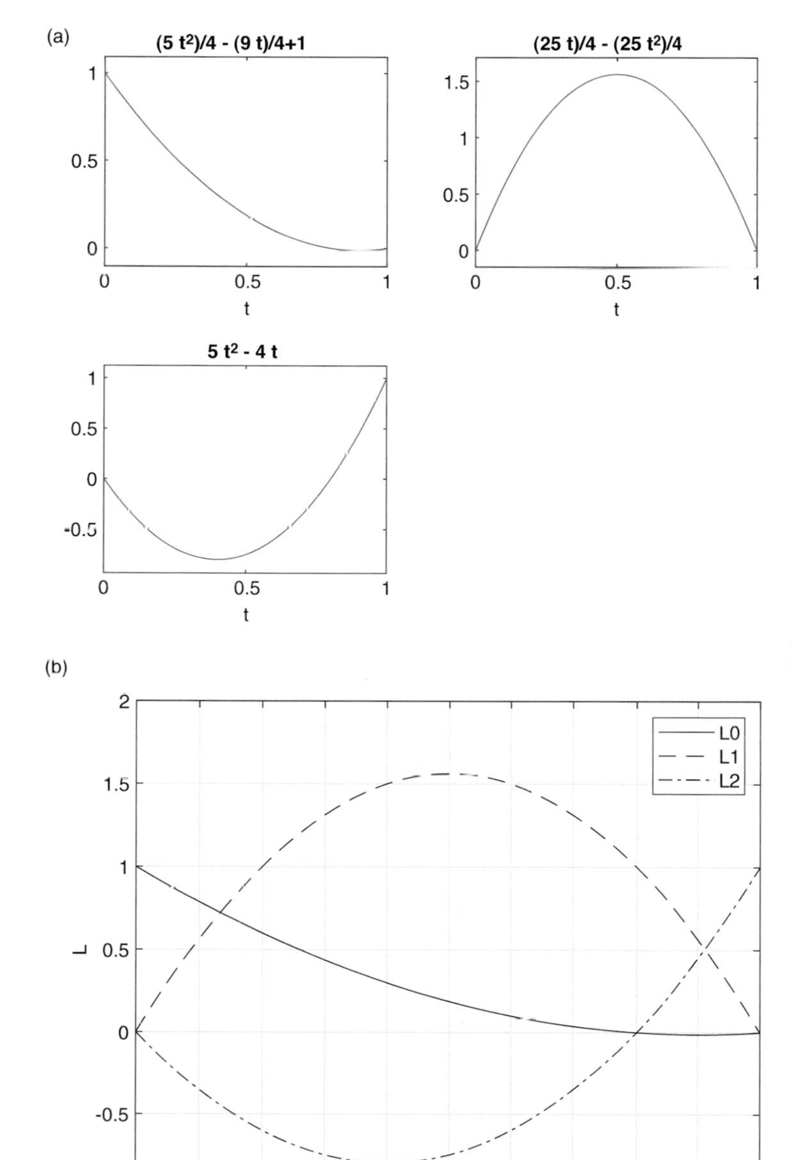

FIGURE 2.3 (a and b) Plots for Example 2.2.

Example 2.3

Find the BFs of a cubic spline having sub-division ratios m = 0.1 and n = 0.7. Hence, show that at each CP only one component of the BFs is non-zero.

From Equation (2.9)

$$L(t)=\begin{bmatrix} 1 & t & t^2 \end{bmatrix}\begin{bmatrix} 1 & 0 & 0 & 0 \\ 1 & 0.1 & 0.01 & 0.001 \\ 1 & 0.7 & 0.49 & 0.343 \\ 1 & 1 & 1 & 1 \end{bmatrix}^{-1}$$

$$=\begin{bmatrix} 1 & t & t^2 \end{bmatrix}\begin{bmatrix} 1 & 0 & 0 & 0 \\ -12.4286 & 12.963 & -0.7937 & 0.2593 \\ 25.7143 & -31.4815 & 8.7302 & -2.963 \\ -14.2857 & 18.5185 & -7.9365 & 3.7037 \end{bmatrix}$$

Simplifying and separating out the component BFs

$$L_0 = 1 - 12.4286t + 25.7143t^2 - 14.2857t^3$$

$$L_1 = 12.963t - 31.4815t^2 + 18.5185t^3$$

$$L_2 = -0.7937t + 8.7302t^2 - 7.9365t^3$$

$$L_3 = 0.2593t - 2.963t^2 + 3.7037t^3$$

At CP 1: Putting $t = 0$, $L_0 = 1$, $L_1 = L_2 = L_3 = 0$
At CP 2: Putting $t = 0.1$, $L_1 = 1$, $L_0 = L_2 = L_3 = 0$
At CP 3: Putting $t = 0.7$, $L_2 = 1$, $L_0 = L_1 = L_3 = 0$
At CP 4: Putting $t = 1$, $L_3 = 1$, $L_0 = L_1 = L_2 = 0$ (Figure 2.4)

MATLAB Code 2.3

```
clear all; clc;
syms t;
m = 0.1; n = 0.7;
C = [1 0 0 0; 1 m m^2 m^3; 1 n n^2 n^3; 1 1 1 1];
B = inv(C);
T = [1 t t^2 t^3];
L = T*B;
fprintf('Blending functions are :\n');
disp(L(1)), disp(L(2)), disp(L(3)), disp(L(4));

%plotting

%Method-1
```

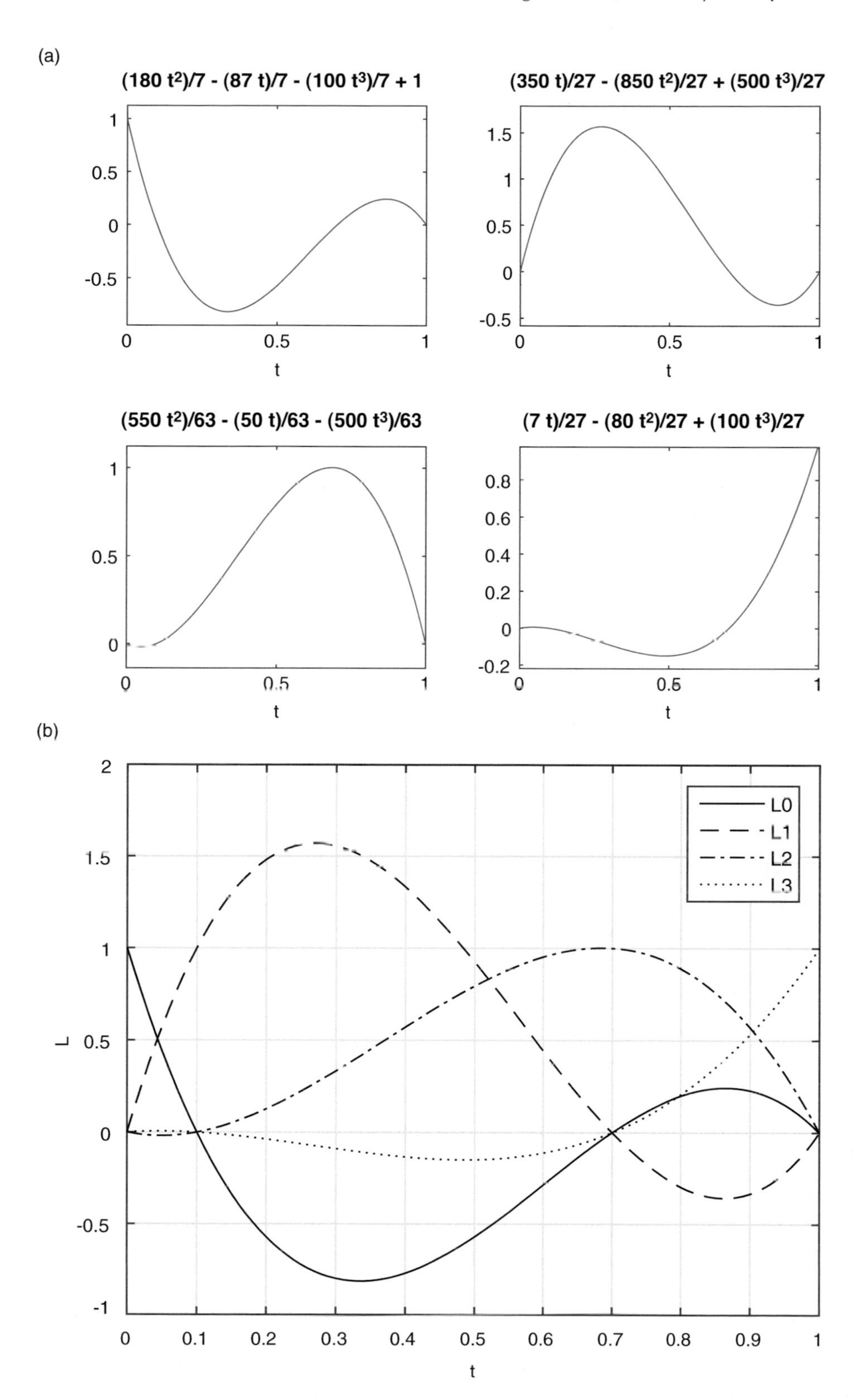

FIGURE 2.4 (a and b) Plots for Example 2.3.

```
figure,
subplot(221), ezplot(L(1), [0,1]);
subplot(222), ezplot(L(2), [0,1]);
subplot(223), ezplot(L(3), [0,1]);
subplot(224), ezplot(L(4), [0,1]);

%Method-2
figure,
tt = 0:.01:1;
L0 = subs(L(1), t, tt);
L1 = subs(L(2), t, tt);
L2 = subs(L(3), t, tt);
L3 = subs(L(4), t, tt);

plot(tt, L0, 'k-', tt, L1, 'k--', tt, L2, 'k-.', tt, L3, 'k:');
grid;
xlabel('t');
ylabel('L');
legend('L0', 'L1', 'L2', 'L3');

fprintf('\nL(0)  :'), disp(subs(L, t, 0));
fprintf('L(0.1)  :'), disp(subs(L, t, 0.1));
fprintf('L(0.7)  :'), disp(subs(L, t, 0.7));
fprintf('L(1)  :'), disp(subs(L, t, 1));
```

The above discussions pave the way for an important conclusion. In general, an approximating spline will not go through any of its CPs, so the location of any point on the curve with respect to the position of the CPs, is determined by the net gravitational pull of all masses together (i.e. the center of mass). However, for interpolating splines the curve actually goes through all of its CPs one after another. Analyzing this behavior from the viewpoint of BFs, leads us to conclude that at each point where the curve actually goes through a CP, the mass at that point must be so great as to nullify the effects of all other masses and force the center of mass to shift to that particular CP. We can therefore say that an interpolating spline can be considered as a special case of an approximating spline where each mass is large enough to nullify the effects of other masses and force the locus of the spline to go through a particular CP. This fact is also evident from the plots of the BFs above. At each CP only a specific BF has value equal to 1 while all the others are reduced to 0. This tells us where the CPs are located by noting where the BF graph has a single 1 and all other zeros. For example for the cubic curve above, the first CP is at $t = 0$, the second CP is at $t = 0.1$, the third CP is at $t = 0.7$ and the fourth CP is at $t = 1$.

2.4 HERMITE SPLINE

A hybrid spline differs from an interpolating spline in that it passes through only a subset of its CPs and hence requires some additional information for its unique characterization. The first hybrid spline that we are going to study is the Hermite spline, named after French

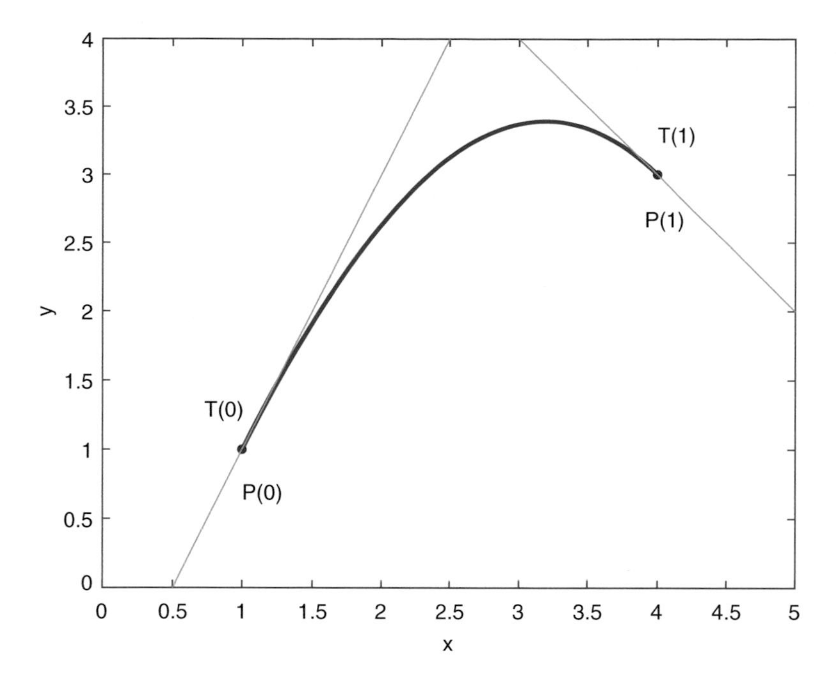

FIGURE 2.5 Hermite spline.

mathematician Charles Hermite (Hearn and Baker, 1996). For a cubic Hermite spline only the start point $P(0)$ i.e. $P(t)$ at $t = 0$, and the end point $P(1)$ i.e. $P(t)$ at $t = 1$, is known. To be uniquely specified two additional pieces of information are also given: the tangent slopes at the start point and end points i.e. $T(0) = P'(0)$ and $T(1) = P'(1)$ (see Figure 2.5).

To find the equation of a cubic Hermite spline, we start with the general equation of a cubic spline:

$$P(t) = a + bt + ct^2 + dt^3 = \begin{bmatrix} 1 & t & t^2 & t^3 \end{bmatrix} \begin{bmatrix} a \\ b \\ c \\ d \end{bmatrix}$$

The derivative of the starting equation is:

$$P'(t) = b + 2ct + 3dt^2 \tag{2.10}$$

Substituting the given conditions in both the starting equation and derivative equation:

$$\begin{aligned} P(0) &= a = P_0 \text{ (say)} \\ P(1) &= a + b + c + d = P_1 \text{ (say)} \\ P'(0) &= b \\ P'(1) &= b + 2c + 3d \end{aligned} \tag{2.11}$$

Rearranging and rewriting the above equations:

$$\begin{aligned} P_0 &= a \\ P_1 &= a+b+c+d \\ P'(0) &= b \\ P'(1) &= b+2c+3d \end{aligned} \tag{2.12}$$

Writing equations in matrix form $G = C{\cdot}A$:

$$\begin{bmatrix} P_0 \\ P_1 \\ P'(0) \\ P'(1) \end{bmatrix} = \begin{bmatrix} 1 & 0 & 0 & 0 \\ 1 & 1 & 1 & 1 \\ 0 & 1 & 0 & 0 \\ 0 & 1 & 2 & 3 \end{bmatrix} \begin{bmatrix} a \\ b \\ c \\ d \end{bmatrix} \tag{2.13}$$

Solving for A:

$$\begin{bmatrix} a \\ b \\ c \\ d \end{bmatrix} = \begin{bmatrix} 1 & 0 & 0 & 0 \\ 0 & 0 & 1 & 0 \\ -3 & 3 & -2 & -1 \\ 2 & -2 & 1 & 1 \end{bmatrix} \begin{bmatrix} P_0 \\ P_1 \\ P'(0) \\ P'(1) \end{bmatrix} \tag{2.14}$$

Substituting in starting equation:

$$P(t) = T \cdot B \cdot G = \begin{bmatrix} 1 & t & t^2 & t^3 \end{bmatrix} \begin{bmatrix} 1 & 0 & 0 & 0 \\ 0 & 0 & 1 & 0 \\ -3 & 3 & -2 & -1 \\ 2 & -2 & 1 & 1 \end{bmatrix} \begin{bmatrix} P_0 \\ P_1 \\ P'(0) \\ P'(1) \end{bmatrix} \tag{2.15}$$

The BFs of a cubic Hermite spline are given by:

$$L(t) = T \cdot B = \begin{bmatrix} 1 & t & t^2 & t^3 \end{bmatrix} \begin{bmatrix} 1 & 0 & 0 & 0 \\ 0 & 0 & 1 & 0 \\ -3 & 3 & -2 & -1 \\ 2 & -2 & 1 & 1 \end{bmatrix} \tag{2.16}$$

Separating out the matrix components:

$$\begin{aligned} L_0 &= 1 - 3t^2 + 2t^3 \\ L_1 &= 3t^2 - 2t^3 \\ L_2 &= t - 2t^2 + t^3 \\ L_3 &= -t^2 + t^3 \end{aligned} \tag{2.17}$$

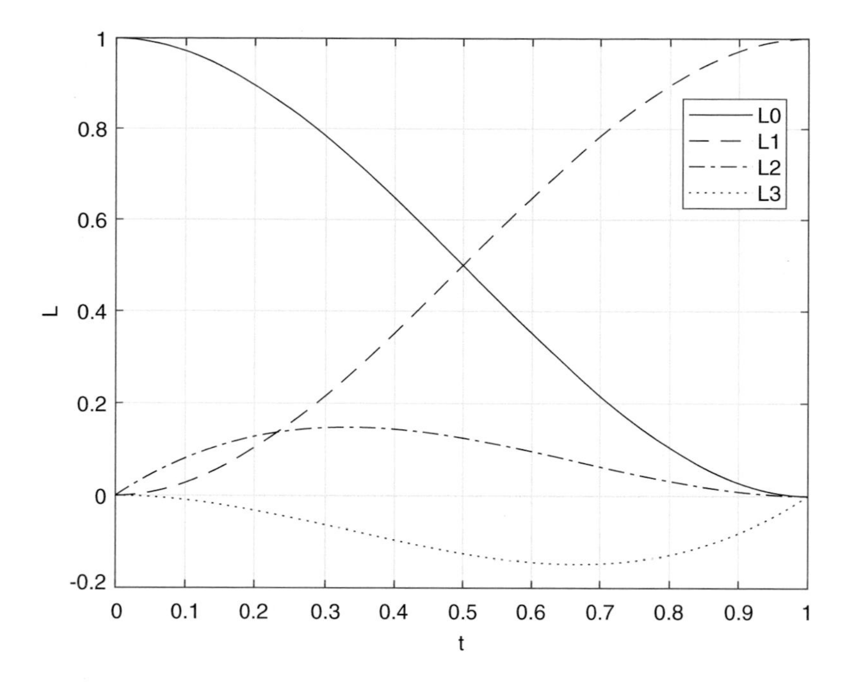

FIGURE 2.6 Blending functions of cubic Hermite spline.

Figure 2.6 shows the plot of the four BFs. One difference with the BFs of a cubic interpolating spline shown in Figure 2.4 can be observed. While for an interpolating spline, the BFs have four points for which only one is non-zero, Figure 2.6 shows that in this case there are only two such points (start and end points) for which this condition is true. This is a reflection of the fact that a hybrid spline passes through a subset of its CPs viz. only the first and last points and not through the intermediate points.

Example 2.4

Find the parametric equation of a cubic Hermite spline through points (1, 1) and (4, 3) and having parametric slopes (3, 6) and (1, –1) at these points.

Here, $P_0 = (1, 1)$, $P_1 = (4, 3)$, $P'(0) = (3, 6)$, $P'(1) = (1, -1)$.

From Equation (2.14)

$$\begin{bmatrix} a \\ b \\ c \\ d \end{bmatrix} = \begin{bmatrix} 1 & 0 & 0 & 0 \\ 0 & 0 & 1 & 0 \\ -3 & 3 & -2 & -1 \\ 2 & -2 & 1 & 1 \end{bmatrix} \begin{bmatrix} 1 & 1 \\ 4 & 3 \\ 3 & 6 \\ 1 & -1 \end{bmatrix} = \begin{bmatrix} 1 & 1 \\ 3 & 6 \\ 2 & -5 \\ -2 & 1 \end{bmatrix}$$

Required parametric equations:

$$x(t) = 1 + 3t + 2t^2 - 2t^3$$

$$y(t) = 1 + 6t - 5t^2 + t^3$$

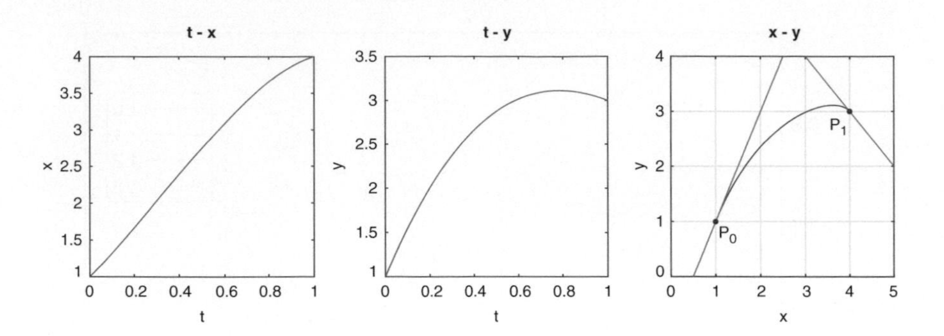

FIGURE 2.7 Plots for Example 2.4.

Verification (Figure 2.7):

$$x(0)=1, y(0)=1, x(1)=4, y(1)=3$$

$$x'(0)=3, y'(0)=6, x'(1)=1, y'(1)=-1$$

MATLAB Code 2.4

```
clear all; clc
syms t;

P0 = [1, 1];
P1 = [4, 3];
T0 = [3, 6];
T1 = [1, -1];
C = [1 0 0 0; 1 1 1 1; 0 1 0 0; 0 1 2 3];
B = inv(C);
X = [P0(1), P1(1)];
Y = [P0(2), P1(2)];
G = [P0; P1; T0; T1];
A = B*G;
T = [1 t t^2 t^3];
P = T*A;
fprintf('Required equations :\n');
x = P(1)
y = P(2)

%plotting
tt = 0:.01:1;
xx = subs(P(1), t, tt);
yy = subs(P(2), t, tt);
subplot(131), plot(tt, xx);
xlabel('t'); ylabel('x'); axis square; title('t - x');
subplot(132), plot(tt, yy);
xlabel('t'); ylabel('y'); axis square; title('t - y');
```

```
subplot(133), plot(xx, yy, 'b-'); hold on;
scatter(X, Y, 20, 'r', 'filled');
xlabel('x'); ylabel('y'); axis square;
axis([0 5 0 4]); grid;
ezplot('(y-1) = 2*(x-1)'); colormap winter;
ezplot('(y-3) = -1*(x-4)');
title('x - y');
text(P0(1), P0(2)-0.5, 'P_0');
text(P1(1), P1(2)-0.5, 'P_1');
hold off;
```

NOTE

colormap: specifies a color scheme using predefined color look-up tables

2.5 CARDINAL SPLINE

A Cardinal spline is similar to a Hermite spline in that it actually passes through two given points P_1 and P_2, and has known gradients at these points, but unlike a Hermite spline, these gradients are not explicitly given. Rather two additional points are given, a previous point P_0 and a subsequent point P_3, and the gradients at P_1 and P_2 are expressed as scalar multiples of the gradients of the line joining P_0 with P_2, and P_1 with P_3, respectively (Hearn and Baker, 1996) (see Figure 2.8).

The line vector connecting P_0 and P_2 is given by the differences of their position vectors i.e. $(P_2 - P_0)$. The gradient at P_1 is a scalar multiple of this line vector i.e. $P_1' = s(P_2 - P_0)$, where s is referred to as the "shape parameter." Likewise, the line vector connecting P_1 and P_3 is given by the differences of their position vectors i.e. $(P_3 - P_1)$. The gradient at P_2 is a scalar multiple

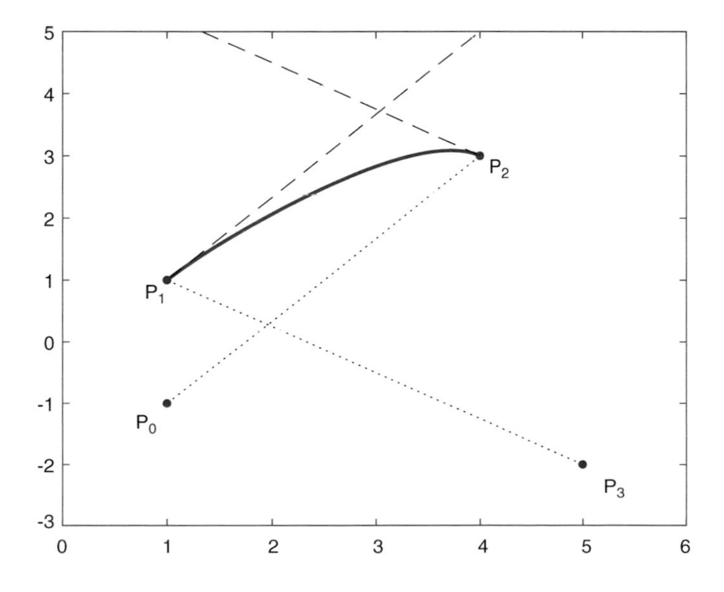

FIGURE 2.8 Cardinal spline.

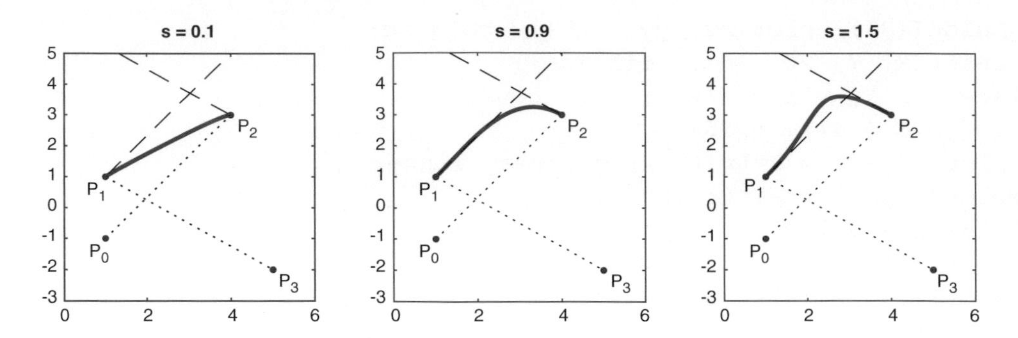

FIGURE 2.9 Effects of varying the shape parameter.

of this line vector i.e. $P_2' = s(P_3 - P_1)$. The value of s typically ranges from 0 to 1. Smaller values of s produce shorter and tighter curves, larger values of s produce longer and looser curves. If $s > 1$, the curve crosses over to the other size of the tangent line (see Figure 2.9).

To derive equation of a cubic Cardinal spline, we start with a general third-degree parametric equation

$$P(t) = a + bt + ct^2 + dt^3 = \begin{bmatrix} 1 & t & t^2 & t^3 \end{bmatrix} \begin{bmatrix} a \\ b \\ c \\ d \end{bmatrix}$$

Calculate derivative of the starting equation

$$P'(t) = b + 2ct + 3dt^2$$

Substituting the starting conditions in the above equations

$$\begin{aligned} P(0) &= a = P_1 \\ P(1) &= a + b + c + d = P_2 \\ P'(0) &= b = P'(1) = s(P_2 - P_0) \\ P'(1) &= b + 2c + 3d = P'(2) = s(P_3 - P_1) \end{aligned} \tag{2.18}$$

Substituting P_1 and P_2 and rearranging:

$$P_0 = a + \left(1 - \frac{1}{s}\right)b + c + d$$

$$P_3 = a + \left(\frac{1}{s}\right)b + \left(\frac{2}{s}\right)c + \left(\frac{3}{s}\right)d$$

Rewriting all four equations in matrix form $G = CA$

$$\begin{bmatrix} P_0 \\ P_1 \\ P_2 \\ P_3 \end{bmatrix} = \begin{bmatrix} 1 & \frac{s-1}{s} & 1 & 1 \\ 1 & 0 & 0 & 0 \\ 1 & 1 & 1 & 1 \\ 1 & \frac{1}{s} & \frac{2}{s} & \frac{3}{s} \end{bmatrix} \begin{bmatrix} a \\ b \\ c \\ d \end{bmatrix} \tag{2.19}$$

Solving for A:

$$\begin{bmatrix} a \\ b \\ c \\ d \end{bmatrix} = \begin{bmatrix} 0 & 1 & 0 & 0 \\ -s & 0 & s & 0 \\ 2s & s-3 & 3-2s & -s \\ -s & 2-s & s-2 & s \end{bmatrix} \begin{bmatrix} P_0 \\ P_1 \\ P_2 \\ P_3 \end{bmatrix} \tag{2.20}$$

Substituting in starting equation:

$$P(t) = T \cdot B \cdot G = \begin{bmatrix} 1 & t & t^2 & t^3 \end{bmatrix} \begin{bmatrix} 0 & 1 & 0 & 0 \\ -s & 0 & s & 0 \\ 2s & s-3 & 3-2s & -s \\ s & 2-s & s-2 & s \end{bmatrix} \begin{bmatrix} P_0 \\ P_1 \\ P_2 \\ P_3 \end{bmatrix} \tag{2.21}$$

The BFs of a cubic Cardinal spline are given by:

$$L(t) = T \cdot B = \begin{bmatrix} 1 & t & t^2 & t^3 \end{bmatrix} \begin{bmatrix} 0 & 1 & 0 & 0 \\ -s & 0 & s & 0 \\ 2s & s-3 & 3-2s & -s \\ -s & 2-s & s-2 & s \end{bmatrix} \tag{2.22}$$

Separating out the matrix components:

$$\begin{aligned} L_0 &= -s \cdot t + 2s \cdot t^2 - s \cdot t^3 \\ L_1 &= 1 + (s-3) \cdot t^2 + (2-s) \cdot t^3 \\ L_2 &= s \cdot t + (3-2s) \cdot t^2 + (s-2) \cdot t^3 \\ L_3 &= -s \cdot t^2 + s \cdot t^3 \end{aligned} \tag{2.23}$$

Example 2.5

Find the BFs of a cubic Cardinal spline with shape parameter $s = 1.5$ and generate a plot to visualize them.

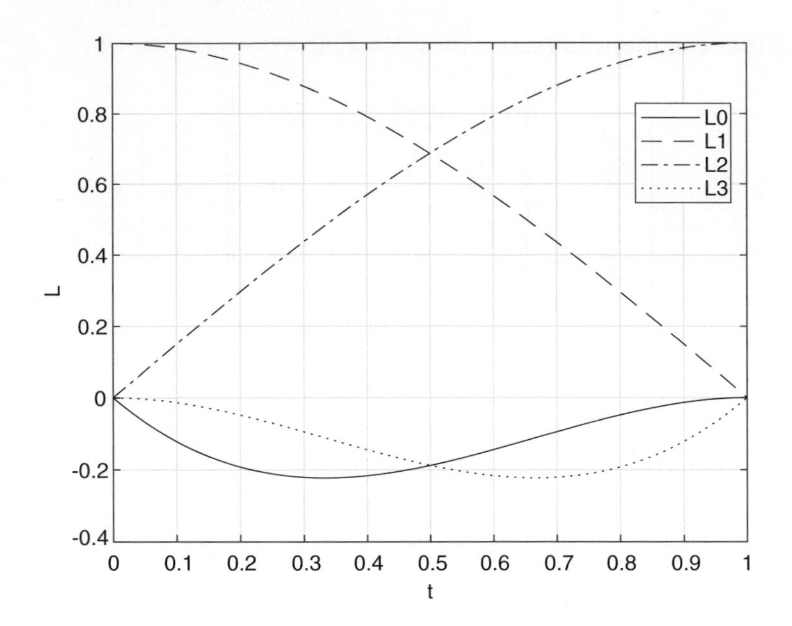

FIGURE 2.10 Plot for Example 2.5.

From Equation (2.23), substituting $s = 1.5$ (Figure 2.10)

$$L_0 = -\left(3t^3\right)/2 + 3t^2 - (3t)/2$$

$$L_1 = t^3/2 - \left(3t^2\right)/2 + 1$$

$$L_2 = -t^3/2 + (3t)/2$$

$$L_3 = \left(3t^3\right)/2 - \left(3t^2\right)/2$$

MATLAB Code 2.5

```
clear all; clc;
syms t;
s = 1.5;
B = [0 1 0 0 ; -s 0 s 0 ; 2*s s-3 3-2*s -s ; -s 2-s s-2 s];
T = [1 t t^2 t^3];
L = T*B;
fprintf('Blending functions are :\n');
disp(L(1)), disp(L(2)), disp(L(3)), disp(L(4));

%plotting
tt = 0:.01:1;
L0 = subs(L(1), t, tt);
L1 = subs(L(2), t, tt);
L2 = subs(L(3), t, tt);
L3 = subs(L(4), t, tt);
```

```
plot(tt, L0, 'k-', tt, L1, 'k--', tt,  L2, 'k-.', tt, L3, 'k:');
xlabel('t'); ylabel('L'); grid;
legend('L0', 'L1', 'L2', 'L3');
```

2.6 CATMULL–ROM SPLINE

A Cardinal spline with value of the shape parameter $s = 0.5$ is referred to as C–R spline named after the American scientist Edwin Catmull and Israeli scientist Raphael Rom. The BFs and basis matrix are computed by substituting $s = 0.5$ in Equation (2.22) to obtain the following:

$$B = \begin{bmatrix} 0 & 1 & 0 & 0 \\ -0.5 & 0 & 0.5 & 0 \\ 1 & -2.5 & 2 & -0.5 \\ -0.5 & 1.5 & -1.5 & 0.5 \end{bmatrix} \tag{2.24}$$

$$\begin{aligned} L_0 &= -0.5t + t^2 - 0.5t^3 \\ L_1 &= 1 - 2.5t^2 + 1.5t^3 \\ L_2 &= 0.5t + 2t^2 - 1.5t^3 \\ L_3 &= -0.5t^2 + 0.5t^3 \end{aligned} \tag{2.25}$$

A plot of the BFs is shown in Figure 2.11.

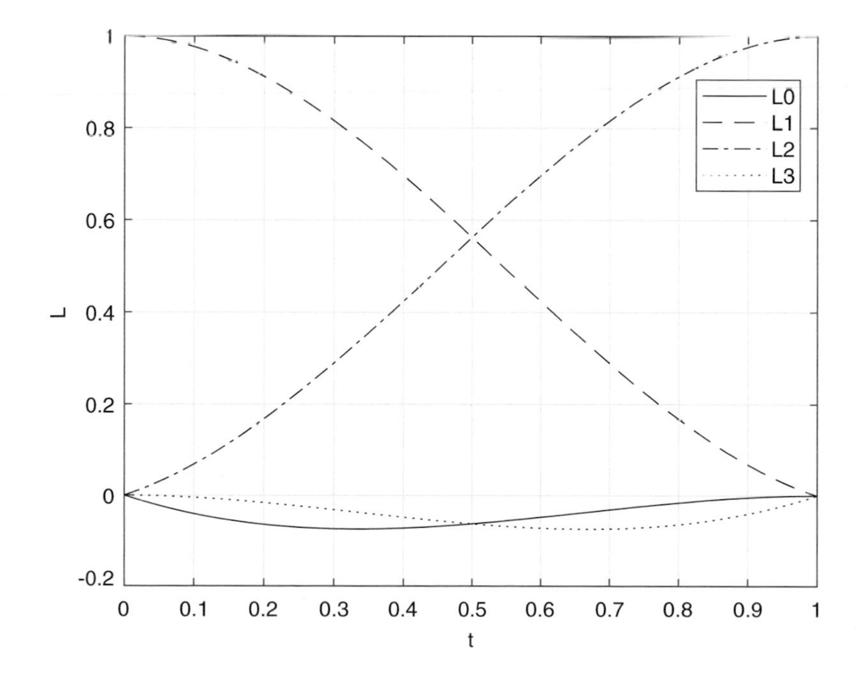

FIGURE 2.11 Blending functions of a C–R spline.

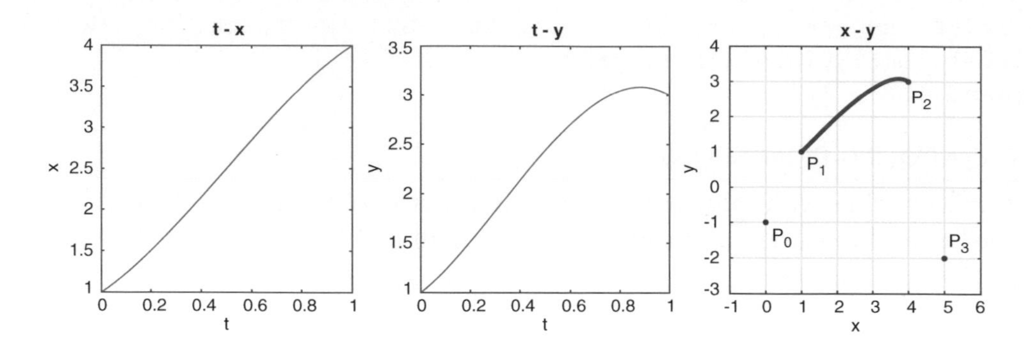

FIGURE 2.12 Plots for Example 2.6.

Example 2.6

Find the parametric equation of a cubic C–R spline associated with the CPs (0, –1), (1, 1), (4, 3), and (5, –2).

From Equations (2.21) and (2.24), substituting the given values

$$P(t)=\begin{bmatrix} 1 & t & t^2 & t^3 \end{bmatrix}\begin{bmatrix} 0 & 1 & 0 & 0 \\ -0.5 & 0 & 0.5 & 0 \\ 1 & -2.5 & 2 & -0.5 \\ -0.5 & 1.5 & -1.5 & 0.5 \end{bmatrix}\begin{bmatrix} 0 & -1 \\ 1 & 1 \\ 4 & 3 \\ 5 & -2 \end{bmatrix}$$

Simplifying and separating out the components

$$x(t)=-2t^3+3t^2+2t+1$$

$$y(t)=-3.5t^3+3.5t^2+2t+1$$

Verification: $x(0)=1, y(0)=1, x(1)=4, y(1)=3$ (Figure 2.12).

MATLAB Code 2.6

```
clear all; clc;
syms t;
P0 = [0, -1]; P1 = [1, 1]; P2 = [4, 3]; P3 = [5, -2]; s = 0.5;
X = [P0(1) P1(1) P2(1) P3(1)]; Y = [P0(2) P1(2) P2(2) P3(2)];
T = [1 t t^2 t^3];
B = [0 1 0 0 ; -s 0 s 0 ; 2*s s-3 3-2*s -s ; -s 2-s s-2 s];
G = [P0; P1; P2; P3];
P = T*B*G;
x = P(1); y = P(2);
fprintf('Required equations are : \n');
x
y
```

```
%plotting
tt = linspace(0, 1);
xx = subs(x, t, tt);
yy = subs(y, t, tt);
subplot(131), plot(tt,xx); title('t - x');
xlabel('t'); ylabel('x'); axis square;
subplot(132), plot(tt,yy);  title('t - y');
xlabel('t'); ylabel('y'); axis square;
subplot(133), plot(xx, yy, 'b-', 'LineWidth', 1.5);
xlabel('x'); ylabel('y'); axis square; title('x - y');
grid; hold on;
scatter(X, Y, 20, 'r', 'filled');
axis([-1 6 -3 4]);
text(P0(1), P0(2)-1, 'P_0');
text(P1(1), P1(2)-1, 'P_1');
text(P2(1), P2(2)-1, 'P_2');
text(P3(1), P3(2)+1, 'P_3');
hold off;
```

2.7 BEZIER SPLINE

Bezier spline, named after French engineer Pierre Bezier, is a popular type of hybrid splines, which satisfies the following characteristics: (a) it passes through the first and last CPs, (b) line joining the first CP and the second CP is tangential to the curve, (c) line joining the last CP and the previous CP is tangential to the curve, and (d) the curve is contained entirely within the convex hull formed by joining all the CPs in sequence (see Figure 2.13).

To satisfy the above conditions the BFs of a Bezier spline is generated by Bernstein polynomials (Hearn and Baker, 1996), named after the Russian mathematician Sergei Bernstein. Bernstein polynomials of degree n is shown below, where ${}^nC_k$ denotes the combination of k items out of n and $0 \le t \le 1$.

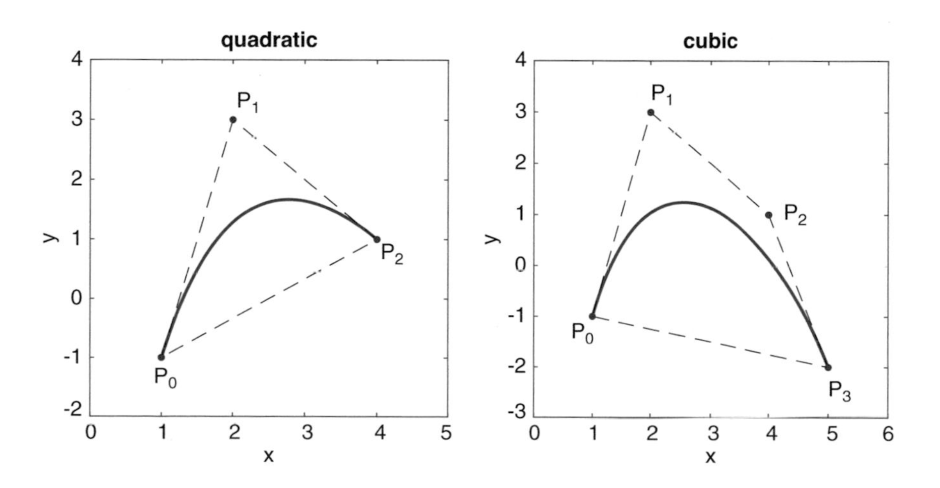

FIGURE 2.13 Bezier splines.

$$L_k = {}^nC_k \cdot (1-t)^{n-k} \cdot t^k$$
$${}^nC_k = \frac{n!}{k!(n-k)!} \tag{2.26}$$

Equation of the curve is $P = L.G$. Here, P_k denote the CPs.

$$P(t) = \sum_{k=0}^{n} L_k P_k = \sum_{k=0}^{n} {}^nC_k \cdot (1-t)^{n-k} \cdot t^k \cdot P_k \tag{2.27}$$

A quadratic Bezier curve is of degree 2 and associated with three CPs P_0, P_1, and P_2. Equation of the curve is derived by putting $n = 2$ in the Bernstein polynomial

$$P(t) = \sum_{k=0}^{2} L_k P_k = \sum_{k=0}^{2} {}^2C_k \cdot (1-t)^{2-k} \cdot t^k \cdot P_k \tag{2.28}$$

Expanding:

$$P(t) = (1-t)^2 \cdot P_0 + 2(1-t) \cdot t \cdot P_1 + t^2 \cdot P_2 \tag{2.29}$$

Separating out the component BFs:

$$L_0 = (1-t)^2 = 1 - 2t + t^2$$
$$L_1 = 2(1-t) \cdot t = 2t - 2t^2 \tag{2.30}$$
$$L_2 = t^2$$

A plot of the blending functions is shown below in Figure 2.14.

Since $L = T.B$ the basis matrix can also be computed trivially:

$$B = \begin{bmatrix} 1 & 0 & 0 \\ -2 & 2 & 0 \\ 1 & -2 & 1 \end{bmatrix} \tag{2.31}$$

Equation of the curve can be expressed in $P = T.B.G$ form

$$P(t) = \begin{bmatrix} 1 & t & t^2 \end{bmatrix} \begin{bmatrix} 1 & 0 & 0 \\ -2 & 2 & 0 \\ 1 & -2 & 1 \end{bmatrix} \begin{bmatrix} P_0 \\ P_1 \\ P_2 \end{bmatrix} \tag{2.32}$$

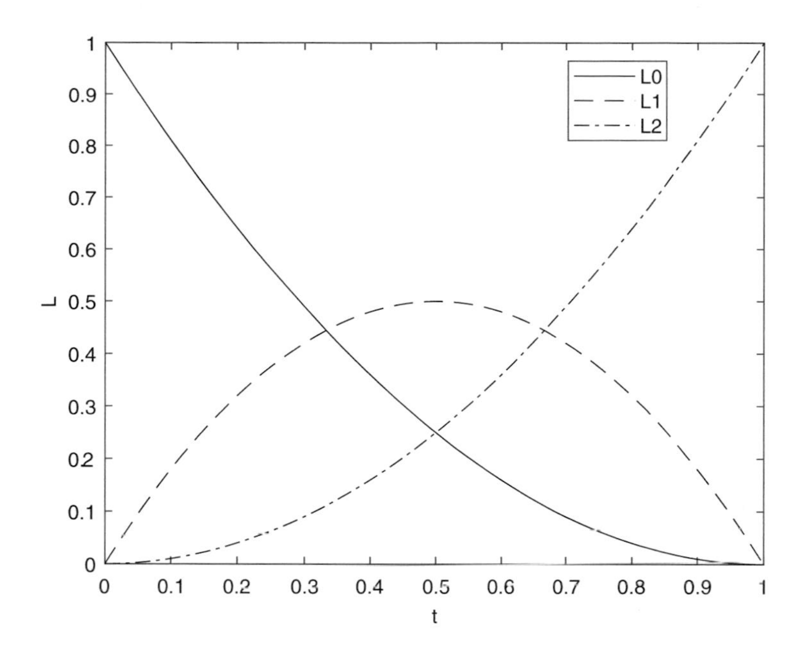

FIGURE 2.14 BFs of a quadratic Bezier spline.

Example 2.7

Find parametric equations of a quadratic Bezier spline with three CPs $P_0(1, -1)$, $P_1(4, 3)$, and $P_2(5, -2)$.

From Equation (2.32), substituting given conditions

$$P(t) = \begin{bmatrix} 1 & t & t^2 \end{bmatrix} \begin{bmatrix} 1 & 0 & 0 \\ -2 & 2 & 0 \\ 1 & -2 & 1 \end{bmatrix} \begin{bmatrix} 1 & -1 \\ 4 & 3 \\ 5 & -2 \end{bmatrix}$$

Separating out the components:

$$x(t) = 1 + 6t - 2t^2$$

$$y(t) = -1 + 8t - 9t^2$$

Verification: $x(0) = 1, x(1) = 5, y(0) = -1, y(1) = -2$ (Figure 2.15).

MATLAB Code 2.7

```
clear all; clc;
syms t;
P0 = [1, -1]; P1 = [4, 3]; P2 = [5, -2];
X = [P0(1) P1(1) P2(1) ];
Y = [P0(2) P1(2) P2(2) ];
```

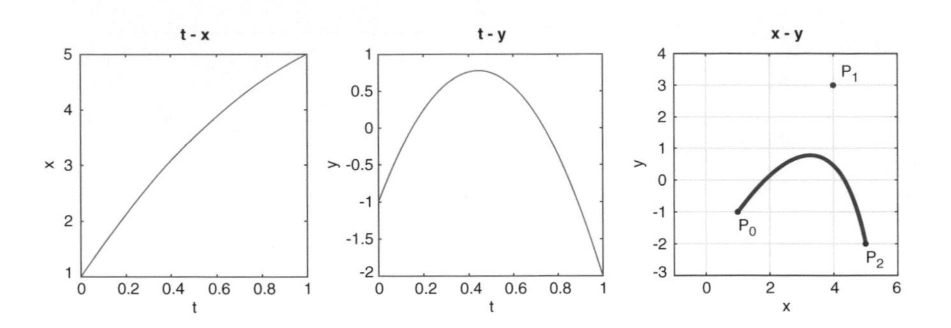

FIGURE 2.15 Plots for Example 2.7.

```
T = [1 t t^2 ];
B = [1 0 0 ; -2 2 0 ; 1 -2 1];
G = [P0; P1; P2];
P = T*B*G;
fprintf('Required equations :\n');
x = P(1)
y = P(2)

%plotting
tt = linspace(0,1);
xx = subs(x, t, tt);
yy = subs(y, t, tt);
subplot(131), plot(tt, xx); title('t - x');
xlabel('t'); ylabel('x'); axis square;
subplot(132), plot(tt, yy);  title('t - y');
xlabel('t'); ylabel('y'); axis square;
subplot(133), plot(xx, yy, 'b-', 'LineWidth', 1.5);
title('x - y');
hold on;
scatter(X, Y, 20, 'r', 'filled');
xlabel('x'); ylabel('y'); axis square;
grid on; d = 0.5;
text(P0(1), P0(2)-d, 'P_0');
text(P1(1)+d, P1(2), 'P_1');
text(P2(1), P2(2)-d, 'P_2');
axis([-1 6 -3 4]);
hold off;
```

A cubic Bezier curve is of degree 3 and associated with four CPs P_0, P_1, P_2, and P_3. Equation of the curve is derived by putting $n = 3$ in the Bernstein polynomial (Foley et al., 1995; Shirley, 2002)

$$P(t)=\sum_{k=0}^{3} L_k P_k = \sum_{k=0}^{3} {}^3C_k \cdot (1-t)^{3-k} \cdot t^k \cdot P_k \tag{2.33}$$

Expanding:

$$P(t) = (1-t)^3 \cdot P_0 + 3\cdot(1-t)^2 \cdot t \cdot P_1 + 3\cdot(1-t)\cdot t^2 \cdot P_2 + t^3 \cdot P_3 \tag{2.34}$$

Separating out the components BFs:

$$\begin{aligned} L_0 &= (1-t)^3 = 1 - 3t + 3t^2 - t^3 \\ L_1 &= 3(1-t)^2 \cdot t = 3t - 6t^2 + 3t^3 \\ L_2 &= 3(1-t)\cdot t^2 = 3t^2 - 3t^3 \\ L_3 &= t^3 \end{aligned} \tag{2.35}$$

A plot of the BFs is shown below in Figure 2.16.

Since $L = T{\cdot}B$, the basis matrix can also be computed trivially:

$$B = \begin{bmatrix} 1 & 0 & 0 & 0 \\ -3 & 3 & 0 & 0 \\ 3 & -6 & 3 & 0 \\ -1 & 3 & -3 & 1 \end{bmatrix} \tag{2.36}$$

Equation of the curve can be expressed in $P = T{\cdot}B{\cdot}G$ form

$$P(t) = \begin{bmatrix} 1 & t & t^2 & t^3 \end{bmatrix} \begin{bmatrix} 1 & 0 & 0 & 0 \\ -3 & 3 & 0 & 0 \\ 3 & -6 & 3 & 0 \\ -1 & 3 & -3 & 1 \end{bmatrix} \begin{bmatrix} P_0 \\ P_1 \\ P_2 \\ P_3 \end{bmatrix} \tag{2.37}$$

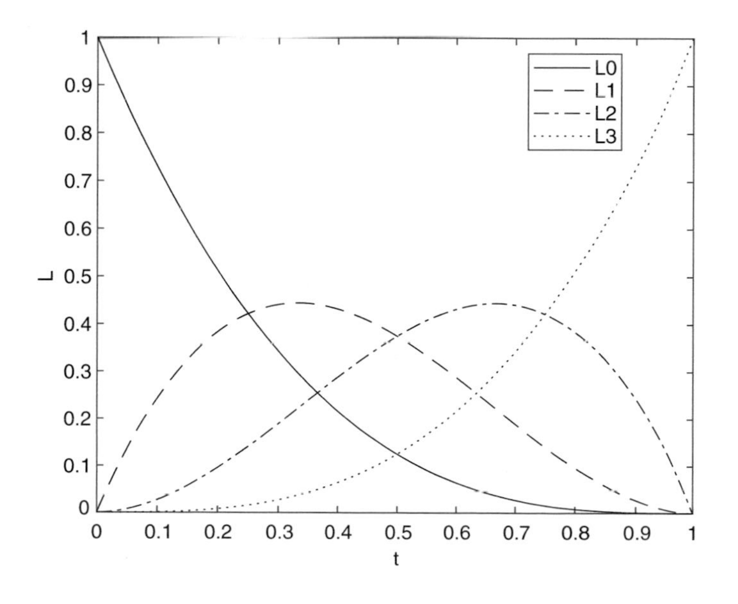

FIGURE 2.16 BFs of cubic Bezier spline.

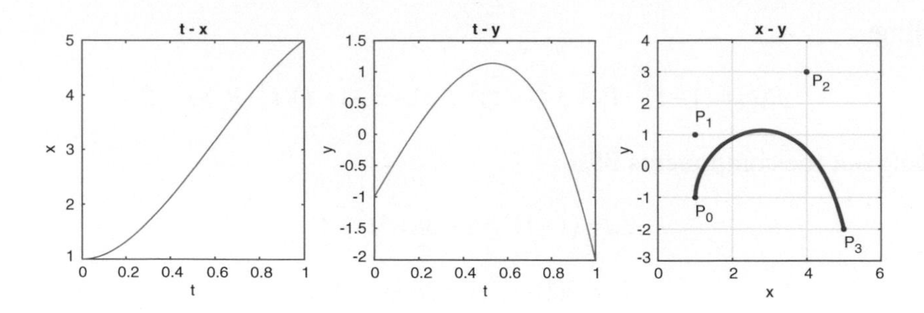

FIGURE 2.17 Plots for Example 2.8.

Example 2.8

Find parametric equations of a cubic Bezier spline associated with four CPs. $P_0(1, -1)$, $P_1(1, 1)$, $P_2(4, 3)$, $P_3(5, -2)$.

From Equation (2.37) substituting given conditions

$$(t)=\begin{bmatrix} 1 & t & t^2 & t^3 \end{bmatrix}\begin{bmatrix} 1 & 0 & 0 & 0 \\ -3 & 3 & 0 & 0 \\ 3 & -6 & 3 & 0 \\ -1 & 3 & -3 & 1 \end{bmatrix}\begin{bmatrix} 1 & -1 \\ 1 & 1 \\ 4 & 3 \\ 5 & -2 \end{bmatrix}$$

Separating out the components:

$$x(t)=1+9t^2-5t^3$$

$$y(t)=-1+6t-7t^3$$

Verification: $x(0)=1, x(1)=5, y(0)=-1, y(1)=-2$ (Figure 2.17).

MATLAB Code 2.8

```
clear all; clc;
syms t;
P0 = [1, -1]; P1 = [1, 1]; P2 = [4, 3]; P3 = [5, -2];
X = [P0(1) P1(1) P2(1) P3(1)];
Y = [P0(2) P1(2) P2(2) P3(2)];
T = [1 t t^2 t^3];
B = [1 0 0 0; -3 3 0 0; 3 -6 3 0; -1 3 -3 1];
G = [P0; P1; P2; P3];
P = T*B*G;
fprintf('Required equations :\n');
x = P(1)
y = P(2)

%plotting
tt = linspace(0,1);
```

```
xx = subs(x, t, tt);
yy = subs(y, t, tt);

subplot(131), plot(tt, xx); title('t - x');
xlabel('t'); ylabel('x'); axis square;
subplot(132), plot(tt, yy);  title('t - y');
xlabel('t'); ylabel('y'); axis square;
subplot(133), plot(xx, yy, 'b-', 'LineWidth', 1.5);
xlabel('x'); ylabel('y'); axis square; hold on; grid;
scatter(X, Y, 20, 'r', 'filled');  title('x - y');
axis([0 6 -3 4]);
d = 0.5;
text(P0(1), P0(2)-d, 'P_0');
text(P1(1), P1(2)+d, 'P_1');
text(P2(1), P2(2)-d, 'P_2');
text(P3(1), P3(2)-d, 'P_3');
hold off;
```

2.8 SPLINE CONVERSIONS

Before ending this chapter, we take a look at conversions of one spline form to another. A spline of one type can be expressed as a spline of another type. Since the spline physically remains the same, its equation is unaltered. However, the basis matrix and CPs change (Hearn and Baker, 1996).

Suppose a spline of type 1 specified by its basis matrix B_1 and CPs in geometry matrix G_1 needs to be converted to a spline of type 2 with basis matrix B_2 and geometry matrix G_1. Then equation of the first spline is $P_1(t) = T{\cdot}B_1{\cdot}G_1$ and equation of the second spline is $P_2(t) = T{\cdot}B_2{\cdot}G_2$. Note, however, that the actual spline physically remains the same and it is only represented in two different manners. The two equations are therefore equivalent, and can be equated:

$$P(t) = T \cdot B_1 \cdot G_1 = T \cdot B_2 \cdot G_2 \tag{2.38}$$

Basis matrices of both splines are known as they are characteristic of the spline type. CPs of the first spline should also be known. CPs of the second spline are found out by solving the following:

$$G_2 = B_2^{-1} \cdot B_1 \cdot G_1 \tag{2.39}$$

Even if for the first curve the CPs are not specifically known but the curve is expressed in the parametric form $\{x(t), y(t)\} = \left\{\left(a_0 + a_1 \cdot t + a_2 \cdot t^2 + a_3 \cdot t^3\right), \left(b_0 + b_1 \cdot t + b_2 \cdot t^2 + b_3 \cdot t^3\right)\right\}$, then we can represent it as:

$$\{x(t), y(t)\} = \begin{bmatrix} 1 & t & t^2 & t^3 \end{bmatrix} \begin{bmatrix} a_0 & b_0 \\ a_1 & b_1 \\ a_2 & b_2 \\ a_3 & b_3 \end{bmatrix} = T \cdot (B \cdot G) \tag{2.40}$$

The second matrix gives the product of B and G and can be directly plugged into Equation (2.39) to calculate G_2.

Example 2.9

Consider a C–R spline through $P_0(1, -1)$, $P_1(1, 4)$, $P_2(4, 4)$, and $P_3(5, -2)$. Convert it to a Bezier spline and find its new CPs. Also compare equations of both the curves.

In this case: $G_{CR} = \begin{bmatrix} 1 & -1 \\ 1 & 4 \\ 4 & 4 \\ 5 & -2 \end{bmatrix}$, $B_{CR} = \begin{bmatrix} 0 & 1 & 0 & 0 \\ -0.5 & 0 & 0.5 & 0 \\ 1 & -2.5 & 2 & -0.5 \\ -0.5 & 1.5 & -1.5 & 0.5 \end{bmatrix}$,

$B_{BZ} = \begin{bmatrix} 1 & 0 & 0 & 0 \\ -3 & 3 & 0 & 0 \\ 3 & -6 & 3 & 0 \\ -1 & 3 & -3 & 1 \end{bmatrix}$

From Equation (2.39)

$$G_{BZ} = B_{BZ}^{-1} \cdot B_{CR} \cdot G_{CR} = \begin{bmatrix} 1 & 4 \\ 1.5 & 4.83 \\ 3.33 & 5 \\ 4 & 4 \end{bmatrix}$$

Thus CPs of the Bezier curve are $Q_0(1,4)$, $Q_1(1.5,4.83)$, $Q_2(3.33,5)$, and $Q_3(4,4)$.

To verify that both of these represent the same physical spline.

Equation of CR spline:

$$P_{CR} = T \cdot B_{CR} \cdot G_{CR} = \begin{bmatrix} 1 & t & t^2 & t^3 \end{bmatrix} \begin{bmatrix} 0 & 1 & 0 & 0 \\ -0.5 & 0 & 0.5 & 0 \\ 1 & -2.5 & 2 & -0.5 \\ -0.5 & 1.5 & -1.5 & 0.5 \end{bmatrix} \begin{bmatrix} 1 & -1 \\ 1 & 4 \\ 4 & 4 \\ 5 & -2 \end{bmatrix}$$

$$= \left[\left(1 + 1.5t + 4t^2 - 2.5t^3\right), \left(4 + 2.5t - 2t^2 - 0.5t^3\right) \right]$$

Equation of Bezier spline:

$$P_{BZ} = T \cdot B_{BZ} \cdot G_{BZ} = \begin{bmatrix} 1 & t & t^2 & t^3 \end{bmatrix} \begin{bmatrix} 1 & 0 & 0 & 0 \\ -3 & 3 & 0 & 0 \\ 3 & -6 & 3 & 0 \\ -1 & 3 & -3 & 1 \end{bmatrix} \begin{bmatrix} 1 & 4 \\ 1.5 & 4.83 \\ 3.33 & 5 \\ 4 & 4 \end{bmatrix}$$

$$= \left[\left(1 + 1.5t + 4t^2 - 2.5t^3\right), \left(4 + 2.5t - 2t^2 - 0.5t^3\right) \right]$$

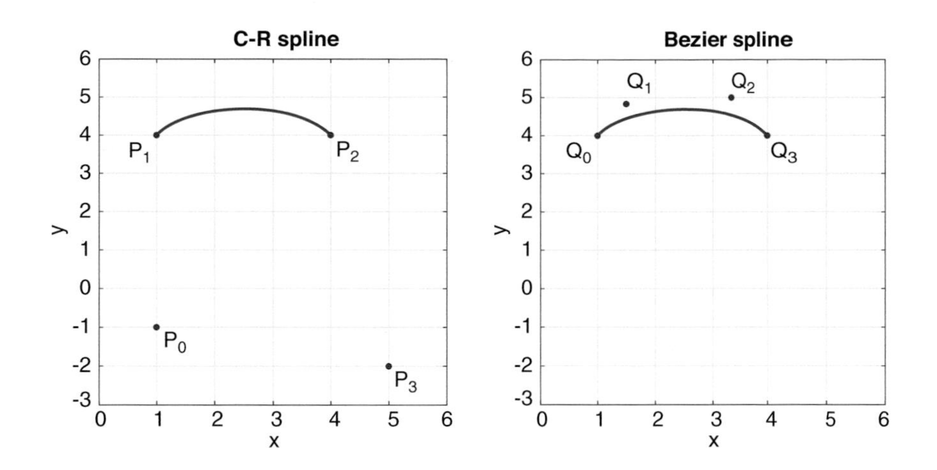

FIGURE 2.18 Plots for Example 2.9.

As can be seen both of these equations point to the same physical spline, only the basis matrix and CPs are different in each case (Figure 2.18).

MATLAB Code 2.9

```
clear all; clc;

B_CR = [0, 1, 0, 0 ; -0.5, 0, 0.5, 0 ; 1, -2.5, 2, -0.5 ; -0.5, 1.5, -1.5, 0.5];
B_BZ = [1, 0, 0, 0 ; -3, 3, 0, 0 ; 3, -6, 3, 0 ; -1, 3, -3, 1];
G_CR = [1, -1 ; 1, 4 ; 4, 4 ; 5, -2];

G_BZ =  inv(B_BZ) * B_CR * G_CR

%verification
syms t;
T = [1, t, t^2, t^3];
PCR = T * B_CR * G_CR
PBZ = T * B_BZ * G_BZ

%plotting
xc0 = G_CR(1,1); yc0 = G_CR(1,2);
xc1 = G_CR(2,1); yc1 = G_CR(2,2);
xc2 = G_CR(3,1); yc2 = G_CR(3,2);
xc3 = G_CR(4,1); yc3 = G_CR(4,2);
Xc = [xc0, xc1, xc2, xc3];
Yc = [yc0, yc1, yc2, yc3];
xb0 = G_BZ(1,1); yb0 = G_BZ(1,2);
xb1 = G_BZ(2,1); yb1 = G_BZ(2,2);
xb2 = G_BZ(3,1); yb2 = G_BZ(3,2);
xb3 = G_BZ(4,1); yb3 = G_BZ(4,2);
Xb = [xb0, xb1, xb2, xb3];
Yb = [yb0, yb1, yb2, yb3];

tt = linspace(1,0);
xcr = subs(PCR(1), t, tt);
```

```
ycr = subs(PCR(2), t, tt);
xbz = subs(PBZ(1), t, tt);
ybz = subs(PBZ(2), t, tt);

subplot(121), plot(xcr, ycr, 'b-', 'LineWidth', 1.5); hold on;
scatter(Xc, Yc, 20, 'r', 'filled'); grid;
axis([0, 6, -3, 6]); title('C-R spline'); axis square;
xlabel('x'); ylabel('y');
d = 0.5;
text(Xc(1), Yc(1)-d, 'P_0');
text(Xc(2), Yc(2)-d, 'P_1');
text(Xc(3), Yc(3)-d, 'P_2');
text(Xc(4), Yc(4)-d, 'P_3');
hold off;

subplot(122), plot(xbz, ybz, 'b-', 'LineWidth', 1.5); hold on;
scatter(Xb, Yb, 20, 'r', 'filled'); grid;
axis([0, 6, -3, 6]); title('Bezier spline'); axis square;
xlabel('x'); ylabel('y');
d = 0.5;
text(Xb(1), Yb(1)-d, 'Q_0');
text(Xb(2), Yb(2)+d, 'Q_1');
text(Xb(3), Yb(3)+d, 'Q_2');
text(Xb(4), Yb(4)-d, 'Q_3');
hold off;
```

NOTE

`axis`: Specifies range of values to display in the order `[xmin, xmax, ymin, ymax]`

2.9 CHAPTER SUMMARY

The following points summarize the topics discussed in this chapter:

- BFs are functions of the parametric variable t used to derive equations of non-interpolating splines.
- BFs are derived by assuming variable masses kept on the CPs, which exert gravitational influence on the spline.
- The actual spline is the locus of the center of mass of these variable masses.
- Equation of a spline can be represented as product of BFs and CP locations.
- A Hermite spline is specified by two CPs and slopes of the curve at these points.
- A Cardinal spline is similar to Hermite spline but slopes are not directly specified.
- Slopes of Cardinal splines are determined by lines joining these points with two additional points.
- A Cardinal spline is also associated with a shape parameter "s," which determine the shape of the curve.
- A C–R spline is a special case of a Cardinal spline with $s = 0.5$.
- A Bezier curve passes through the first and last CPs and is entirely contained within its convex hull.
- BFs of Bezier splines can be derived from Bernstein polynomials of the same degree.
- A spline of one type can be converted to a spline of another type having a different set of CPs.

2.10 REVIEW QUESTIONS

1. What are meant by BFs?
2. How are BFs related to basis matrix?
3. For an interpolating spline, under what conditions are BFs 0 and 1?
4. Justify that interpolating splines are a special case of non-interpolating splines.
5. What are the boundary conditions for specifying Hermite splines?
6. How are Cardinal splines different from a Hermite spline?
7. What does the shape parameter of Cardinal splines determine?
8. Why is a C–R spline a special case of Cardinal spline?
9. Why are BFs of Bezier splines derived from Bernstein polynomials?
10. How is a spline of one type converted to a spline of another type?

2.11 PRACTICE PROBLEMS

1. For a cubic Hermite spline $P(t)=1+bt-2at^2-bt^3$ having equal slopes at start and end points, find a relation between constants a and b.

2. Find equations of a Cardinal spline through (1, –1), (1, 1), (4, 3), and (5, –2) with shape parameter (1) $s=0.1$ and (2) $s=0.9$.

3. A cubic curve has the following BFs, where a, b, and c are the constants: $L_0=1-at+bt^2-2ct^3$, $L_1=2bt-ct^2+at^3$, $L_2=a-2t+3t^2$, $L_3=-3-at^2-ct^3$. Find its basis matrix.

4. A cubic interpolating spline with CPs (3, –4), (2, 3), (–2, –3), and (1, 0) and subdivision ratios 0.3 and 0.5 are converted into a Cardinal spline with shape factor 0.7. Find its CPs and equation of the curves.

5. A quadratic Bezier curve is associated with following CPs: (2, 2), (0, 0), and (–4, 4). Find its equation.

6. Find parametric equations of a cubic Bezier spline having CPs (3, –4), (2, 3), (0, 0), and (–2, –3).

7. Show that for a quadratic Bezier curve with CPs P_0, P_1, and P_2, the following expression holds true: $P(0)+P'(1)-2(P_2-P_0)$.

8. Show that for a cubic Bezier curve with CPs P_0, P_1, P_2, and P_3 the following expressions are true: $P'(0)=3(P_1-P_0)$ and $P'(1)=3(P_3-P_2)$.

9. Convert the parametric curve $\left(1+2t+3t^2, 4+5t+6t^2\right)$ to a quadratic Bezier curve and find its CPs.

10. Convert the cubic Bezier spline having CPs (3, –4), (2, 3), (0, 0), and (–2, –3) to a cubic C–R spline and find its CPs.

CHAPTER 3

Approximating Splines

3.1 INTRODUCTION

In the previous chapters, we have discussed how interpolating splines and hybrid splines are characterized using control points (CPs) and blending functions (BFs). Bezier splines in particular are quite popular and extensively used tools in graphics packages for modeling splines. However, Bezier splines have two major drawbacks, which have been the motivation for designing approximating splines. The first drawback is that the number of CPs is dependent on the degree of the curve i.e. it is not possible to increase the number of CPs for smoother control without increasing the degree of the polynomial and hence computational complexity. The second drawback is that there is no provision of local control i.e. shifting the location of a single CP will change the shape of the entire spline instead of a local portion of the spline. In graphics local, control is usually desirable since it enables small adjustments in the spline to be made (Hearn and Baker, 1996).

To overcome, these drawbacks a new type of spline called B-spline has been proposed. B-spline stands for "basis spline" and these are truly approximating i.e. in general they do not go through any of the CPs but under special conditions they may be forced to do so. Essentially, B-splines consist of multiple curve segments with continuity at join points. This enables local control i.e. when a CP is moved only a set of local curve segments are changed instead of the entire curve. The BFs of B-splines are calculated using an algorithm known as Cox de Boor algorithm, named after the German mathematician Carl-Wilhelm Reinhold de Boor. Values of the parametric variable t at the join points are stored in a vector called the "knot vector (KV)." If the knot values are equally spaced, then the resulting spline is called uniform B-spline; otherwise, it is referred to as non-uniform. B-splines are called open uniform when KV values are repeated.

The BFs and equation of B-splines are not single entities but a collection of expressions for each segment. For example, if there are four segments, designated as A, B, C, and D then each BF B has four sub-components BA, BB, BC, and BD and represented as B = {BA, BB, BC, BD}. If there are say three CPs P0, P1, and P2 then there would be three

BFs B0, B1, and B2 each with four sub-components i.e. B0A, … , B0D, B1A, … , B1D, and B2A, … , B2D. Usually the BFs are represented one per line as:

B0 = {B0A, B0B, B0C, B0D}

B1 = {B1A, B1B, B1C, B1D}

B2 = {B2A, B2B, B2C, B2D}

Similarly, the curve equation P also would have four sub-components P = {PA, PB, PC, PD}. We have seen before that equation of a curve can be expressed as products of CPs and BFs, hence the sub-components can be written as: PA = P0.B0A + P1.B1A + P2.B2A, PB = P0.B0B + P1.B1B + P2.B2B and so on. Note that the sub-components are valid for different ranges of t and hence cannot be added up but needs to be represented as a matrix of values. Typically segment A is valid over the range $t_0 \le t < t_1$, B over $t_1 \le t < t_2$ and so on. Since the curve expressions can be quite large, in most cases the equations are written vertically instead of horizontally, with the range of values mentioned for each segment, as shown below:

$$P = \begin{cases} P_A = P_{0.B0A} + P_{1.B1A} + P_{2.B2A}(t_0 \le t < t_1) \text{ [segment } A] \\ P_B = P_{0.B0B} + P_{1.B1B} + P_{2.B2B}(t_1 \le t < t_2) \text{ [segment } B] \\ P_C = P_{0.B0C} + P_{1.B1C} + P_{2.B2C}(t_2 \le t < t_3) \text{ [segment } C] \end{cases}$$

The details about computations of the B and P values are illustrated in the following sections.

3.2 LINEAR UNIFORM *B*-SPLINE

A *B*-spline has two defining parameters: d which is related to the degree of the spline and n which is related to the number of CPs. The degree of the spline is actually $(d - 1)$ and the number of CPs is $(n + 1)$. The other related parameters are derived as explained below.

For a linear *B*-spline, we start with $d = 2$ and $n = 2$

Then degree of the curve: $d - 1 = 1$

Number of CP: $n + 1 = 3$

Number of BF: $n + 1 = 3$

Number of curve segments: $d + n = 4$

Number of elements in the KV: $d + n + 1 = 5$

Let the curve segments be designated as A, B, C, and D and the CPs be P_0, P_1, and P_2. Let the elements in the KV be designated as = $\{t_k\}$, where k cycles over the values {0, 1, 2, 3, 4}.

In this case, the KV is $T=[t_0, t_1, t_2, t_3, t_4]$. Let the BFs be designated in the form $B_{k,d}$. Since there are three CPs, the BFs of the curve are given by $B_{0,2}$, $B_{1,2}$, and $B_{2,2}$ and higher values of k are not relevant here. The equation of the curve is given by:

$$P(t)=P_0 \cdot B_{0,2}+P_1 \cdot B_{1,2}+P_2 \cdot B_{2,2} \tag{3.1}$$

Figure 3.1 is a schematic diagram which indicates how the three CPs P_0, P_1, and P_2 are associated with the three BFs $B_{0,2}$, $B_{1,2}$, and $B_{2,2}$, the four segments A, B, C, and D and the five element KV $[t_0, t_1, t_2, t_3, t_4]$. Irrespective of the actual locations of the CPs, their influences over specific segments remain the same. CP P_0 exerts influence over segments A and B, P_1 over segment B and C, and P_2 over C and D. This provides the local control property characteristic of B-splines: changing the location of one CP changes only two segments of the spline while the remaining parts of the spline remains unchanged. The dashed lines in the figure indicate the range of influence. In the remaining part of this section, we provide verification of this fact.

The expressions of the BFs are calculated using an algorithm known as Cox de Boor algorithm, (Hearn and Baker, 1996) which is represented below:

$$\begin{gathered} B_{k,1}=1, \quad \text{if } t_k \leq t \leq t_{k+1}, \quad \text{else } 0 \\ B_{k,d}=\left(\frac{t-t_k}{t_{k+d-1}-t_k}\right) \cdot B_{k,d-1}+\left(\frac{t_{k+d}-t}{t_{k+d}-t_{k+1}}\right) \cdot B_{k+1,d-1} \end{gathered} \tag{3.2}$$

The first line states that the first-order terms, with $d=1$, of the BFs are equal to 1 if value of t for a particular segment lies within a specific range, else 0. The second line dictates how the higher order terms, with $d>1$, are calculated based on first-order terms. The algorithm will be illustrated later. One point to be noted is that the algorithm is essentially recursive in nature as the higher order terms, pertaining to degree d, are calculated based on the lower order terms, pertaining to degree $d-1$. Thus even though the equation of the curve requires only second-order terms $B_{k,2}$ as shown in Equation (3.1), the first-order terms $B_{k,1}$ would need to be computed first.

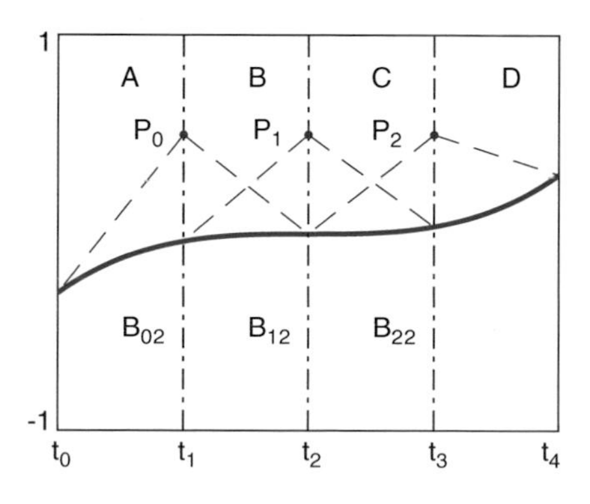

FIGURE 3.1 Linear uniform B-spline with three CPs.

To find the BF and curve equation each segment has to be analyzed separately. But first, we need numerical values for the KV. Let us make a simplifying assumption that KV values are as follows: $t_0 = 0, t_1 = 1, t_2 = 2, t_3 = 3, t_4 = 4$. Thus the KV becomes $T = [0, 1, 2, 3, 4]$. In due course, we will justify this assumption.

Segment A: $t_0 \le t < t_1$

$$B_{0,1} = 1, B_{1,1} = 0, B_{2,1} = 0, B_{3,1} = 0, B_{4,1} = 0$$

$$B_{0,2} = \left(\frac{t - t_0}{t_1 - t_0}\right) \cdot B_{0,1} + \left(\frac{t_2 - t}{t_2 - t_1}\right) \cdot B_{1,1} = t$$

$$B_{1,2} = \left(\frac{t - t_1}{t_2 - t_1}\right) \cdot B_{1,1} + \left(\frac{t_3 - t}{t_3 - t_2}\right) \cdot B_{2,1} = 0$$

$$B_{2,2} = \left(\frac{t - t_2}{t_3 - t_2}\right) \cdot B_{2,1} + \left(\frac{t_4 - t}{t_4 - t_3}\right) \cdot B_{3,1} = 0$$

Segment B: $t_1 \le t < t_2$

$$B_{0,1} = 0,\ B_{1,1} = 1, B_{2,1} = 0,\ B_{3,1} = 0,\ B_{4,1} = 0$$

$$B_{0,2} = \left(\frac{t - t_0}{t_1 - t_0}\right) \cdot B_{0,1} + \left(\frac{t_2 - t}{t_2 - t_1}\right) \cdot B_{1,1} = 2 - t$$

$$B_{1,2} = \left(\frac{t - t_1}{t_2 - t_1}\right) \cdot B_{1,1} + \left(\frac{t_3 - t}{t_3 - t_2}\right) \cdot B_{2,1} = t - 1$$

$$B_{2,2} = \left(\frac{t - t_2}{t_3 - t_2}\right) \cdot B_{2,1} + \left(\frac{t_4 - t}{t_4 - t_3}\right) \cdot B_{3,1} = 0$$

Segment C: $t_2 \le t < t_3$

$$B_{0,1} = 0,\ B_{1,1} = 0,\ B_{2,1} = 1,\ B_{3,1} = 0,\ B_{4,1} = 0$$

$$B_{0,2} = \left(\frac{t - t_0}{t_1 - t_0}\right) \cdot B_{0,1} + \left(\frac{t_2 - t}{t_2 - t_1}\right) \cdot B_{1,1} = 0$$

$$B_{1,2} = \left(\frac{t - t_1}{t_2 - t_1}\right) \cdot B_{1,1} + \left(\frac{t_3 - t}{t_3 - t_2}\right) \cdot B_{2,1} = 3 - t$$

$$B_{2,2} = \left(\frac{t - t_2}{t_3 - t_2}\right) \cdot B_{2,1} + \left(\frac{t_4 - t}{t_4 - t_3}\right) \cdot B_{3,1} = t - 2$$

Segment D: $t_3 \le t < t_4$

$$B_{0,1} = 0,\ B_{1,1} = 0,\ B_{2,1} = 0,\ B_{3,1} = 1,\ B_{4,1} = 0$$

$$B_{0,2} = \left(\frac{t - t_0}{t_1 - t_0}\right) \cdot B_{0,1} + \left(\frac{t_2 - t}{t_2 - t_1}\right) \cdot B_{1,1} = 0$$

$$B_{1,2} = \left(\frac{t - t_1}{t_2 - t_1}\right) \cdot B_{1,1} + \left(\frac{t_3 - t}{t_3 - t_2}\right) \cdot B_{2,1} = 0$$

$$B_{2,2} = \left(\frac{t - t_2}{t_3 - t_2}\right) \cdot B_{2,1} + \left(\frac{t_4 - t}{t_4 - t_3}\right) \cdot B_{3,1} = 4 - t$$

Note that each segment has a specific value of k but we still cycle k over other possible values while calculating the BFs because we want to find out the influence of other CPs onto that segment. For example, for segment $A, = 0$, but we calculate $B_{1,2}$ and $B_{2,2}$ to find out what is the influence of the second and third CP on segment A. In this case, we see that both $B_{1,2}$ and $B_{2,2}$ are zeros, which signifies that segment A is only controlled by the first CP P_0 and is not influenced by the other CPs P_1 and P_2. See Equation (3.1). Similarly for segment B, $k = 1$ but we see that both $B_{0,2}$ and $B_{1,2}$ are non-zeros, which implies that segment B is influenced by two CPs P_0 and P_1.

Now as per Equation (3.1), the BFs required to determine curve equation are $B_{0,2}$, $B_{1,2}$, and $B_{3,2}$. These BFs, however, have different values for different segments. So all these values need to be collected together specifying the segment for which they are valid. We use a separate subscript A, B, C, and D to denote the relevant segment.

$$\begin{aligned} B_{0,2} &= \{B_{0,2A}, B_{0,2B}, B_{0,2C}, B_{0,2D}\} \\ B_{1,2} &= \{B_{1,2A}, B_{1,2B}, B_{1,2C}, B_{1,2D}\} \\ B_{2,2} &= \{B_{2,2A}, B_{2,2B}, B_{2,2C}, B_{2,2D}\} \end{aligned} \tag{3.3}$$

Table 3.1 summarizes the segment-wise calculation of the BF values. As mentioned before a KV of $T = [0, 1, 2, 3, 4]$ is assumed.

Substituting the above values in Equation (3.3) we get:

$$\begin{aligned} B_{0,2} &= \{t, 2 - t, 0, 0\} \\ B_{1,2} &= \{0, t - 1, 3 - t, 0\} \\ B_{2,2} &= \{0, 0, t - 2, 4 - t\} \end{aligned} \tag{3.4}$$

Equation (3.4) specifies the BFs of a linear uniform B-spline having three CPs (represented by three vertical rows) and four CPs (represented by four element vectors per row).

TABLE 3.1 Computation of BFs of Linear Uniform B-Spline

Segment	t	$B_{k,1}$	$B_{k,2}$
A	$t_0 \le t < t_1$	$B_{0,1} = 1$ $B_{1,1} = 0$ $B_{2,1} = 0$ $B_{3,1} = 0$	$B_{0,2} = t$ $B_{1,2} = 0$ $B_{2,2} = 0$
B	$t_1 \le t < t_2$	$B_{0,1} = 0$ $B_{1,1} = 1$ $B_{2,1} = 0$ $B_{3,1} = 0$	$B_{0,2} = 2 - t$ $B_{1,2} = t - 1$ $B_{2,2} = 0$
C	$t_2 \le t < t_3$	$B_{0,1} = 0$ $B_{1,1} = 0$ $B_{2,1} = 1$ $B_{3,1} = 0$	$B_{0,2} = 0$ $B_{1,2} = 3 - t$ $B_{2,2} = t - 2$
D	$t_3 \le t < t_4$	$B_{0,1} = 0$ $B_{1,1} = 0$ $B_{2,1} = 0$ $B_{3,1} = 1$	$B_{0,2} = 0$ $B_{1,2} = 0$ $B_{2,2} = 4 - t$

An alternative way of representation of the BFs is shown in Equation (3.5) where only the non-zero values are indicated along with the segment name and range for t for which they are valid.

$$
\begin{aligned}
B_{0,2} &= \begin{cases} t & (0 \le t < 1)A \\ 2 - t & (1 \le t < 2)B \end{cases} \\
B_{1,2} &= \begin{cases} t - 1 & (1 \le t < 2)B \\ 3 - t & (2 \le t < 3)C \end{cases} \\
B_{2,2} &= \begin{cases} t - 2 & (2 \le t < 3)C \\ 4 - t & (3 \le t < 4)D \end{cases}
\end{aligned}
\tag{3.5}
$$

Equation (3.5) shows that $B_{0,2}$ and hence, P_0 influences segments A and B, $B_{1,2}$ and P_1 influences segments B and C, and $B_{2,2}$ and P_2 influences segments C and D, a fact that has been indicated by the dashed lines in Figure 3.1.

The plot of the BFs is shown in Figure 3.2. Each BF has the same shape but shifted toward the right by 1 with respect to the previous one. Thus each BF can be obtained from the previous one by substituting t with $(t - 1)$ as is evident from Equation (3.5). Also as shown in Equation (3.4), each BF has four sub-divisions out of which two are non-zeros. The first curve for $B_{0,2}$ has non-zero parts for segments A $(0 \le t < 1)$ and B $(1 \le t < 2)$, the second curve for $B_{1,2}$ has non-zero parts for segments B $(1 \le t < 2)$ and C $(2 \le t < 3)$, the third curve for $B_{2,2}$ has non-zero parts for segments C $(2 \le t < 3)$ and D $(3 \le t < 4)$. Since the BFs are associated with the CPs, this again provides an indication of the local control property

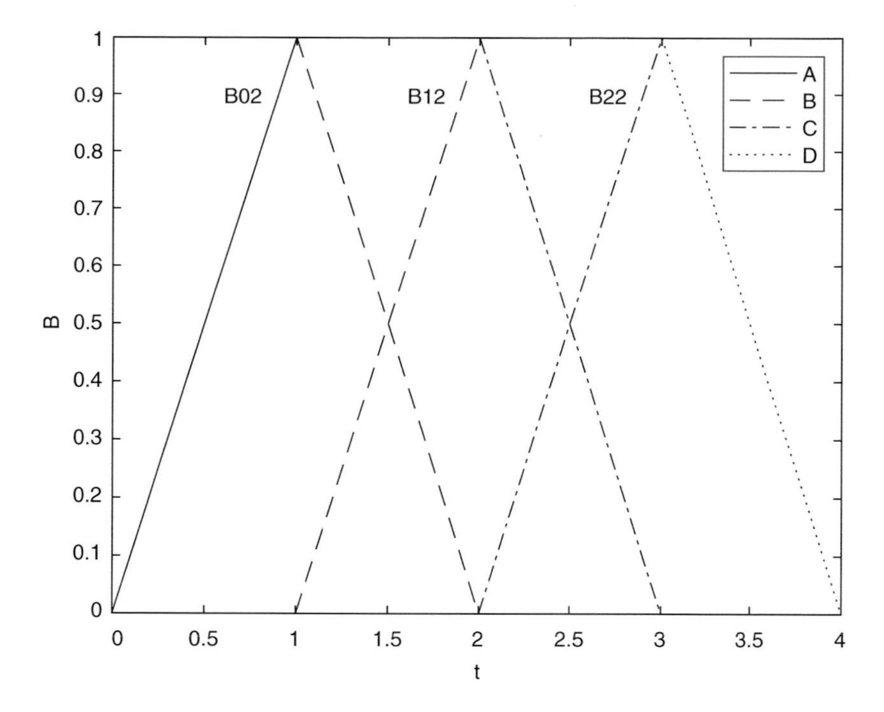

FIGURE 3.2 BFs of a linear uniform B-spline.

of the spline i.e. the first CP has influence over the first two segments A and B, the second CP has influence over B and C, and so on. This means that if the first CP is changed, it will affect only the two of the four segments while the rest of the spline will remain unchanged. This is in contrast with the interpolating and hybrid curves where each BF is valid over the entire range of t.

The equation of the spline is a collection of the equations of its four segments:

$$P(t) = \begin{cases} P_A & (0 \le t < 1) \\ P_B & (1 \le t < 2) \\ P_C & (2 \le t < 3) \\ P_D & (3 \le t < 4) \end{cases} \tag{3.6}$$

where

$$\begin{aligned} P_A &= P_0 \cdot B_{0,2A} + P_1 \cdot B_{1,2A} + P_2 \cdot B_{2,2A} \\ P_B &= P_0 \cdot B_{0,2B} + P_1 \cdot B_{1,2B} + P_2 \cdot B_{2,2B} \\ P_C &= P_0 \cdot B_{0,2C} + P_1 \cdot B_{1,2C} + P_2 \cdot B_{2,2C} \\ P_D &= P_0 \cdot B_{0,2D} + P_1 \cdot B_{1,2D} + P_2 \cdot B_{2,2D} \end{aligned} \tag{3.7}$$

Substituting values of the BFs from Table 3.1 into Equation (3.6) we get:

$$P(t)=\begin{cases} P_0 \cdot t & (0 \le t < 1) \\ P_0 \cdot (2-t) + P_1 \cdot (t-1) & (1 \le t < 2) \\ P_1 \cdot (3-t) + P_2 \cdot (t-2) & (2 \le t < 3) \\ P_2 \cdot (4-t) & (3 \le t < 4) \end{cases} \tag{3.8}$$

Equation (3.8) represents the equation of a linear uniform *B*-spline with three CPs and four segments. The four parts are the equations of the four segments.

Example 3.1

Find the equation of a linear uniform B-spline having CPs (2, –3), (5, 5), and (8, –1). Also write a program to plot the BFs.

From Equation (3.8) substituting the values of the given CPs:

$$x(t)=\begin{cases} 2t & (0 \le t < 1) \\ 3t-1 & (1 \le t < 2) \\ 3t-1 & (2 \le t < 3) \\ -8t+32 & (3 \le t < 4) \end{cases}$$

$$y(t)=\begin{cases} -3t & (0 \le t < 1) \\ 8t-11 & (1 \le t < 2) \\ -6t+17 & (2 \le t < 3) \\ t-4 & (3 \le t < 4) \end{cases}$$

MATLAB® Code 3.1

```
clear all; format compact; clc;

t0 = 0; t1 = 1; t2 = 2; t3 = 3; t4 = 4;
T = [t0, t1, t2, t3, t4];

syms t P0 P1 P2;

%Segment A
B01 = 1; B11 = 0; B21 = 0; B31 = 0; B41 = 0; B51 = 0; B61 = 0;
S1 = (t - t0)/(t1 - t0); S2 = (t2 - t)/(t2 - t1); B02 = S1*B01 + S2*B11; B02A = B02;
S1 = (t - t1)/(t2 - t1); S2 = (t3 - t)/(t3 - t2); B12 = S1*B11 + S2*B21; B12A = B12;
S1 = (t - t2)/(t3 - t2); S2 = (t4 - t)/(t4 - t3); B22 = S1*B21 + S2*B31; B22A = B22;

%Segment B
B01 = 0; B11 = 1; B21 = 0; B31 = 0; B41 = 0; B51 = 0; B61 = 0;
S1 = (t - t0)/(t1 - t0); S2 = (t2 - t)/(t2 - t1); B02 = S1*B01 + S2*B11; B02B = B02;
S1 = (t - t1)/(t2 - t1); S2 = (t3 - t)/(t3 - t2); B12 = S1*B11 + S2*B21; B12B = B12;
S1 = (t - t2)/(t3 - t2); S2 = (t4 - t)/(t4 - t3); B22 = S1*B21 + S2*B31; B22B = B22;

%Segment C
B01 = 0; B11 = 0; B21 = 1; B31 = 0; B41 = 0; B51 = 0; B61 = 0;
S1 = (t - t0)/(t1 - t0); S2 = (t2 - t)/(t2 - t1); B02 = S1*B01 + S2*B11; B02C = B02;
S1 = (t - t1)/(t2 - t1); S2 = (t3 - t)/(t3 - t2); B12 = S1*B11 + S2*B21; B12C = B12;
S1 = (t - t2)/(t3 - t2); S2 = (t4 - t)/(t4 - t3); B22 = S1*B21 + S2*B31; B22C = B22;

%Segment D
B01 = 0; B11 = 0; B21 = 0; B31 = 1; B41 = 0; B51 = 0; B61 = 0;
S1 = (t - t0)/(t1 - t0); S2 = (t2 - t)/(t2 - t1); B02 = S1*B01 + S2*B11; B02D = B02;
S1 = (t - t1)/(t2 - t1); S2 = (t3 - t)/(t3 - t2); B12 = S1*B11 + S2*B21; B12D = B12;
S1 = (t - t2)/(t3 - t2); S2 = (t4 - t)/(t4 - t3); B22 = S1*B21 + S2*B31; B22D = B22;
```

```
fprintf('Blending functions :\n');

B02 = [B02A, B02B, B02C, B02D]
B12 = [B12A, B12B, B12C, B12D]
B22 = [B22A, B22B, B22C, B22D]

fprintf('\n');

fprintf('General Equation of Curve :\n');

P = P0*B02 + P1*B12 + P2*B22

fprintf('\n');

x0 = 2; x1 = 5; x2 = 8;
y0 = -3; y1 = 5; y2 = -1;

fprintf('Actual Equation  :\n');

x = subs(P, ([P0, P1, P2]), ([x0, x1, x2]))
y = subs(P, ([P0, P1, P2]), ([y0, y1, y2]))

%plotting BFs

tta = linspace(t0, t1);
ttb = linspace(t1, t2);
ttc = linspace(t2, t3);
ttd = linspace(t3, t4);

B02aa = subs(B02A, t, tta);
```

```
B02bb = subs(B02B, t, ttb);
B02cc = subs(B02C, t, ttc);
B02dd = subs(B02D, t, ttd);

B12aa = subs(B12A, t, tta);
B12bb = subs(B12B, t, ttb);
B12cc = subs(B12C, t, ttc);
B12dd = subs(B12D, t, ttd);

B22aa = subs(B22A, t, tta);
B22bb = subs(B22B, t, ttb);
B22cc = subs(B22C, t, ttc);
B22dd = subs(B22D, t, ttd);

figure
plot(tta, B02aa, 'k-', ttb,B02bb, 'k--', ttc,B02cc, 'k-.', ttd,B02dd, 'k:'); hold on;
plot(tta, B12aa, 'k-', ttb,B12bb, 'k--', ttc,B12cc, 'k-.', ttd,B12dd, 'k:');
plot(tta, B22aa, 'k-', ttb,B22bb, 'k--', ttc,B22cc, 'k-.', ttd,B22dd, 'k:'); hold off;
xlabel('t'); ylabel('B');
legend('A', 'B', 'C', 'D');
text(0.6, 0.9, 'B02'); text(1.6, 0.9, 'B12'); text(2.6, 0.9, 'B22');
```

NOTE

See Figure 3.2 for a plot of the BFs.

3.3 CHANGING NUMBER OF CONTROL POINTS

One of the objectives of designing *B*-splines is to have local control, a fact we have seen justified in the previous section since various CPs affect only specific curve segments instead of the entire curve. The other objective was to make the number of CPs independent of the degree of the curve. To investigate this point let us now increase the number of CPs by 1 while keeping the degree same as before and find out whether such a combination produces a valid curve equation.

For this case, we start with $d = 2$ and $n = 3$

Then degree of the curve: $d - 1 = 1$

Number of CPs: $n + 1 = 4$

Number of curve segments: $d + n = 5$

Let the curve segments be designated as *A*, *B*, *C*, *D*, and *E* (see Figure 3.3) and the CPs be P_0, P_1, P_2, and P_3.

Number of elements in the KV: $d + n + 1 = 6$

Let the elements in the KV be designated as $T = \{t_k\}$, where k cycles over the values {0, 1, 2, 3, 4, 5}. In this case, $T = [t_0, t_1, t_2, t_3, t_4, t_5, t_6]$.

Number of BFs is same as the number of CPs i.e. 4. Let the BFs be designated as $B_{k,d}$. In this case, the BFs are $B_{0,2}$, $B_{1,2}$, $B_{2,2}$, and $B_{3,2}$. Higher values of k are not relevant and hence need not be calculated. Equation of the curve is given by:

$$P(t) = P_0 \cdot B_{0,2} + P_1 \cdot B_{1,2} + P_2 \cdot B_{2,2} + P_3 \cdot B_{3,2} \tag{3.9}$$

If the process outlined in the previous section is followed, the reader will be able to verify that this does indeed produce a valid *B*-spline. It is therefore left as an exercise. The final BFs and the equation of the curve are given below as a reference. Compare with Equations (3.5) and (3.8) to see the difference.

$$B_{0,2} = \begin{cases} t & (0 \le t < 1) \\ 2 - t & (1 \le t < 2) \end{cases}$$

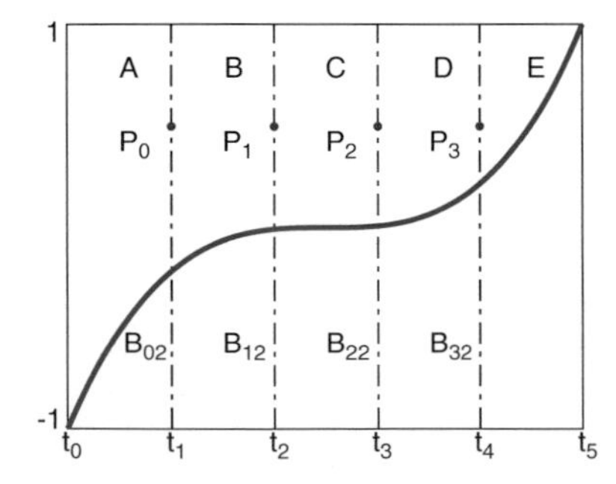

FIGURE 3.3 Linear uniform *B*-spline with four CPs.

$$B_{1,2} = \begin{cases} t-1 & (1 \le t < 2) \\ 3-t & (2 \le t < 3) \end{cases}$$

$$B_{2,2} = \begin{cases} t-2 & (2 \le t < 3) \\ 4-t & (3 \le t < 4) \end{cases}$$

$$B_{3,2} = \begin{cases} t-3 & (3 \le t < 4) \\ 5-t & (4 \le t < 5) \end{cases}$$

$$P(t) = \begin{cases} P_0 \cdot t & (0 \le t < 1) \\ P_0 \cdot (2-t) + P_1 \cdot (t-1) & (1 \le t < 2) \\ P_1 \cdot (3-t) + P_2 \cdot (t-2) & (2 \le t < 3) \\ P_2 \cdot (4-t) + P_3 \cdot (t-3) & (3 \le t < 4) \\ P_3 \cdot (5-t) & (4 \le t < 5) \end{cases}$$

3.4 QUADRATIC UNIFORM *B*-SPLINE

For generating a quadratic *B*-spline, we need to start with $d = 3$ and $n = 3$

Then degree of the curve: $d - 1 = 2$

Number of CPs: $n + 1 = 4$

Number of BFs: $n + 1 = 4$

Number of curve segments: $d + n = 6$

Number of elements in the KV: $d + n + 1 = 7$

Let the curve segments be *A*, *B*, *C*, *D*, *E*, and *F* and CPs be P_0, P_1, P_2, and P_3 (see Figure 3.4).

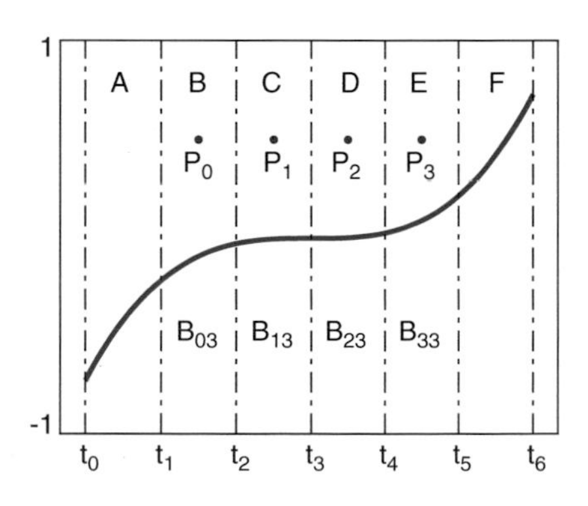

FIGURE 3.4 Quadratic uniform *B*-spline with four CPs.

Let the KV be $T = \{t_k\}$ for $k = \{0, 1, 2, 3, 4, 5, 6\}$. In this case, $T = [t_0, t_1, t_2, t_3, t_4, t_5, t_6]$.
Let the BFs be $B_{0,3}$, $B_{1,3}$, $B_{2,3}$, and $B_{3,3}$
Equation of the curve is:

$$P(t) = P_0 \cdot B_{0,3} + P_1 \cdot B_{1,3} + P_2 \cdot B_{2,3} + P_3 \cdot B_{3,3} \tag{3.10}$$

As before, we assume the KV to be $T = [0, 1, 2, 3, 4, 5, 6]$. The first-order terms $B_{0,3}$, $B_{0,1}$, $B_{1,1}$, $B_{2,1}$, $B_{3,1}$, and $B_{4,1}$ will be either 0 or 1 as per the first condition of the Cox de Boor algorithm. According to the second condition of the algorithm, the second-order terms are computed as follows:

$$\begin{aligned}
B_{0,2} &= (t-0)\cdot B_{0,1} + (2-t)\cdot B_{1,1} \\
B_{1,2} &= (t-1)\cdot B_{1,1} + (3-t)\cdot B_{2,1} \\
B_{2,2} &= (t-2)\cdot B_{2,1} + (4-t)\cdot B_{3,1} \\
B_{3,2} &= (t-3)\cdot B_{3,1} + (5-t)\cdot B_{4,1} \\
B_{4,2} &= (t-4)\cdot B_{4,1} + (6-t)\cdot B_{5,1}
\end{aligned} \tag{3.11}$$

This time, there are also third-order terms, which are calculated from second-order terms:

$$\begin{aligned}
B_{0,3} &= \left(\frac{1}{2}\right)(t-0)\cdot B_{0,2} + \left(\frac{1}{2}\right)(3-t)\cdot B_{1,2} \\
B_{1,3} &= \left(\frac{1}{2}\right)(t-1)\cdot B_{1,2} + \left(\frac{1}{2}\right)(4-t)\cdot B_{2,2} \\
B_{2,3} &= \left(\frac{1}{2}\right)(t-2)\cdot B_{2,2} + \left(\frac{1}{2}\right)(5-t)\cdot B_{3,2} \\
B_{3,3} &= \left(\frac{1}{2}\right)(t-3)\cdot B_{3,2} + \left(\frac{1}{2}\right)(6-t)\cdot B_{4,2}
\end{aligned} \tag{3.12}$$

Since there are six segments each BF consists of six sub-components:

$$\begin{aligned}
B_{0,3} &= \left\{B_{0,3A}, B_{0,3B}, B_{0,3C}, B_{0,3D}, B_{0,3E}, B_{0,3F}\right\} \\
B_{1,3} &= \left\{B_{1,3A}, B_{1,3B}, B_{1,3C}, B_{1,3D}, B_{1,3E}, B_{1,3F}\right\} \\
B_{2,3} &= \left\{B_{2,3A}, B_{2,3B}, B_{2,3C}, B_{2,3D}, B_{2,3E}, B_{2,3F}\right\} \\
B_{3,3} &= \left\{B_{3,3A}, B_{3,3B}, B_{3,3C}, B_{3,3D}, B_{3,3E}, B_{3,3F}\right\}
\end{aligned} \tag{3.13}$$

Table 3.2 summarizes the calculation of the BF values.

TABLE 3.2 Computation of BFs of Quadratic Uniform B-Spline

Segment	t	$B_{k,1}$	$B_{k,2}$	$B_{k,3}$
A	$0 \le t < 1$	$B_{0,1} = 1$ $B_{1,1} = 0$ $B_{2,1} = 0$ $B_{3,1} = 0$ $B_{4,1} = 0$ $B_{5,1} = 0$	$B_{0,2} = t$ $B_{1,2} = 0$ $B_{2,2} = 0$ $B_{3,2} = 0$ $B_{4,2} = 0$	$B_{0,3} = \left(\frac{1}{2}\right)t^2$ $B_{1,3} = 0$ $B_{2,3} = 0$ $B_{3,3} = 0$
B	$1 \le t < 2$	$B_{0,1} = 0$ $B_{1,1} = 1$ $B_{2,1} = 0$ $B_{3,1} = 0$ $B_{4,1} = 0$ $B_{5,1} = 0$	$B_{0,2} = 2 - t$ $B_{1,2} = t - 1$ $B_{2,2} = 0$ $B_{3,2} = 0$ $B_{4,2} = 0$	$B_{0,3} = -t^2 + 3t - \frac{3}{2}$ $B_{1,3} = \left(\frac{1}{2}\right)(t-1)^2$ $B_{2,3} = 0$ $B_{3,3} = 0$
C	$2 \le t < 3$	$B_{0,1} = 0$ $B_{1,1} = 0$ $B_{2,1} = 1$ $B_{3,1} = 0$ $B_{4,1} = 0$ $B_{5,1} = 0$	$B_{0,2} = 0$ $B_{1,2} = 3 - t$ $B_{2,2} = t - 2$ $B_{3,2} = 0$ $B_{4,2} = 0$	$B_{0,3} = \left(\frac{1}{2}\right)(t-3)^2$ $B_{1,3} = -t^2 + 5t - \frac{11}{2}$ $B_{2,3} = \left(\frac{1}{2}\right)(t-2)^2$ $B_{3,3} = 0$
D	$3 \le t < 4$	$B_{0,1} = 0$ $B_{1,1} = 0$ $B_{2,1} = 0$ $B_{3,1} = 1$ $B_{4,1} = 0$ $B_{5,1} = 0$	$B_{0,2} = 0$ $B_{1,2} = 0$ $B_{2,2} = 4 - t$ $B_{3,2} = t - 3$ $B_{4,2} = 0$	$B_{0,3} = 0$ $B_{1,3} = \left(\frac{1}{2}\right)(t-4)^2$ $B_{2,3} = -t^2 + 7t - \frac{23}{2}$ $B_{3,3} = \left(\frac{1}{2}\right)(t-3)^2$
E	$4 \le t < 5$	$B_{0,1} = 0$ $B_{1,1} = 0$ $B_{2,1} = 0$ $B_{3,1} = 0$ $B_{4,1} = 1$ $B_{5,1} = 0$	$B_{0,2} = 0$ $B_{1,2} = 0$ $B_{2,2} = 0$ $B_{3,2} = 5 - t$ $B_{4,2} = t - 4$	$B_{0,3} = 0$ $B_{1,3} = 0$ $B_{2,3} = \left(\frac{1}{2}\right)(t-5)^2$ $B_{3,3} = -t^2 + 9t - \frac{39}{2}$
F	$5 \le t < 6$	$B_{0,1} = 0$ $B_{1,1} = 0$ $B_{2,1} = 0$ $B_{3,1} = 0$ $B_{4,1} = 0$ $B_{5,1} = 1$	$B_{0,2} = 0$ $B_{1,2} = 0$ $B_{2,2} = 0$ $B_{3,2} = 0$ $B_{4,2} = 6 - t$	$B_{0,3} = 0$ $B_{1,3} = 0$ $B_{2,3} = 0$ $B_{3,3} = \left(\frac{1}{2}\right)(t-6)^2$

Substituting the above values in Equation (3.13) we get:

$$
\begin{aligned}
B_{0,3} &= \begin{cases} \left(\frac{1}{2}\right)t^2 & (0 \le t < 1)\ A \\ -t^2 + 3t - \frac{3}{2} & (1 \le t < 2)\ B \\ \left(\frac{1}{2}\right)(t-3)^2 & (2 \le t < 3)\ C \end{cases} \\
B_{1,3} &= \begin{cases} \left(\frac{1}{2}\right)(t-1)^2 & (1 \le t < 2)\ B \\ -t^2 + 5t - \frac{11}{2} & (2 \le t < 3)\ C \\ \left(\frac{1}{2}\right)(t-4)^2 & (3 \le t < 4)\ D \end{cases} \\
B_{2,3} &= \begin{cases} \left(\frac{1}{2}\right)(t-2)^2 & (2 \le t < 3)\ C \\ -t^2 + 7t - \frac{23}{2} & (3 \le t < 4)\ D \\ \left(\frac{1}{2}\right)(t-5)^2 & (4 \le t < 5)\ E \end{cases} \\
B_{3,3} &= \begin{cases} \left(\frac{1}{2}\right)(t-3)^2 & (3 \le t < 4)\ D \\ -t^2 + 9t - \frac{39}{2} & (4 \le t < 5)\ E \\ \left(\frac{1}{2}\right)(t-6)^2 & (5 \le t < 6)\ F \end{cases}
\end{aligned} \tag{3.14}
$$

Equation (3.14) represents the BFs of a quadratic uniform *B*-spline with four CPs and six segments. The plot of the BFs is shown in Figure 3.5. Each BF has the same shape but shifted towards the right by 1 with respect to the previous one. Thus each BF can be

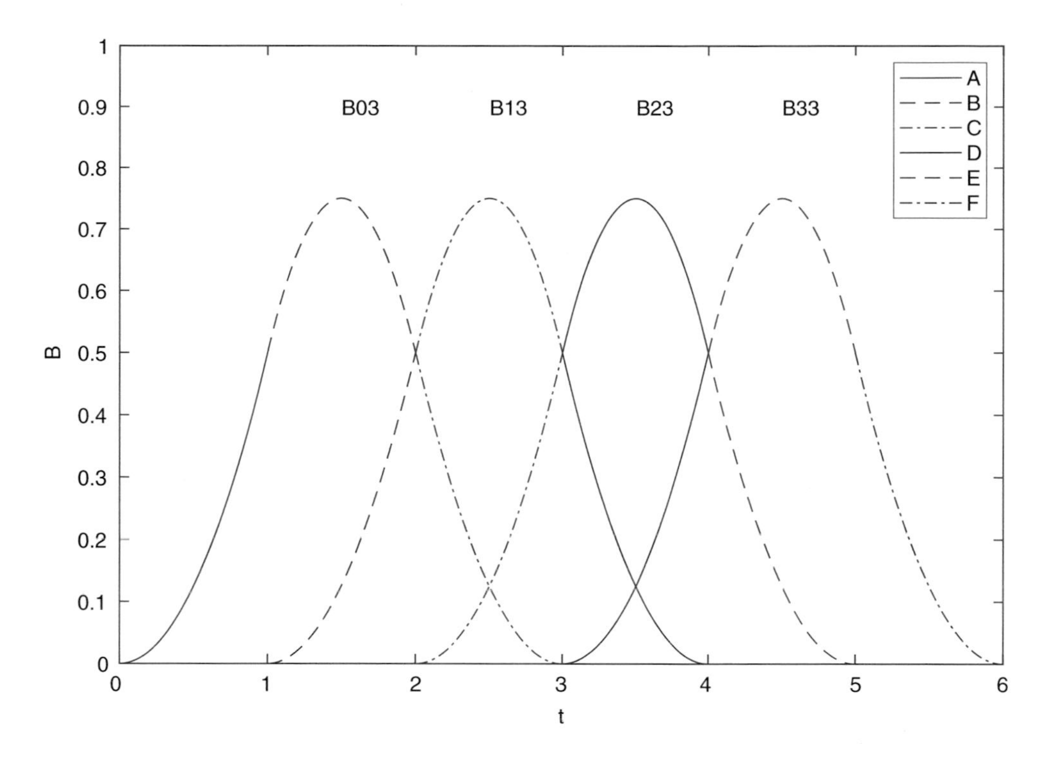

FIGURE 3.5 BFs of quadratic uniform B-spline.

obtained from the previous one by substituting t with $(t-1)$. Each BF has six sub-divisions out of which three are non-zeros. The first curve for $B_{0,3}$ has non-zero parts for segments $A(0 \le t < 1), B(1 \le t < 2), C(2 \le t < 3)$, the second curve for $B_{1,3}$ has non-zero parts for segments $B(1 \le t < 2), C(2 \le t < 3), D(3 \le t < 4)$, the third curve for $B_{2,3}$ has non-zero parts for segments $C(2 \le t < 3), D(3 \le t < 4), E(4 \le t < 5)$, and the fourth curve for $B_{3,3}$ has non-zero parts for segments $D(3 \le t < 4), E(4 \le t < 5), F(5 \le t < 6)$.

The equation of the spline is a collection of the equations of its six segments:

$$P(t) = \begin{cases} P_A & (0 \le t < 1) \\ P_B & (1 \le t < 2) \\ P_C & (2 \le t < 3) \\ P_D & (3 \le t < 4) \\ P_E & (4 \le t < 5) \\ P_F & (5 \le t < 6) \end{cases} \tag{3.15}$$

where

$$\begin{aligned}
P_A &= P_0 \cdot B_{0,3A} + P_1 \cdot B_{1,3A} + P_2 \cdot B_{2,3A} + P_3 \cdot B_{3,3A} \\
P_B &= P_0 \cdot B_{0,3B} + P_1 \cdot B_{1,3B} + P_2 \cdot B_{2,3B} + P_3 \cdot B_{3,3B} \\
P_C &= P_0 \cdot B_{0,3C} + P_1 \cdot B_{1,3C} + P_2 \cdot B_{2,3C} + P_3 \cdot B_{3,3C} \\
P_D &= P_0 \cdot B_{0,3D} + P_1 \cdot B_{1,3D} + P_2 \cdot B_{2,3D} + P_3 \cdot B_{3,3D} \\
P_E &= P_0 \cdot B_{0,3E} + P_1 \cdot B_{1,3E} + P_2 \cdot B_{2,3E} + P_3 \cdot B_{3,3E} \\
P_F &= P_0 \cdot B_{0,3F} + P_1 \cdot B_{1,3F} + P_2 \cdot B_{2,3F} + P_3 \cdot B_{3,3F}
\end{aligned} \tag{3.16}$$

Substituting values of the BFs from Table 3.2 into Equation (3.15) and specifying the range for which they are valid we get:

$$P(t) = \begin{cases}
P_0 \cdot \left(\frac{1}{2}\right) t^2 & (0 \le t < 1) \\
P_0 \cdot \left(-t^2 + 3t - \frac{3}{2}\right) + P_1 \cdot \left(\frac{1}{2}\right)(t-1)^2 & (1 \le t < 2) \\
P_0 \cdot \left(\frac{1}{2}\right)(t-3)^2 + P_1 \cdot \left(-t^2 + 5t - \frac{11}{2}\right) + P_2 \cdot \left(\frac{1}{2}\right)(t-2)^2 & (2 \le t < 3) \\
P_1 \cdot \left(\frac{1}{2}\right)(t-4)^2 + P_2 \cdot \left(-t^2 + 7t - \frac{23}{2}\right) + P_3 \cdot \left(\frac{1}{2}\right)(t-3)^2 & (3 \le t < 4) \\
P_2 \cdot \left(\frac{1}{2}\right)(t-5)^2 + P_3 \cdot \left(-t^2 + 9t - \frac{39}{2}\right) & (4 \le t < 5) \\
P_3 \cdot \left(\frac{1}{2}\right)(t-6)^2 & (5 \le t < 6)
\end{cases} \tag{3.17}$$

Equation (3.17) represents the equation of a quadratic uniform B-spline with four CPs and six segments. The six parts of the equation represent the six segments of the curve.

Example 3.2

Find the equation of a uniform quadratic B-spline having CPs $P_0(1, 2)$, $P_1(4, 1)$, $P_2(6, 5)$, and $P_3(8, -1)$. Also write a program to plot the BFs and the actual curve.

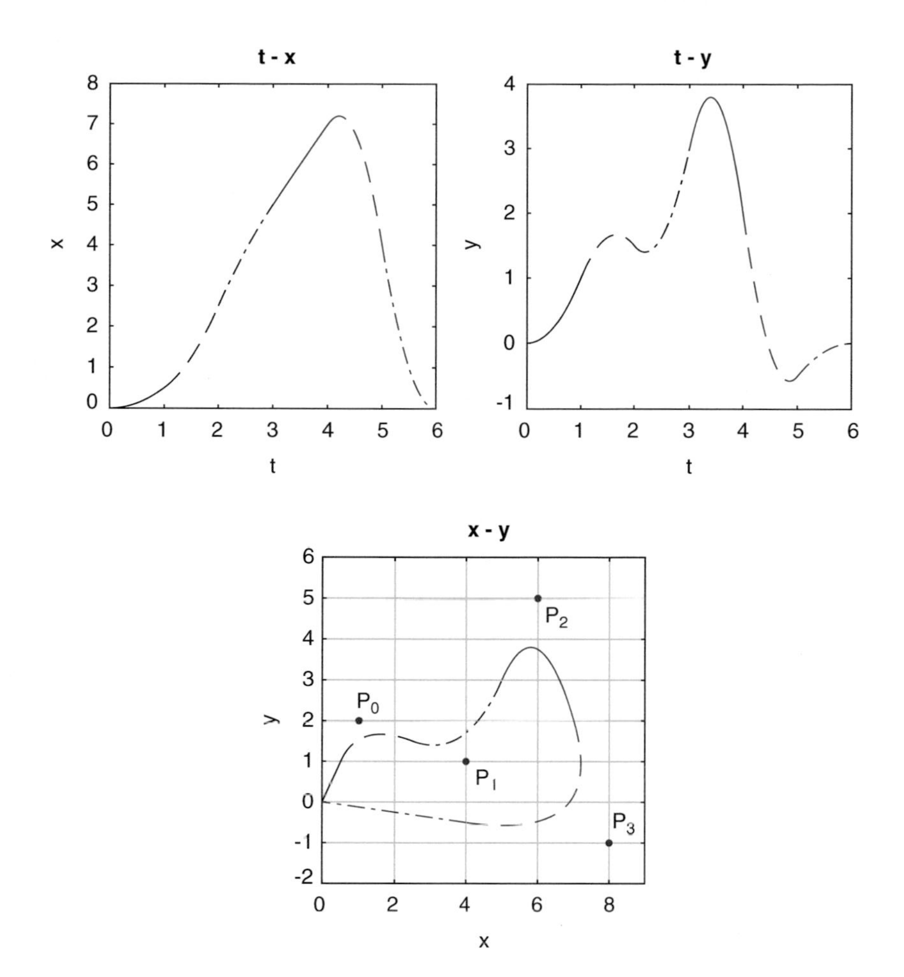

FIGURE 3.6 Plots for Example 3.2.

From Equation (3.17), substituting the values of the given CPs (Figure 3.6):

$$x(t)=\begin{cases} \dfrac{t^2}{2} & (0\le t<1) \\ t^2-t+\dfrac{1}{2} & (1\le t<2) \\ \dfrac{t^2}{2}+5t-\dfrac{11}{2} & (2\le t<3) \\ 2t-1 & (3\le t<4) \\ -5t^2+42t-81 & (4\le t<5) \\ 4(t-6)^2 & (5\le t<6) \end{cases} \qquad y(t)=\begin{cases} t^2 & (0\le t<1) \\ -\dfrac{3t^2}{2}+5t-\dfrac{5}{2} & (1\le t<2) \\ \dfrac{5t^2}{2}-11t+\dfrac{27}{2} & (2\le t<3) \\ -5t^2+34t-54 & (3\le t<4) \\ \dfrac{7t^2}{2}-34t+82 & (4\le t<5) \\ -\dfrac{1}{2}(t-6)^2 & (5\le t<6) \end{cases}$$

MATLAB Code 3.2

```
clear all; format compact; clc;

t0 = 0; t1 = 1; t2 = 2; t3 = 3; t4 = 4; t5 = 5; t6 = 6;
T = [t0, t1, t2, t3, t4, t5, t6];

syms t P0 P1 P2 P3;

%Segment A
B01 = 1; B11 = 0; B21 = 0; B31 = 0; B41 = 0; B51 = 0; B61 = 0;
S1 = (t - t0)/(t1 - t0); S2 = (t2 - t)/(t2 - t1); B02 = S1*B01 + S2*B11; B02A = B02;
S1 = (t - t1)/(t2 - t1); S2 = (t3 - t)/(t3 - t2); B12 = S1*B11 + S2*B21; B12A = B12;
S1 = (t - t2)/(t3 - t2); S2 = (t4 - t)/(t4 - t3); B22 = S1*B21 + S2*B31; B22A = B22;
S1 = (t - t3)/(t4 - t3); S2 = (t5 - t)/(t5 - t4); B32 = S1*B31 + S2*B41; B32A = B32;
S1 = (t - t4)/(t5 - t4); S2 = (t6 - t)/(t6 - t5); B42 = S1*B41 + S2*B51; B42A = B42;
S1 = (t - t0)/(t2 - t0); S2 = (t3 - t)/(t3 - t1); B03 = S1*B02 + S2*B12; B03A = B03;
S1 = (t - t1)/(t3 - t1); S2 = (t4 - t)/(t4 - t2); B13 = S1*B12 + S2*B22; B13A = B13;
S1 = (t - t2)/(t4 - t2); S2 = (t5 - t)/(t5 - t3); B23 = S1*B22 + S2*B32; B23A = B23;
S1 = (t - t3)/(t5 - t3); S2 = (t6 - t)/(t6 - t4); B33 = S1*B32 + S2*B42; B33A = B33;

%Segment B
B01 = 0; B11 = 1; B21 = 0; B31 = 0; B41 = 0; B51 = 0; B61 = 0;
S1 = (t - t0)/(t1 - t0); S2 = (t2 - t)/(t2 - t1); B02 = S1*B01 + S2*B11; B02B = B02;
S1 = (t - t1)/(t2 - t1); S2 = (t3 - t)/(t3 - t2); B12 = S1*B11 + S2*B21; B12B = B12;
S1 = (t - t2)/(t3 - t2); S2 = (t4 - t)/(t4 - t3); B22 = S1*B21 + S2*B31; B22B = B22;
S1 = (t - t3)/(t4 - t3); S2 = (t5 - t)/(t5 - t4); B32 = S1*B31 + S2*B41; B32B = B32;
S1 = (t - t4)/(t5 - t4); S2 = (t6 - t)/(t6 - t5); B42 = S1*B41 + S2*B51; B42B = B42;
S1 = (t - t0)/(t2 - t0); S2 = (t3 - t)/(t3 - t1); B03 = S1*B02 + S2*B12; B03B = B03;
S1 = (t - t1)/(t3 - t1); S2 = (t4 - t)/(t4 - t2); B13 = S1*B12 + S2*B22; B13B = B13;
S1 = (t - t2)/(t4 - t2); S2 = (t5 - t)/(t5 - t3); B23 = S1*B22 + S2*B32; B23B = B23;
S1 = (t - t3)/(t5 - t3); S2 = (t6 - t)/(t6 - t4); B33 = S1*B32 + S2*B42; B33B = B33;
```

```
%Segment C
B01 = 0; B11 = 0; B21 = 1; B31 = 0; B41 = 0; B51 = 0; B61 = 0;
S1 = (t - t0)/(t1 - t0); S2 = (t2 - t)/(t2 - t1); B02 = S1*B01 + S2*B11; B02C = B02;
S1 = (t - t1)/(t2 - t1); S2 = (t3 - t)/(t3 - t2); B12 = S1*B11 + S2*B21; B12C = B12;
S1 = (t - t2)/(t3 - t2); S2 = (t4 - t)/(t4 - t3); B22 = S1*B21 + S2*B31; B22C = B22;
S1 = (t - t3)/(t4 - t3); S2 = (t5 - t)/(t5 - t4); B32 = S1*B31 + S2*B41; B32C = B32;
S1 = (t - t4)/(t5 - t4); S2 = (t6 - t)/(t6 - t5); B42 = S1*B41 + S2*B51; B42C = B42;
S1 = (t - t0)/(t2 - t0); S2 = (t3 - t)/(t3 - t1); B03 = S1*B02 + S2*B12; B03C = B03;
S1 = (t - t1)/(t3 - t1); S2 = (t4 - t)/(t4 - t2); B13 = S1*B12 + S2*B22; B13C = B13;
S1 = (t - t2)/(t4 - t2); S2 = (t5 - t)/(t5 - t3); B23 = S1*B22 + S2*B32; B23C = B23;
S1 = (t - t3)/(t5 - t3); S2 = (t6 - t)/(t6 - t4); B33 = S1*B32 + S2*B42; B33C = B33;

%Segment D
B01 = 0; B11 = 0; B21 = 0; B31 = 1; B41 = 0; B51 = 0; B61 = 0;
S1 = (t - t0)/(t1 - t0); S2 = (t2 - t)/(t2 - t1); B02 = S1*B01 + S2*B11; B02D = B02;
S1 = (t - t1)/(t2 - t1); S2 = (t3 - t)/(t3 - t2); B12 = S1*B11 + S2*B21; B12D = B12;
S1 = (t - t2)/(t3 - t2); S2 = (t4 - t)/(t4 - t3); B22 = S1*B21 + S2*B31; B22D = B22;
S1 = (t - t3)/(t4 - t3); S2 = (t5 - t)/(t5 - t4); B32 = S1*B31 + S2*B41; B32D = B32;
S1 = (t - t4)/(t5 - t4); S2 = (t6 - t)/(t6 - t5); B42 = S1*B41 + S2*B51; B42D = B42;
S1 = (t - t0)/(t2 - t0); S2 = (t3 - t)/(t3 - t1); B03 = S1*B02 + S2*B12; B03D = B03;
S1 = (t - t1)/(t3 - t1); S2 = (t4 - t)/(t4 - t2); B13 = S1*B12 + S2*B22; B13D = B13;
S1 = (t - t2)/(t4 - t2); S2 = (t5 - t)/(t5 - t3); B23 = S1*B22 + S2*B32; B23D = B23;
S1 = (t - t3)/(t5 - t3); S2 = (t6 - t)/(t6 - t4); B33 = S1*B32 + S2*B42; B33D = B33;

%Segment E
B01 = 0; B11 = 0; B21 = 0; B31 = 0; B41 = 1; B51 = 0; B61 = 0;
S1 = (t - t0)/(t1 - t0); S2 = (t2 - t)/(t2 - t1); B02 = S1*B01 + S2*B11; B02E = B02;
S1 = (t - t1)/(t2 - t1); S2 = (t3 - t)/(t3 - t2); B12 = S1*B11 + S2*B21; B12E = B12;
S1 = (t - t2)/(t3 - t2); S2 = (t4 - t)/(t4 - t3); B22 = S1*B21 + S2*B31; B22E = B22;
S1 = (t - t3)/(t4 - t3); S2 = (t5 - t)/(t5 - t4); B32 = S1*B31 + S2*B41; B32E = B32;
S1 = (t - t4)/(t5 - t4); S2 = (t6 - t)/(t6 - t5); B42 = S1*B41 + S2*B51; B42E = B42;
S1 = (t - t0)/(t2 - t0); S2 = (t3 - t)/(t3 - t1); B03 = S1*B02 + S2*B12; B03E = B03;
```

```
S1 = (t - t1)/(t3 - t1); S2 = (t4 - t)/(t4 - t2); B13 = S1*B12 + S2*B22; B13E = B13;
S1 = (t - t2)/(t4 - t2); S2 = (t5 - t)/(t5 - t3); B23 = S1*B22 + S2*B32; B23E = B23;
S1 = (t - t3)/(t5 - t3); S2 = (t6 - t)/(t6 - t4); B33 = S1*B32 + S2*B42; B33E = B33;

%Segment F
B01 = 0; B11 = 0; B21 = 0; B31 = 0; B41 = 0; B51 = 1; B61 = 0;
S1 = (t - t0)/(t1 - t0); S2 = (t2 - t)/(t2 - t1); B02 = S1*B01 + S2*B11; B02F = B02;
S1 = (t - t1)/(t2 - t1); S2 = (t3 - t)/(t3 - t2); B12 = S1*B11 + S2*B21; B12F = B12;
S1 = (t - t2)/(t3 - t2); S2 = (t4 - t)/(t4 - t3); B22 = S1*B21 + S2*B31; B22F = B22;
S1 = (t - t3)/(t4 - t3); S2 = (t5 - t)/(t5 - t4); B32 = S1*B31 + S2*B41; B32F = B32;
S1 = (t - t4)/(t5 - t4); S2 = (t6 - t)/(t6 - t5); B42 = S1*B41 + S2*B51; B42F = B42;
S1 = (t - t0)/(t2 - t0); S2 = (t3 - t)/(t3 - t1); B03 = S1*B02 + S2*B12; B03F = B03;
S1 = (t - t1)/(t3 - t1); S2 = (t4 - t)/(t4 - t2); B13 = S1*B12 + S2*B22; B13F = B13;
S1 = (t - t2)/(t4 - t2); S2 = (t5 - t)/(t5 - t3); B23 = S1*B22 + S2*B32; B23F = B23;
S1 = (t - t3)/(t5 - t3); S2 = (t6 - t)/(t6 - t4); B33 = S1*B32 + S2*B42; B33F = B33;

fprintf('Blending functions :\n');

B03 = [B03A, B03B, B03C, B03D, B03E, B03F]; B03 = simplify(B03)
B13 = [B13A, B13B, B13C, B13D, B13E, B13F]; B13 = simplify(B13)
B23 = [B23A, B23B, B23C, B23D, B23E, B23F]; B23 = simplify(B23)
B33 = [B33A, B33B, B33C, B33D, B33E, B33F]; B33 = simplify(B33)

fprintf('\n');

fprintf('General Equation of Curve :\n');

P = P0*B03 + P1*B13 + P2*B23 + P3*B33

fprintf('\n');

x0 = 1; x1 = 4; x2 = 6; x3 = 8;
y0 = 2; y1 = 1; y2 = 5; y3 = -1;

fprintf('Actual Equation  :\n');
```

```
x = subs(P, ([P0, P1, P2, P3]), ([x0, x1, x2, x3])); x = simplify(x)
y = subs(P, ([P0, P1, P2, P3]), ([y0, y1, y2, y3])); y = simplify(y)

%plotting BFs

tta = linspace(t0, t1);
ttb = linspace(t1, t2);
ttc = linspace(t2, t3);
ttd = linspace(t3, t4);
tte = linspace(t4, t5);
ttf = linspace(t5, t6);

B03aa = subs(B03A, t, tta);
B03bb = subs(B03B, t, ttb);
B03cc = subs(B03C, t, ttc);
B03dd = subs(B03D, t, ttd);
B03ee = subs(B03E, t, tte);
B03ff = subs(B03F, t, ttf);

B13aa = subs(B13A, t, tta);
B13bb = subs(B13B, t, ttb);
B13cc = subs(B13C, t, ttc);
B13dd = subs(B13D, t, ttd);
B13ee = subs(B13E, t, tte);
B13ff = subs(B13F, t, ttf);

B23aa = subs(B23A, t, tta);
B23bb = subs(B23B, t, ttb);
B23cc = subs(B23C, t, ttc);
B23dd = subs(B23D, t, ttd);
B23ee = subs(B23E, t, tte);
B23ff = subs(B23F, t, ttf);

B33aa = subs(B33A, t, tta);
```

```
B33bb = subs(B33B, t, ttb);
B33cc = subs(B33C, t, ttc);
B33dd = subs(B33D, t, ttd);
B33ee = subs(B33E, t, tte);
B33ff = subs(B33F, t, ttf);

figure,
plot(tta, B03aa, 'k-', ttb, B03bb, 'k--', ttc, B03cc, 'k-.', ...
    ttd, B03dd, 'b-', tte, B03ee, 'b--', ttf, B03ff, 'b-.');
hold on;
plot(tta, B13aa, 'k-', ttb, B13bb, 'k--', ttc, B13cc, 'k-.', ...
    ttd, B13dd, 'b-', tte, B13ee, 'b--', ttf, B13ff, 'b-.');
plot(tta, B23aa, 'k-', ttb, B23bb, 'k--', ttc, B23cc, 'k-.', ...
    ttd, B23dd, 'b-', tte, B23ee, 'b--', ttf, B23ff, 'b-.');
plot(tta, B33aa, 'k-', ttb, B33bb, 'k--', ttc, B33cc, 'k-.', ...
    ttd, B33dd, 'b-', tte, B33ee, 'b--', ttf, B33ff, 'b-.');
xlabel ('t'); ylabel('B'); title('B03 - B13 - B23 - B33');
legend('A', 'B', 'C', 'D', 'E', 'F');
hold off;

%plotting curve

xa = x(1); ya = y(1);
xb = x(2); yb = y(2);
xc = x(3); yc = y(3);
xd = x(4); yd = y(4);
xe = x(5); ye = y(5);
xf = x(6); yf = y(6);
xaa = subs(xa, t, tta); yaa = subs(ya, t, tta);
xbb = subs(xb, t, ttb); ybb = subs(yb, t, ttb);
xcc = subs(xc, t, ttc); ycc = subs(yc, t, ttc);
xdd = subs(xd, t, ttd); ydd = subs(yd, t, ttd);
```

```
xee = subs(xe, t, tte); yee = subs(ye, t, tte);
xff = subs(xf, t, ttf); yff = subs(yf, t, ttf);

X = [x0, x1, x2, x3]; Y = [y0, y1, y2, y3];

figure
subplot(131),
plot(tta, xaa, 'k-', ttb, xbb,  'k--', ttc, xcc,  'k-.', ...
    ttd, xdd,  'b-', tte, xee,  'b--', ttf, xff, 'b-.');
xlabel ('t'); ylabel('x'); axis square; title('t - x');
subplot(132),
plot(tta, yaa, 'k-', ttb, ybb,  'k--', ttc, ycc,  'k-.', ...
    ttd, ydd,  'b-', tte, yee,  'b--', ttf, yff, 'b-.')
xlabel ('t'); ylabel('y'); axis square; title('t - y');
subplot(133),
plot(xaa, yaa, 'k-', xbb, ybb,  'k--', xcc, ycc,  'k-.', ...
    xdd, ydd,  'b-', xee, yee,  'b--', xff, yff, 'b-.'); hold on;
scatter(X, Y, 20, 'r', 'filled');
xlabel ('x'); ylabel('y'); axis square; grid; title('x - y');
axis([0 9 -2 6]);
d = 0.5;
text(x0, y0+d, 'P_0');
text(x1, y1-d, 'P_1');
text(x2, y2-d, 'P_2');
text(x3, y3+d, 'P_3');
hold off;
```

NOTE

...: continues the current command or function call onto the next line
`simplify`: simplifies an equation by resolving all intersections and nestings

3.5 JUSTIFICATION FOR KNOT-VECTOR VALUES

Now, it is time to provide a justification for choosing KV values of [0, 1, 2, 3, …] that we have been using so far. Let us first choose some other set of values and observe the result we get. So let us assume that for a uniform quadratic B-spline the KV is changed to $T = [0, 5, 10, 15, 20, 25, 30]$. We can choose arbitrarily any set of values, the only constraint being that gaps between the values should be uniform, as is the requirement for a uniform B-spline. So in this case we choose a KV with gaps 5 times the earlier values. If we proceed along the same steps as in Section 3.4 and calculate the BFs we get the result shown below:

$$
\begin{aligned}
B_{0,3} &= \begin{cases} \left(\dfrac{1}{50}\right)t^2 & (0 \le t < 1)A \\ -\dfrac{t^2}{25} + \dfrac{3t}{5} - \dfrac{3}{2} & (1 \le t < 2)B \\ \left(\dfrac{1}{50}\right)(t-15)^2 & (2 \le t < 3)C \end{cases} \\
B_{1,3} &= \begin{cases} \left(\dfrac{1}{50}\right)(t-5)^2 & (1 \le t < 2)B \\ -\dfrac{t^2}{25} + t - \dfrac{11}{2} & (2 \le t < 3)C \\ \left(\dfrac{1}{50}\right)(t-20)^2 & (3 \le t < 4)D \end{cases} \\
B_{2,3} &= \begin{cases} \left(\dfrac{1}{50}\right)(t-10)^2 & (2 \le t < 3)C \\ -\dfrac{t^2}{25} + \dfrac{7t}{5} - \dfrac{23}{2} & (3 \le t < 4)D \\ \left(\dfrac{1}{50}\right)(t-25)^2 & (4 \le t < 5)E \end{cases} \\
B_{3,3} &= \begin{cases} \left(\dfrac{1}{50}\right)(t-15)^2 & (3 \le t < 4)D \\ -\dfrac{t^2}{25} + \dfrac{9t}{5} - \dfrac{39}{2} & (4 \le t < 5)E \\ \left(\dfrac{1}{50}\right)(t-30)^2 & (5 \le t < 6)F \end{cases}
\end{aligned} \tag{3.18}
$$

Comparing with Equation (3.14) tells us that t has been replaced by $\frac{t}{5}$ i.e. all t values have been scaled by a factor of 5. So now, to get the same B values require the t values to be 5 times larger than earlier. This is reflected in the BF plot in Figure 3.7, which shows the t-axis to be expanded by 5 times. Compare with the BF plot in Figure 3.5.

On plugging the new KV into Example 3.2, we can observe the effects on the parametric curves. The $x(t)$ and $y(t)$ values shown below are seen to be affected similar to the BFs i.e. the values have been scaled by 5.

$$
x(t)=\begin{cases}
\dfrac{t^2}{50} & (0\le t<1)\\[4pt]
\dfrac{t^2}{25}-\dfrac{t}{5}+\dfrac{1}{2} & (1\le t<2)\\[4pt]
\dfrac{t^2}{50}+t-\dfrac{11}{2} & (2\le t<3)\\[4pt]
\dfrac{2t}{5}-1 & (3\le t<4)\\[4pt]
-\dfrac{t^2}{5}+\dfrac{42t}{5}-81 & (4\le t<5)\\[4pt]
\dfrac{4}{25}(t-30)^2 & (5\le t<6)
\end{cases}
\qquad
y(t)=\begin{cases}
\dfrac{t^2}{25} & (0\le t<1)\\[4pt]
-\dfrac{3t^2}{50}+t-\dfrac{5}{2} & (1\le t<2)\\[4pt]
\dfrac{t^2}{10}-\dfrac{11t}{5}+\dfrac{27}{2} & (2\le t<3)\\[4pt]
-\dfrac{t^2}{5}+\dfrac{34t}{5}-54 & (3\le t<4)\\[4pt]
\dfrac{7t^2}{50}-\dfrac{34t}{5}+82 & (4\le t<5)\\[4pt]
-\dfrac{1}{50}(t-30)^2 & (5\le t<6)
\end{cases}
$$

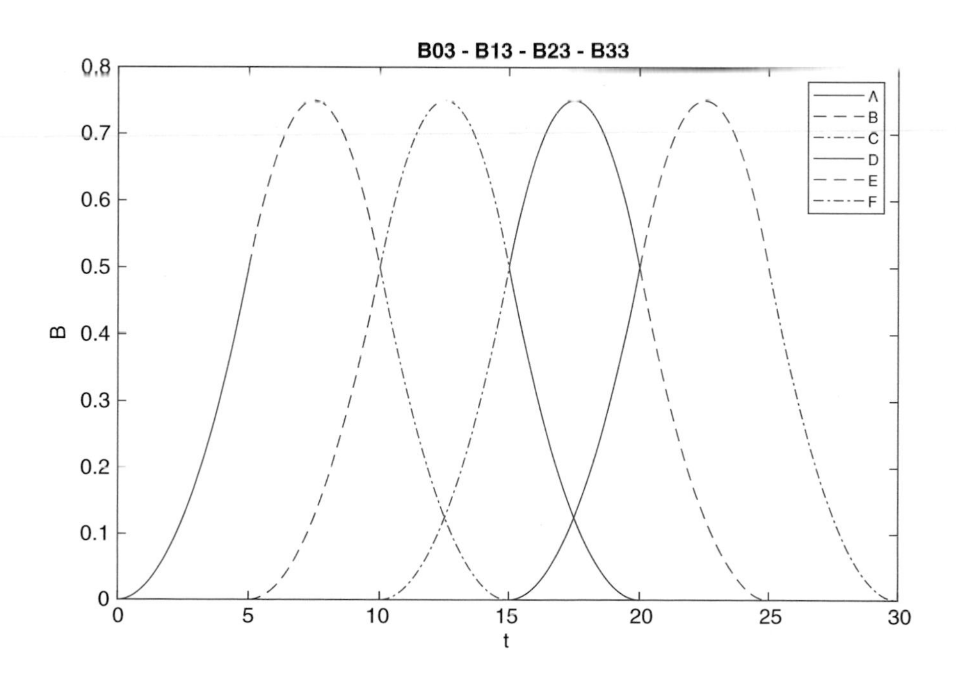

FIGURE 3.7 Change in BFs with change in KV.

On plotting the actual curves, the $x-t$ and $y-t$ graphs are seen to be similarly affected i.e. the t values have changed by 5 times. See Figure 3.8. The different line types indicate the segment intervals. However, since the x and y values have remained unchanged, the $x-y$ plot remains exactly the same as before as it is unaffected by any change in t values so long as it affects the x and y values uniformly by the same amount. Compare with Figure 3.6. For example, from Figure 3.6 for $t=2, x=2.5, y=1.5$ and from Figure 3.8 for $t=10, x=2.5, y=1.5$, which implies that the x vs. y point remains unaffected. This is true for all the points and so the $x-y$ plot of the curve remains same as before. Obviously, this is true for any scaling factor by which t is changed.

Readers are encouraged to generate the plots and verify the results by using different KV values. Then can easily do so by using MATLAB Code 3.2 and simply changing the KV in the second line of the program.

This leads us to conclude that the actual curve in the spatial domain is independent of the KV values since it does not depend on the scaling factor of the t values. Hence, it is customary to choose the smallest t values viz. 0, 1, 2, ... to reduce the complexity of the equations but in reality, we can choose any values for the KV and would get the same result.

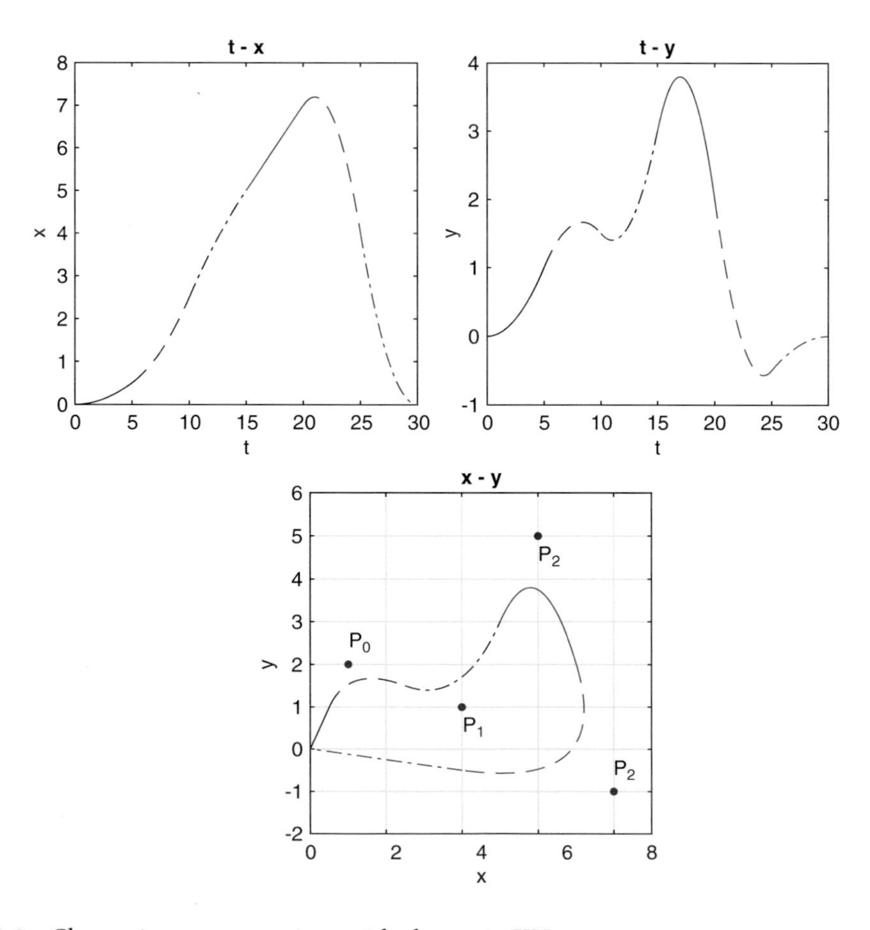

FIGURE 3.8 Change in curve equations with change in KV.

3.6 QUADRATIC OPEN-UNIFORM *B*-SPLINE

In an open uniform spline, the KV is uniform except at the ends where it is repeated d times where $(d-1)$ is the degree of the curve. Repeated values are referred to as multiplicity. Multiplicity implies that denominators of the Cox de Boor terms become zero in many cases. Hence, an important assumption needs to be made here: division by zero is treated as zero.

Consider an open-uniform quadratic *B*-spline with $d = 3$ and $n = 3$. Let the KV be chosen as $T = \{1, 1, 1, 2, 3, 3, 3\}$. As before, performing a segment-wise analysis the results obtained are tabulated in Table 3.3.

Substituting the above values in Equation (3.13) we get:

$$
\begin{aligned}
B_{0,3} &= \begin{cases} (t-2)^2 & (2 \le t < 3)\,C \end{cases} \\
B_{1,3} &= \begin{cases} -\dfrac{3}{2}t^2 + 5t - \dfrac{7}{2} & (2 \le t < 3)\,C \\ \left(\dfrac{1}{2}\right)(t-3)^2 & (3 \le t < 4)\,D \end{cases} \\
B_{2,3} &= \begin{cases} \left(\dfrac{1}{2}\right)(t-1)^2 & (2 \le t < 3)\,C \\ -\dfrac{3}{2}t^2 + 7t - \dfrac{15}{2} & (3 \le t < 4)\,D \end{cases} \\
B_{3,3} &= \begin{cases} (t-2)^2 & (3 \le t < 4)\,D \end{cases}
\end{aligned}
\tag{3.19}
$$

Equation (3.19) represents the BFs of a quadratic open-uniform *B*-spline with four CPs. The plot of the BFs is shown in Figure 3.9. Only the portion pertaining to segments $C(2 \le t < 3)$ and $D(3 \le t < 4)$ are present.

The curve equation is obtained by substituting the BF values into Equation (3.15):

$$
P(t) = \begin{cases}
0 & (0 \le t < 1) \\
0 & (1 \le t < 2) \\
P_0 \cdot (t-2)^2 + P_1 \cdot \left(-\dfrac{3}{2}t^2 + 5t - \dfrac{7}{2}\right) + P_2 \cdot \left(\dfrac{1}{2}\right)(t-1)^2 & (2 \le t < 3) \\
P_1 \cdot \left(\dfrac{1}{2}\right)(t-3)^2 + P_2 \cdot \left(-\dfrac{3}{2}t^2 + 7t - \dfrac{15}{2}\right) + P_3 \cdot (t-2)^2 & (3 \le t < 4) \\
0 & (4 \le t < 5) \\
0 & (5 \le t < 6)
\end{cases}
\tag{3.20}
$$

TABLE 3.3 Computation of BFs of Quadratic Open-Uniform *B*-Spline

Segment	t	$B_{k,1}$	$B_{k,2}$	$B_{k,3}$
A	$0 \le t < 1$	$B_{0,1}=1$ $B_{1,1}=0$ $B_{2,1}=0$ $B_{3,1}=0$ $B_{4,1}=0$ $B_{5,1}=0$	$B_{0,2}=0$ $B_{1,2}=0$ $B_{2,2}=0$ $B_{3,2}=0$ $B_{4,2}=0$	$B_{0,3}=0$ $B_{1,3}=0$ $B_{2,3}=0$ $B_{3,3}=0$
B	$1 \le t < 2$	$B_{0,1}=0$ $B_{1,1}=1$ $B_{2,1}=0$ $B_{3,1}=0$ $B_{4,1}=0$ $B_{5,1}=0$	$B_{0,2}=0$ $B_{1,2}=0$ $B_{2,2}=0$ $B_{3,2}=0$ $B_{4,2}=0$	$B_{0,3}=0$ $B_{1,3}=0$ $B_{2,3}=0$ $B_{3,3}=0$
C	$2 \le t < 3$	$B_{0,1}=0$ $B_{1,1}=0$ $B_{2,1}=1$ $B_{3,1}=0$ $B_{4,1}=0$ $B_{5,1}=0$	$B_{0,2}=0$ $B_{1,2}=2-t$ $B_{2,2}=t-1$ $B_{3,2}=0$ $B_{4,2}=0$	$B_{0,3}=(t-2)^2$ $B_{1,3}=-\frac{3}{2}t^2+5t-\frac{7}{2}$ $B_{2,3}=\left(\frac{1}{2}\right)(t-1)^2$ $B_{3,3}=0$
D	$3 \le t < 4$	$B_{0,1}=0$ $B_{1,1}=0$ $B_{2,1}=0$ $B_{3,1}=1$ $B_{4,1}=0$ $B_{5,1}=0$	$B_{0,2}=0$ $B_{1,2}=0$ $B_{2,2}=3-t$ $B_{3,2}=t-2$ $B_{4,2}=0$	$B_{0,3}=0$ $B_{1,3}=\left(\frac{1}{2}\right)(t-3)^2$ $B_{2,3}=-\frac{3}{2}t^2+7t-\frac{15}{2}$ $B_{3,3}=(t-2)^2$
E	$4 \le t < 5$	$B_{0,1}=0$ $B_{1,1}=0$ $B_{2,1}=0$ $B_{3,1}=0$ $B_{4,1}=1$ $B_{5,1}=0$	$B_{0,2}=0$ $B_{1,2}=0$ $B_{2,2}=0$ $B_{3,2}=0$ $B_{4,2}=0$	$B_{0,3}=0$ $B_{1,3}=0$ $B_{2,3}=0$ $B_{3,3}=0$
F	$5 \le t < 6$	$B_{0,1}=0$ $B_{1,1}=0$ $B_{2,1}=0$ $B_{3,1}=0$ $B_{4,1}=0$ $B_{5,1}=1$	$B_{0,2}=0$ $B_{1,2}=0$ $B_{2,2}=0$ $B_{3,2}=0$ $B_{4,2}=0$	$B_{0,3}=0$ $B_{1,3}=0$ $B_{2,3}=0$ $B_{3,3}=0$

A plot of the curve in Figure 3.10 shows the effect of multiplicity on the spline: it has forced the curve to actually go through the first and last CPs. This has created a spline with only two segments *C* and *D*, while the other segments *A*, *B*, *E*, and *F* are non-existent. We can, therefore, conclude that an approximating spline can behave like an interpolating or hybrid spline if the KV has repeated values.

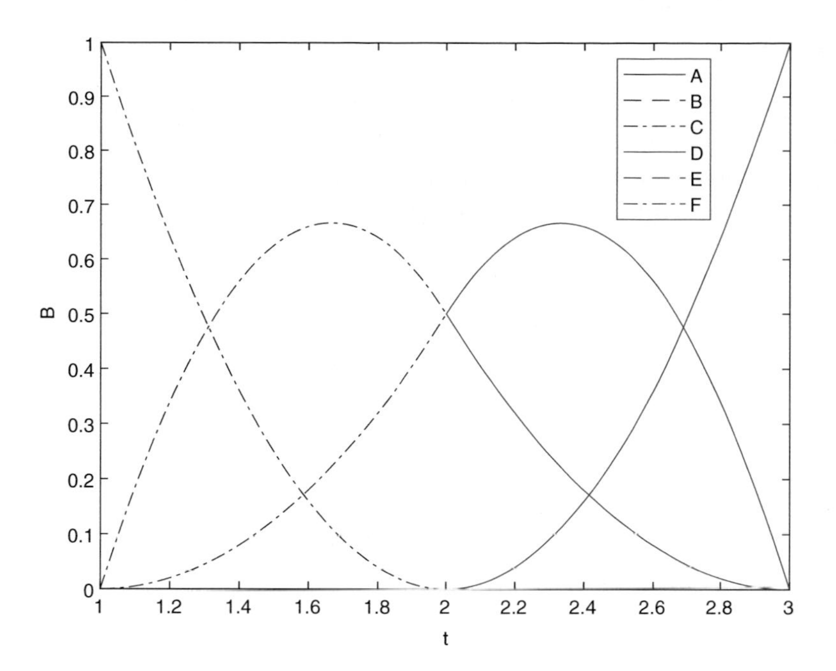

FIGURE 3.9 BFs of a quadratic open-uniform B-spline.

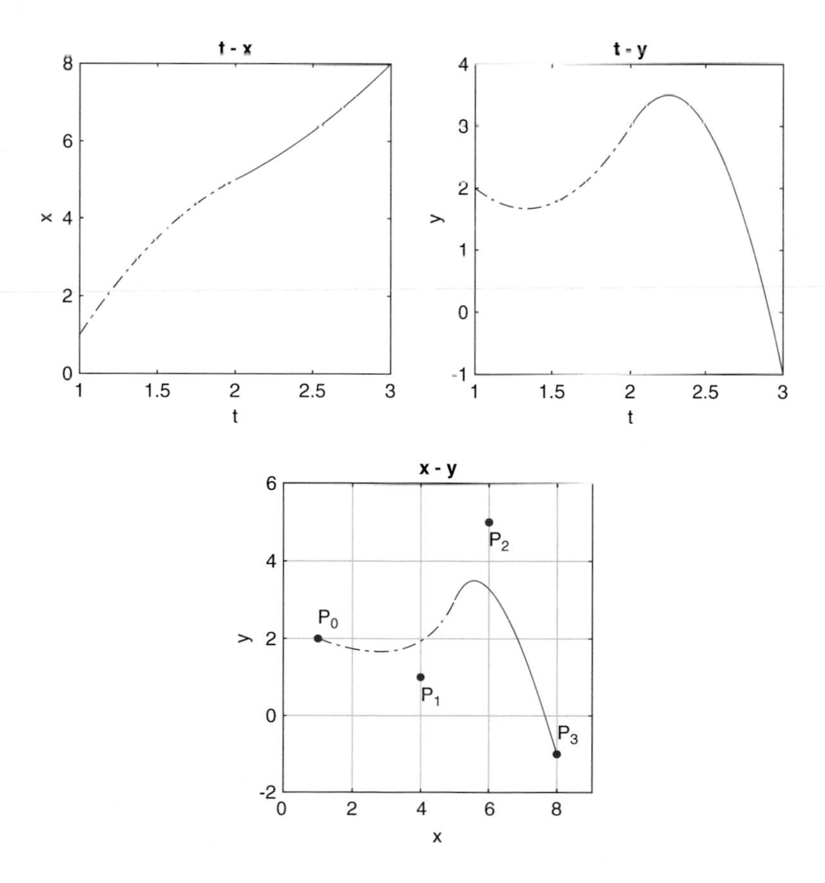

FIGURE 3.10 Open-uniform quadratic B-spline curve.

Example 3.3

Find the equation of an open-uniform quadratic B-spline having CPs (1, 2), (4, 1), (6, 5), and (8, –1) assuming a KV T = [1, 1, 1, 2, 3, 3, 3]. Also write a program to plot the BFs and the actual curve.

From Equation (3.20) substituting the given CPs we get:

$$x(t)=\begin{cases} 0 & (0 \le t < 1) \\ 0 & (1 \le t < 2) \\ -2t^2+10t-7 & (2 \le t < 3) \\ -t^2-2t+5 & (3 \le t < 4) \\ 0 & (4 \le t < 5) \\ 0 & (5 \le t < 6) \end{cases} \qquad y(t)=\begin{cases} 0 & (0 \le t < 1) \\ 0 & (1 \le t < 2) \\ 3t^2-8t+7 & (2 \le t < 3) \\ -8t^2+36t-37 & (3 \le t < 4) \\ 0 & (4 \le t < 5) \\ 0 & (5 \le t < 6) \end{cases}$$

MATLAB Code 3.3

The code will be almost exactly same as MATLAB Code 3.2 except for an additional check to avoid a divide-by-zero condition before calculating each BF. If a divide-by-zero condition exists, then the expression is to replaced by 0; otherwise, it should be calculated in the normal process. This is illustrated as follows.

```
D =(t1 - t0); if D == 0, S1 = 0; else S1 = (t - t0)/D; end;
D =(t2 - t1); if D == 0, S2 = 0; else S2 = (t - t0)/D; end;
B02 = S1*B01 + S2*B11; B02A = B02;
```

3.7 QUADRATIC NON-UNIFORM *B*-SPLINE

In a non-uniform spline, the KV is not uniform i.e. the gaps between the knot elements are different. This makes the BFs unsymmetric and they tend to cluster together where the gaps are smaller. This has the effect of drawing the curve towards the corresponding CPs. Figure 3.11 shows the BFs of a non-uniform quadratic *B*-spline. The amount of non-symmetry would depend upon the gaps in the KV.

Example 3.4

Find the equation of a non-uniform quadratic B-spline having CPs (1, 2), (4, 1), (6, 5), and (8, –1) assuming a KV T = [0, 1, 3, 5, 15, 18, 20].

Using a segment-wise analysis:

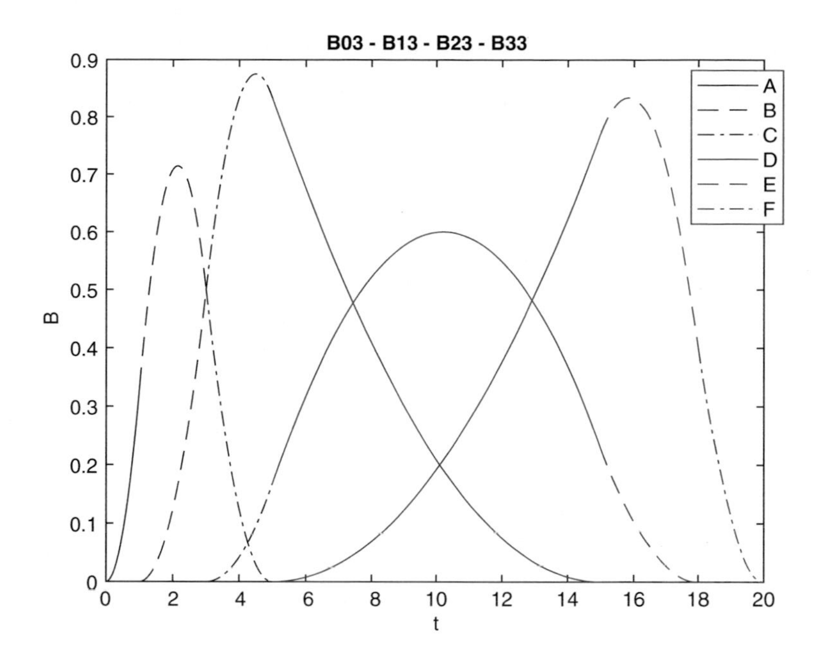

FIGURE 3.11 BFs of a quadratic non-uniform *B*-spline.

$$x(t)=\begin{cases} \dfrac{t^2}{3} & (0\le t<1) \\ \dfrac{5t^2}{24}+\dfrac{t}{4}-\dfrac{1}{8} & (1\le t<2) \\ -\dfrac{7t^2}{24}+\dfrac{13t}{4}-\dfrac{37}{8} & (2\le t<3) \\ -\dfrac{t^2}{780}+\dfrac{9t}{26}+\dfrac{137}{52} & (3\le t<4) \\ -\dfrac{38t^2}{65}+\dfrac{232t}{13}-\dfrac{1,672}{13} & (4\le t<5) \\ \dfrac{4(t-20)^2}{5} & (5\le t<6) \end{cases} \qquad y(t)=\begin{cases} \dfrac{2t^2}{3} & (0\le t<1) \\ -\dfrac{11t^2}{24}+\dfrac{9t}{4}-\dfrac{9}{8} & (1\le t<2) \\ \dfrac{7t^2}{24}-\dfrac{9t}{4}+\dfrac{45}{8} & (2\le t<3) \\ -\dfrac{31t^2}{390}+\dfrac{19t}{13}-\dfrac{95}{26} & (3\le t<4) \\ \dfrac{43t^2}{195}-\dfrac{98t}{13}+\dfrac{830}{13} & (4\le t<5) \\ -\dfrac{(t-20)^2}{10} & (5\le t<6) \end{cases}$$

MATLAB Code 3.4

Same as MATLAB Code 3.2 with a change in the KV.

3.8 CUBIC UNIFORM *B*-SPLINE

A cubic uniform *B*-spline is treated in much the same manner as a quadratic uniform *B*-spline with additional layer of complexity arising out of the fourth-order BF terms generated by the Cox de Boor algorithm.

For generating a cubic *B*-spline, we need to start with $d = 4$ and $n = 4$

Then degree of the curve: $d - 1 = 3$

Number of CPs: $n + 1 = 5$

Number of curve segments: $d + n = 8$

Number of elements in the KV: $d + n + 1 = 9$

Let the curve segments be *A*, *B*, *C*, *D*, *E*, *F*, *G*, and *H* and CPs be P_0, P_1, P_2, P_3, and P_4 (see Figure 3.12).

Let the KV be $T = \{t_k\}$ for $k = \{0, 1, 2, 3, 4, 5, 6, 7, 8\}$. In this case, $T = [t_0, t_1, t_2, t_3, t_4, t_5, t_6, t_7, t_8]$.

Let the BFs are $B_{0,4}$, $B_{1,4}$, $B_{2,4}$, $B_{3,4}$, and $B_{4,4.}$

Equation of the curve is given by:

$$P(t) = P_0 \cdot B_{0,4} + P_1 \cdot B_{1,4} + P_2 \cdot B_{2,4} + P_3 \cdot B_{3,4} + P_4 \cdot B_{4,4} \tag{3.21}$$

As before, we assume the KV to be $T = [0, 1, 2, 3, 4, 5, 6, 7, 8]$. The first-order terms $B_{0,1}, B_{1,1}, B_{2,1}, B_{3,1}, B_{4,1}, B_{5,1}, B_{6,1}, B_{7,1}$, and $B_{8,1}$ will be either 0 or 1 as per the first condition of the Cox de Boor algorithm. The second-order terms are computed as follows:

$$B_{0,2} = (t-0) \cdot B_{0,1} + (2-t) \cdot B_{1,1}$$

$$B_{1,2} = (t-1) \cdot B_{1,1} + (3-t) \cdot B_{2,1}$$

$$B_{2,2} = (t-2) \cdot B_{2,1} + (4-t) \cdot B_{3,1}$$

$$B_{3,2} = (t-3) \cdot B_{3,1} + (5-t) \cdot B_{4,1}$$

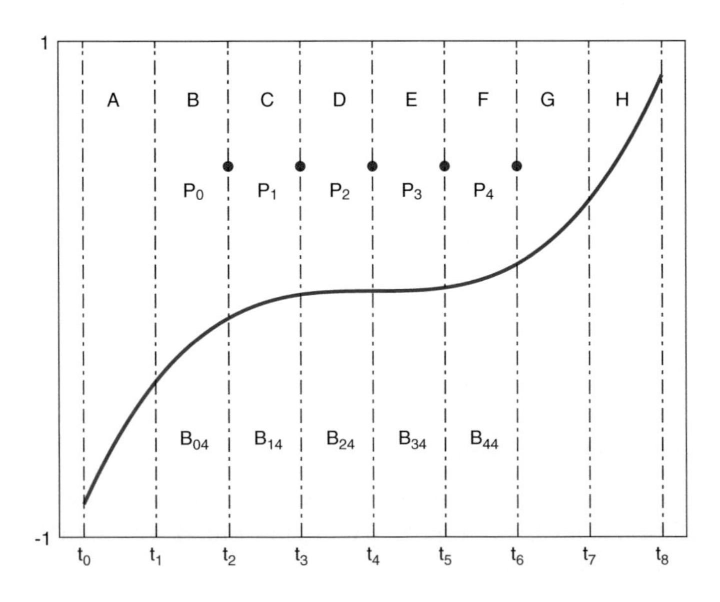

FIGURE 3.12 Cubic uniform *B*-spline with five CPs.

$$\begin{aligned}B_{4,2} &= (t-4)\cdot B_{4,1} + (6-t)\cdot B_{5,1}\\ B_{5,2} &= (t-5)\cdot B_{5,1} + (7-t)\cdot B_{6,1}\\ B_{6,2} &= (t-6)\cdot B_{6,1} + (8-t)\cdot B_{7,1}\\ B_{7,2} &= (t-7)\cdot B_{7,1} + (9-t)\cdot B_{8,1}\end{aligned} \tag{3.22}$$

The third-order terms are calculated from second-order terms:

$$\begin{aligned}B_{0,3} &= \left(\frac{1}{2}\right)(t-0)\cdot B_{0,2} + \left(\frac{1}{2}\right)(3-t)\cdot B_{1,2}\\ B_{1,3} &= \left(\frac{1}{2}\right)(t-1)\cdot B_{1,2} + \left(\frac{1}{2}\right)(4-t)\cdot B_{2,2}\\ B_{2,3} &= \left(\frac{1}{2}\right)(t-2)\cdot B_{2,2} + \left(\frac{1}{2}\right)(5-t)\cdot B_{3,2}\\ B_{3,3} &= \left(\frac{1}{2}\right)(t-3)\cdot B_{3,2} + \left(\frac{1}{2}\right)(6-t)\cdot B_{4,2}\\ B_{4,3} &= \left(\frac{1}{2}\right)(t-4)\cdot B_{4,2} + \left(\frac{1}{2}\right)(7-t)\cdot B_{5,2}\\ B_{5,3} &= \left(\frac{1}{2}\right)(t-5)\cdot B_{5,2} + \left(\frac{1}{2}\right)(8-t)\cdot B_{6,2}\\ B_{6,3} &= \left(\frac{1}{2}\right)(t-6)\cdot B_{6,2} + \left(\frac{1}{2}\right)(9-t)\cdot B_{7,2}\end{aligned} \tag{3.23}$$

The fourth-order terms are calculated from the third-order terms:

$$\begin{aligned}B_{0,4} &= \left(\frac{1}{3}\right)(t-0)\cdot B_{0,3} + \left(\frac{1}{3}\right)(4-t)\cdot B_{1,3}\\ B_{1,4} &= \left(\frac{1}{3}\right)(t-1)\cdot B_{1,3} + \left(\frac{1}{3}\right)(5-t)\cdot B_{2,3}\\ B_{2,4} &= \left(\frac{1}{3}\right)(t-2)\cdot B_{2,3} + \left(\frac{1}{3}\right)(6-t)\cdot B_{3,3}\\ B_{3,4} &= \left(\frac{1}{3}\right)(t-3)\cdot B_{3,3} + \left(\frac{1}{3}\right)(7-t)\cdot B_{4,3}\\ B_{4,4} &= \left(\frac{1}{3}\right)(t-4)\cdot B_{4,3} + \left(\frac{1}{3}\right)(8-t)\cdot B_{5,3}\end{aligned} \tag{3.24}$$

Since there are eight segments each BF consists of eight sub-components:

$$\begin{aligned}
B_{0,4} &= \{B_{0,4A}, B_{0,4B}, B_{0,4C}, B_{0,4D}, B_{0,4E}, B_{0,4F}, B_{0,4G}, B_{0,4H}\} \\
B_{1,4} &= \{B_{1,4A}, B_{1,4B}, B_{1,4C}, B_{1,4D}, B_{1,4E}, B_{1,4F}, B_{1,4G}, B_{1,4H}\} \\
B_{2,4} &= \{B_{2,4A}, B_{2,4B}, B_{2,4C}, B_{2,4D}, B_{2,4E}, B_{2,4F}, B_{2,4G}, B_{2,4H}\} \\
B_{3,4} &= \{B_{3,4A}, B_{3,4B}, B_{3,4C}, B_{3,4D}, B_{3,4E}, B_{3,4F}, B_{3,4G}, B_{3,4H}\} \\
B_{4,4} &= \{B_{4,4A}, B_{4,4B}, B_{4,4C}, B_{4,4D}, B_{4,4E}, B_{4,4F}, B_{4,4G}, B_{4,4H}\}
\end{aligned} \tag{3.25}$$

Table 3.4 summarizes the calculation of the BF values.

TABLE 3.4 Computation of BFs of Cubic Uniform *B*-Spline

Segment	t	$B_{k,1}$	$B_{k,2}$	$B_{k,3}$	$B_{k,4}$
A	$0 \le t < 1$	$B_{0,1}=1$ $B_{1,1}=0$ $B_{2,1}=0$ $B_{3,1}=0$ $B_{4,1}=0$ $B_{5,1}=0$ $B_{6,1}=0$ $B_{7,1}=0$	$B_{0,2}=t$ $B_{1,2}=0$ $B_{2,2}=0$ $B_{3,2}=0$ $B_{4,2}=0$	$B_{0,3}=\left(\frac{1}{2}\right)t^2$ $B_{1,3}=0$ $B_{2,3}=0$ $B_{3,3}=0$	$B_{0,4}=\left(\frac{1}{6}\right)t^3$
B	$1 \le t < 2$	$B_{0,1}=0$ $B_{1,1}=1$ $B_{2,1}=0$ $B_{3,1}=0$ $B_{4,1}=0$ $B_{5,1}=0$ $B_{6,1}=0$ $B_{7,1}=0$	$B_{0,2}=2-t$ $B_{1,2}=t-1$ $B_{2,2}=0$ $B_{3,2}=0$ $B_{4,2}=0$	$B_{0,3}=-t^2+3t-\frac{3}{2}$ $B_{1,3}=\left(\frac{1}{2}\right)(t-1)^2$ $B_{2,3}=0$ $B_{3,3}=0$	$B_{0,4}=-\frac{t^3}{3}+2t^2-2t+\frac{2}{3}$ $B_{1,4}=\left(\frac{1}{6}\right)(t-1)^3$
C	$2 \le t < 3$	$B_{0,1}=0$ $B_{1,1}=0$ $B_{2,1}=1$ $B_{3,1}=0$ $B_{4,1}=0$ $B_{5,1}=0$ $B_{6,1}=0$ $B_{7,1}=0$	$B_{0,2}=0$ $B_{1,2}=3-t$ $B_{2,2}=t-2$ $B_{3,2}=0$ $B_{4,2}=0$	$B_{0,3}=\left(\frac{1}{2}\right)(t-3)^2$ $B_{1,3}=-t^2+5t-\frac{11}{2}$ $B_{2,3}=\left(\frac{1}{2}\right)(t-2)^2$ $B_{3,3}=0$	$B_{0,4}=-\frac{t^3}{2}-4t^2+10t-\frac{22}{3}$ $B_{1,4}=-\frac{t^3}{2}+\frac{7t^2}{2}-\frac{15t}{2}+\frac{31}{6}$ $B_{2,4}=\left(\frac{1}{6}\right)(t-2)^3$
D	$3 \le t < 4$	$B_{0,1}=0$ $B_{1,1}=0$ $B_{2,1}=0$ $B_{3,1}=1$ $B_{4,1}=0$ $B_{5,1}=0$ $B_{6,1}=0$ $B_{7,1}=0$	$B_{0,2}=0$ $B_{1,2}=0$ $B_{2,2}=4-t$ $B_{3,2}=t-3$ $B_{4,2}=0$	$B_{0,3}=0$ $B_{1,3}=\left(\frac{1}{2}\right)(t-4)^2$ $B_{2,3}=-t^2+7t-\frac{23}{2}$ $B_{3,3}=\left(\frac{1}{2}\right)(t-3)^2$	$B_{0,4}=\left(\frac{1}{6}\right)(t-4)^3$ $B_{1,4}=-\frac{t^3}{2}-\frac{11t^2}{2}+\frac{39t}{2}-\frac{131}{6}$ $B_{2,4}=-\frac{t^3}{2}+\frac{5t^2}{2}-16t+\frac{50}{3}$ $B_{3,4}=\left(\frac{1}{6}\right)(t-3)^3$

(*Continued*)

TABLE 3.4 (*Continued*) Computation of BFs of Cubic Uniform *B*-Spline

Segment	t	$B_{k,1}$	$B_{k,2}$	$B_{k,3}$	$B_{k,4}$
E	$4 \le t < 5$	$B_{0,1}=0$ $B_{1,1}=0$ $B_{2,1}=0$ $B_{3,1}=0$ $B_{4,1}=1$ $B_{5,1}=0$ $B_{6,1}=0$ $B_{7,1}=0$	$B_{0,2}=0$ $B_{1,2}=0$ $B_{2,2}=0$ $B_{3,2}=5-t$ $B_{4,2}=t-4$	$B_{0,3}=0$ $B_{1,3}=0$ $B_{2,3}=\left(\frac{1}{2}\right)(t-5)^2$ $B_{3,3}=-t^2+9t-\frac{39}{2}$ $B_{4,3}=\left(\frac{1}{2}\right)(t-4)^2$	$B_{1,4}=\left(\frac{1}{6}\right)(t-5)^3$ $B_{2,4}=\frac{t^3}{2}-\frac{7t^2}{2}+32t-\frac{142}{3}$ $B_{3,4}=-\frac{t^3}{2}+\frac{13t^2}{2}-\frac{55t}{2}+\frac{229}{6}$ $B_{4,4}=\left(\frac{1}{6}\right)(t-4)^3$
F	$5 \le t < 6$	$B_{0,1}=0$ $B_{1,1}=0$ $B_{2,1}=0$ $B_{3,1}=0$ $B_{4,1}=0$ $B_{5,1}=1$ $B_{6,1}=0$ $B_{7,1}=0$	$B_{0,2}=0$ $B_{1,2}=0$ $B_{2,2}=0$ $B_{3,2}=0$ $B_{4,2}=6-t$ $B_{5,2}=t-5$	$B_{0,3}=0$ $B_{1,3}=0$ $B_{2,3}=0$ $B_{3,3}=\left(\frac{1}{2}\right)(t-6)^2$ $B_{4,3}=-t^2+11t-\frac{59}{2}$ $B_{5,3}=\left(\frac{1}{2}\right)(t-5)^2$	$B_{2,4}=\left(\frac{1}{6}\right)(t-6)^3$ $B_{3,4}=\frac{t^3}{2}-\frac{17t^2}{2}+\frac{95t}{2}-\frac{521}{6}$ $B_{4,4}=-\frac{t^3}{2}+8t^2-42t+\frac{218}{3}$
G	$5 \le t < 6$	$B_{0,1}=0$ $B_{1,1}=0$ $B_{2,1}=0$ $B_{3,1}=0$ $B_{4,1}=0$ $B_{5,1}=0$ $B_{6,1}=1$ $B_{7,1}=0$	$B_{0,2}=0$ $B_{1,2}=0$ $B_{2,2}=0$ $B_{3,2}=0$ $B_{4,2}=0$ $B_{5,2}=7-t$ $B_{6,2}=t-6$	$B_{0,3}=0$ $B_{1,3}=0$ $B_{2,3}=0$ $B_{3,3}=0$ $B_{4,3}=\left(\frac{1}{2}\right)(t-7)^2$ $B_{5,3}=-t^2+13t-\frac{83}{2}$ $B_{6,3}=\left(\frac{1}{2}\right)(t-6)^2$	$B_{3,4}=\left(\frac{1}{6}\right)(t-7)^3$ $B_{4,4}=\frac{t^3}{2}-10t^2+66t-\frac{430}{3}$
H	$5 \le t < 6$	$B_{0,1}=0$ $B_{1,1}=0$ $B_{2,1}=0$ $B_{3,1}=0$ $B_{4,1}=0$ $B_{5,1}=0$ $B_{6,1}=0$ $B_{7,1}=1$	$B_{0,2}=0$ $B_{1,2}=0$ $B_{2,2}=0$ $B_{3,2}=0$ $B_{4,2}=0$ $B_{5,2}=0$ $B_{6,2}=8-t$ $B_{7,2}=t-7$	$B_{0,3}=0$ $B_{1,3}=0$ $B_{2,3}=0$ $B_{3,3}=0$ $B_{5,3}=\left(\frac{1}{2}\right)(t-8)^2$ $B_{6,3}=-\frac{t^2}{2}+7t-24$	$B_{4,4}=\left(\frac{1}{6}\right)(t-8)^3$

Substituting the above values in Equation (3.25) we get:

$$B_{0,4} = \begin{cases} \left(\frac{1}{6}\right)t^3 & (0 \le t < 1)A \\ -\frac{t^3}{3} + 2t^2 - 2t + \frac{2}{3} & (1 \le t < 2)B \\ -\frac{t^3}{2} - 4t^2 + 10t - \frac{22}{3} & (2 \le t < 3)C \\ -\left(\frac{1}{6}\right)(t-4)^3 & (3 \le t < 4)D \end{cases}$$

$$B_{1,4} = \begin{cases} \left(\frac{1}{6}\right)(t-1)^3 & (1 \le t < 2)A \\ -\frac{t^3}{2} + \frac{7t^2}{2} - \frac{15t}{2} + \frac{31}{6} & (2 \le t < 3)B \\ -\frac{t^3}{2} - \frac{11t^2}{2} + \frac{39t}{2} - \frac{131}{6} & (3 \le t < 4)C \\ -\left(\frac{1}{6}\right)(t-5)^3 & (4 \le t < 5)D \end{cases}$$

$$B_{2,4} = \begin{cases} \left(\frac{1}{6}\right)(t-2)^3 & (2 \le t < 3)A \\ -\frac{t^3}{2} + \frac{5t^2}{2} - 16t + \frac{50}{3} & (3 \le t < 4)B \\ \frac{t^3}{2} - \frac{7t^2}{2} + 32t - \frac{142}{3} & (4 \le t < 5)C \\ -\left(\frac{1}{6}\right)(t-6)^3 & (5 \le t < 6)D \end{cases}$$

$$B_{3,4} = \begin{cases} \left(\frac{1}{6}\right)(t-3)^3 & (3 \le t < 4) A \\ -\frac{t^3}{2} + \frac{13t^2}{2} - \frac{55t}{2} + \frac{229}{6} & (4 \le t < 5) B \\ \frac{t^3}{2} - \frac{17t^2}{2} + \frac{95t}{2} - \frac{521}{6} & (5 \le t < 6) C \\ -\left(\frac{1}{6}\right)(t-7)^3 & (6 \le t < 7) D \end{cases}$$

$$B_{4,4} = \begin{cases} \left(\frac{1}{6}\right)(t-4)^3 & (4 \le t < 5) A \\ -\frac{t^3}{2} + 8t^2 - 42t + \frac{218}{3} & (5 \le t < 6) B \\ \frac{t^3}{2} - 10t^2 + 66t - \frac{430}{3} & (6 \le t < 7) C \\ -\left(\frac{1}{6}\right)(t-8)^3 & (7 \le t < 8) D \end{cases} \tag{3.26}$$

Equation (3.26) represents the BFs of a cubic uniform *B*-spline with five CPs and eight segments. The plot of the BFs is shown in Figure 3.13. Each BF has the same shape but shifted toward the right by 1 with respect to the previous one. Thus, each BF can be obtained from the previous one by substituting t with $(t - 1)$. As shown in Equation (3.26), each BF has eight sub-divisions out of which four are non-zeros. The first curve for *B*04 has non-zero parts for segments A $(0 \le t < 1)$, B $(1 \le t < 2)$, C $(2 \le t < 3)$, and D $(3 \le t < 4)$, the second curve for *B*14 has non-zero parts for segments B $(1 \le t < 2)$, C $(2 \le t < 3)$, D $(3 \le t < 4)$, and E $(4 \le t < 5)$, the third curve for *B*24 has non-zero parts for segments C $(2 \le t < 3)$, D $(3 \le t < 4)$, E $(4 \le t < 5)$, and F $(5 \le t < 6)$, the fourth curve for *B*34 has non-zero parts for segments D $(3 \le t < 4)$, E $(4 \le t < 5)$, F $(5 \le t < 6)$, and G $(6 \le t < 7)$, and the fourth curve for *B*44 has non-zero parts for segments E $(4 \le t < 5)$, F $(5 \le t < 6)$, G $(6 \le t < 7)$, and H $(7 \le t < 8)$. Since the BFs are associated with the CPs, this provides an indication of the local control property of the cubic spline. The first CP has influence over the first four segments A, B, C, and D, the second CP has influence over B, C, D, and E, and so on. This means that if the first CP is changed, it will affect only the first four segments while the rest of the spline will remain unchanged.

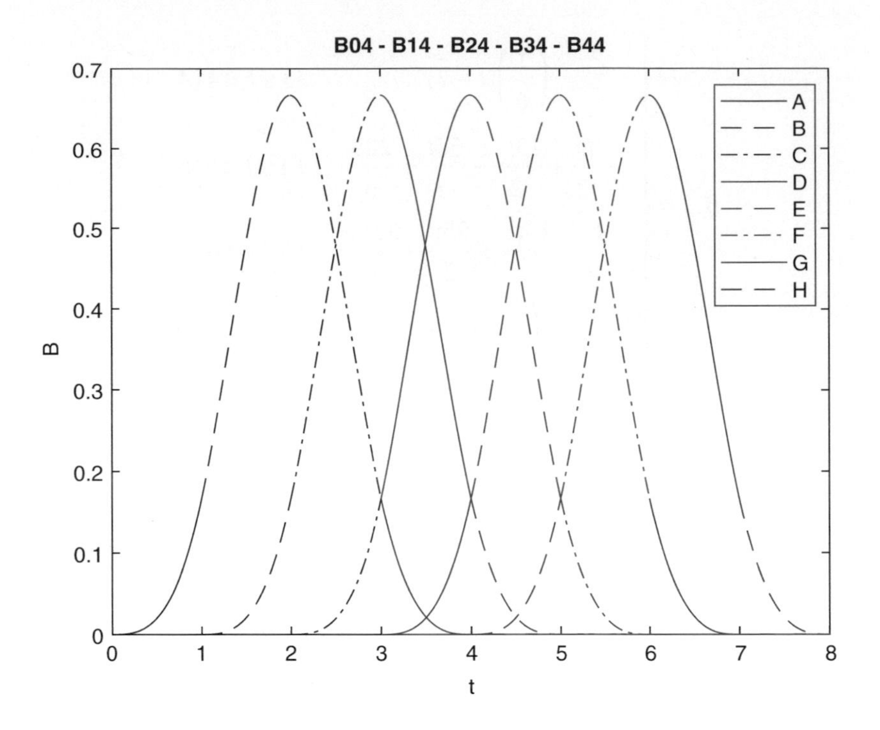

FIGURE 3.13 BFs of cubic uniform B-spline with five CPs.

The equation of the spline is a collection of the equations of its eight segments:

$$P(t)=\begin{cases} P_A & (0 \le t < 1) \\ P_B & (1 \le t < 2) \\ P_C & (2 \le t < 3) \\ P_D & (3 \le t < 4) \\ P_E & (4 \le t < 5) \\ P_F & (5 \le t < 6) \\ P_G & (6 \le t < 7) \\ P_H & (7 \le t < 8) \end{cases} \tag{3.27}$$

where

$$P_A = P_0 \cdot B_{0,4A} + P_1 \cdot B_{1,4A} + P_2 \cdot B_{2,4A} + P_3 \cdot B_{3,4A} + P_4 \cdot B_{4,4A}$$

$$P_B = P_0 \cdot B_{0,4B} + P_1 \cdot B_{1,4B} + P_2 \cdot B_{2,4B} + P_3 \cdot B_{3,4B} + P_4 \cdot B_{4,4B}$$

$$P_C = P_0 \cdot B_{0,4C} + P_1 \cdot B_{1,4C} + P_2 \cdot B_{2,4C} + P_3 \cdot B_{3,4C} + P_4 \cdot B_{4,4C}$$

$$P_D = P_0 \cdot B_{0,4D} + P_1 \cdot B_{1,4D} + P_2 \cdot B_{2,4D} + P_3 \cdot B_{3,4D} + P_4 \cdot B_{4,4D}$$

$$\begin{aligned}
P_E &= P_0 \cdot B_{0,4E} + P_1 \cdot B_{1,4E} + P_2 \cdot B_{2,4E} + P_3 \cdot B_{3,4E} + P_4 \cdot B_{4,4E} \\
P_F &= P_0 \cdot B_{0,4F} + P_1 \cdot B_{1,4F} + P_2 \cdot B_{2,4F} + P_3 \cdot B_{3,4F} + P_4 \cdot B_{4,4F} \\
P_G &= P_0 \cdot B_{0,4G} + P_1 \cdot B_{1,4G} + P_2 \cdot B_{2,4G} + P_3 \cdot B_{3,4G} + P_4 \cdot B_{4,4G} \\
P_H &= P_0 \cdot B_{0,4H} + P_1 \cdot B_{1,4H} + P_2 \cdot B_{2,4H} + P_3 \cdot B_{3,4H} + P_4 \cdot B_{4,4H}
\end{aligned} \tag{3.28}$$

Substituting values of the BFs from Table 3.4 into Equation (3.28) we get:

$$P(t) = \begin{cases}
P_0 \cdot \left(\frac{1}{6}\right) t^3 \\
P_0 \cdot \left(-\frac{t^3}{3} + 2t^2 - 2t + \frac{2}{3}\right) + P_1 \cdot \left(\frac{1}{6}\right)(t-1)^3 \\
P_0 \cdot \left(-\frac{t^3}{2} - 4t^2 + 10t - \frac{22}{3}\right) + P_1 \cdot \left(-\frac{t^3}{2} + \frac{7t^2}{2} - \frac{15t}{2} + \frac{31}{6}\right) + P_2 \cdot \left(\frac{1}{6}\right)(t-2)^3 \\
P_0 \cdot \left(\frac{1}{6}\right)(4-t)^3 + P_1 \cdot \left(-\frac{t^3}{2} - \frac{11t^2}{2} + \frac{39t}{2} - \frac{131}{6}\right) + P_2 \cdot \left(-\frac{t^3}{2} + \frac{5t^2}{2} - 16t + \frac{50}{3}\right) + P_3 \cdot \left(\frac{1}{2}\right)(t-3)^2 \\
P_1 \cdot \left(\frac{1}{6}\right)(5-t)^3 + P_2 \cdot \left(\frac{t^3}{2} - \frac{7t^2}{2} + 32t - \frac{142}{3}\right) + P_3 \cdot \left(-\frac{t^3}{2} + \frac{13t^2}{2} - \frac{55t}{2} + \frac{229}{6}\right) + P_4 \cdot \left(\frac{1}{6}\right)(t-4)^3 \\
P_2 \cdot \left(\frac{1}{6}\right)(6-t)^3 + P_3 \cdot \left(\frac{t^3}{2} - \frac{17t^2}{2} + \frac{95t}{2} - \frac{521}{6}\right) + P_4 \cdot \left(-\frac{t^3}{2} + 8t^2 - 42t + \frac{218}{3}\right) \\
P_3 \cdot \left(\frac{1}{6}\right)(7-t)^3 + P_4 \cdot \left(\frac{t^3}{2} - 10t^2 + 66t - \frac{430}{3}\right) \\
P_4 \cdot \left(\frac{1}{6}\right)(8-t)^3
\end{cases} \tag{3.29}$$

Equation (3.29) represents the curve equation of a cubic uniform B-spline with five CPs. The eight sub-components of the equations represents the parts for the eight segments.

Example 3.5

Find the equation of a uniform cubic B-spline having CPs (–1, 0), (0, 1), (1, 0), (0, –1), and (–0.5, –0.5). Also write a program to plot its BFs and the actual curve.

From Equation (3.29), substituting the values of the given CPs we get (Figure 3.14):

$$x(t)=\begin{cases} \dfrac{t^3}{6} & (0 \le t < 1) \\ \dfrac{t^3}{2}-2t^2+2t-\dfrac{2}{3} & (1 \le t < 2) \\ -\dfrac{t^3}{3}+3t^2-8t+6 & (2 \le t < 3) \\ -\dfrac{t^3}{3}+3t^2-8t+6 & (3 \le t < 4) \\ \dfrac{5t^3}{12}-6t^2+28t-42 & (4 \le t < 5) \\ \dfrac{t^3}{12}-2t^2+3t-\dfrac{1}{3} & (5 \le t < 6) \\ -\dfrac{t^3}{4}+5t^2-33t+\dfrac{215}{3} & (6 \le t < 7) \\ \dfrac{(t-8)^3}{12} & (7 \le t < 8) \end{cases} \qquad y(t)=\begin{cases} 0 & (0 \le t < 1) \\ \dfrac{(t-1)^3}{6} & (1 \le t < 2) \\ -\dfrac{t^3}{2}+\dfrac{7t^2}{2}-\dfrac{15t}{2}+\dfrac{31}{6} & (2 \le t < 3) \\ \dfrac{t^3}{3}-4t^2+15t-\dfrac{52}{3} & (3 \le t < 4) \\ \dfrac{t^3}{4}-3t^2+11t-12 & (4 \le t < 5) \\ -\dfrac{t^3}{4}+\dfrac{9t^2}{2}-\dfrac{53t}{2}+\dfrac{101}{2} & (5 \le t < 6) \\ -\dfrac{t^3}{12}+\dfrac{3t^2}{2}-\dfrac{17t}{2}+\dfrac{29}{2} & (6 \le t < 7) \\ \dfrac{(t-8)^3}{12} & (7 \le t < 8) \end{cases}$$

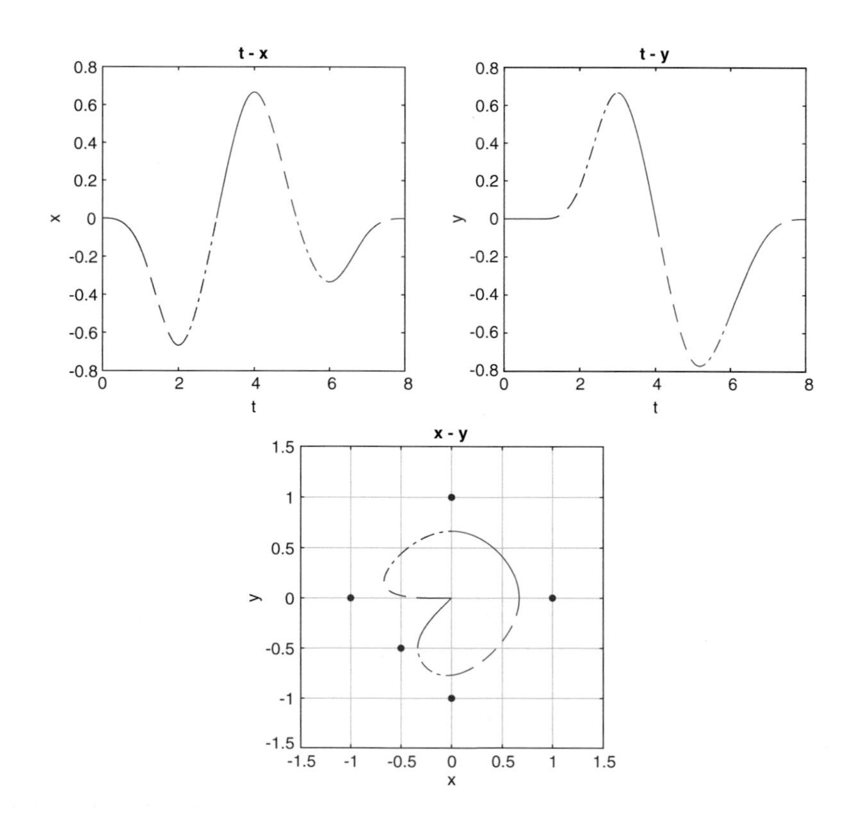

FIGURE 3.14 Plots for Example 3.5.

MATLAB Code 3.5

```
clear all; format compact; clc;

x0 = -1;    x1 = 0;     x2 = 1;     x3 = 0;     x4 = -0.5;
y0 = 0;     y1 = 1;     y2 = 0;     y3 = -1;    y4 = -0.5;

t0 = 0; t1 = 1; t2 = 2; t3 = 3; t4 = 4; t5 = 5; t6 = 6; t7 = 7; t8 = 8;
T = [t0, t1, t2, t3, t4, t5, t6, t7, t8];

syms t P0 P1 P2 P3 P4;
syms B01 B11 B21 B31 B41 B51 B61 B71;

B02 = ((t - t0)/(t1 - t0))*B01 + ((t2 - t)/(t2 - t1))*B11;
B12 = ((t - t1)/(t2 - t1))*B11 + ((t3 - t)/(t3 - t2))*B21;
B22 = ((t - t2)/(t3 - t2))*B21 + ((t4 - t)/(t4 - t3))*B31;
B32 = ((t - t3)/(t4 - t3))*B31 + ((t5 - t)/(t5 - t4))*B41;
B42 = ((t - t4)/(t5 - t4))*B41 + ((t6 - t)/(t6 - t5))*B51;
B52 = ((t - t5)/(t6 - t5))*B51 + ((t7 - t)/(t7 - t6))*B61;
B62 = ((t - t6)/(t7 - t6))*B61 + ((t8 - t)/(t8 - t7))*B71;
B72 = ((t - t7)/(t8 - t7))*B71 + 0;

B03 = ((t - t0)/(t2 - t0))*B02 + ((t3 - t)/(t3 - t1))*B12;
B13 = ((t - t1)/(t3 - t1))*B12 + ((t4 - t)/(t4 - t2))*B22;
B23 = ((t - t2)/(t4 - t2))*B22 + ((t5 - t)/(t5 - t3))*B32;
B33 = ((t - t3)/(t5 - t3))*B32 + ((t6 - t)/(t6 - t4))*B42;
B43 = ((t - t4)/(t6 - t4))*B42 + ((t7 - t)/(t7 - t5))*B52;
B53 = ((t - t5)/(t7 - t5))*B52 + ((t8 - t)/(t8 - t6))*B62;
B63 = ((t - t6)/(t8 - t6))*B62 + 0;

B04 = ((t - t0)/(t3 - t0))*B03 + ((t4 - t)/(t4 - t1))*B13;
B14 = ((t - t1)/(t4 - t1))*B13 + ((t5 - t)/(t5 - t2))*B23;
```

```
B24 = ((t - t2)/(t5 - t2))*B23 + ((t6 - t)/(t6 - t3))*B33;
B34 = ((t - t3)/(t6 - t3))*B33 + ((t7 - t)/(t7 - t4))*B43;
B44 = ((t - t4)/(t7 - t4))*B43 + ((t8 - t)/(t8 - t5))*B53;
B54 = ((t - t5)/(t8 - t5))*B53 + 0;

%Segment A
B02A = subs(B02, {B01,B11,B21,B31,B41,B51,B61,B71},{1,0,0,0,0,0,0,0});
B12A = subs(B12, {B01,B11,B21,B31,B41,B51,B61,B71},{1,0,0,0,0,0,0,0});
B22A = subs(B22, {B01,B11,B21,B31,B41,B51,B61,B71},{1,0,0,0,0,0,0,0});
B32A = subs(B32, {B01,B11,B21,B31,B41,B51,B61,B71},{1,0,0,0,0,0,0,0});
B42A = subs(B42, {B01,B11,B21,B31,B41,B51,B61,B71},{1,0,0,0,0,0,0,0});
B52A = subs(B52, {B01,B11,B21,B31,B41,B51,B61,B71},{1,0,0,0,0,0,0,0});
B62A = subs(B62, {B01,B11,B21,B31,B41,B51,B61,B71},{1,0,0,0,0,0,0,0});
B72A = subs(B72, {B01,B11,B21,B31,B41,B51,B61,B71},{1,0,0,0,0,0,0,0});
B03A = subs(B03, {B01,B11,B21,B31,B41,B51,B61,B71},{1,0,0,0,0,0,0,0});
B13A = subs(B13, {B01,B11,B21,B31,B41,B51,B61,B71},{1,0,0,0,0,0,0,0});
B23A = subs(B23, {B01,B11,B21,B31,B41,B51,B61,B71},{1,0,0,0,0,0,0,0});
B33A = subs(B33, {B01,B11,B21,B31,B41,B51,B61,B71},{1,0,0,0,0,0,0,0});
B43A = subs(B43, {B01,B11,B21,B31,B41,B51,B61,B71},{1,0,0,0,0,0,0,0});
B53A = subs(B53, {B01,B11,B21,B31,B41,B51,B61,B71},{1,0,0,0,0,0,0,0});
B63A = subs(B63, {B01,B11,B21,B31,B41,B51,B61,B71},{1,0,0,0,0,0,0,0});
B04A = subs(B04, {B01,B11,B21,B31,B41,B51,B61,B71},{1,0,0,0,0,0,0,0});
B14A = subs(B14, {B01,B11,B21,B31,B41,B51,B61,B71},{1,0,0,0,0,0,0,0});
B24A = subs(B24, {B01,B11,B21,B31,B41,B51,B61,B71},{1,0,0,0,0,0,0,0});
B34A = subs(B34, {B01,B11,B21,B31,B41,B51,B61,B71},{1,0,0,0,0,0,0,0});
B44A = subs(B44, {B01,B11,B21,B31,B41,B51,B61,B71},{1,0,0,0,0,0,0,0});
B54A = subs(B54, {B01,B11,B21,B31,B41,B51,B61,B71},{1,0,0,0,0,0,0,0});

%Segment B
B02B = subs(B02, {B01,B11,B21,B31,B41,B51,B61,B71},{0,1,0,0,0,0,0,0});
B12B = subs(B12, {B01,B11,B21,B31,B41,B51,B61,B71},{0,1,0,0,0,0,0,0});
```

```
B22B = subs(B22, {B01,B11,B21,B31,B41,B51,B61,B71},{0,1,0,0,0,0,0,0});
B32B = subs(B32, {B01,B11,B21,B31,B41,B51,B61,B71},{0,1,0,0,0,0,0,0});
B42B = subs(B42, {B01,B11,B21,B31,B41,B51,B61,B71},{0,1,0,0,0,0,0,0});
B52B = subs(B52, {B01,B11,B21,B31,B41,B51,B61,B71},{0,1,0,0,0,0,0,0});
B62B = subs(B62, {B01,B11,B21,B31,B41,B51,B61,B71},{0,1,0,0,0,0,0,0});
B72B = subs(B72, {B01,B11,B21,B31,B41,B51,B61,B71},{0,1,0,0,0,0,0,0});
B03B = subs(B03, {B01,B11,B21,B31,B41,B51,B61,B71},{0,1,0,0,0,0,0,0});
B13B = subs(B13, {B01,B11,B21,B31,B41,B51,B61,B71},{0,1,0,0,0,0,0,0});
B23B = subs(B23, {B01,B11,B21,B31,B41,B51,B61,B71},{0,1,0,0,0,0,0,0});
B33B = subs(B33, {B01,B11,B21,B31,B41,B51,B61,B71},{0,1,0,0,0,0,0,0});
B43B = subs(B43, {B01,B11,B21,B31,B41,B51,B61,B71},{0,1,0,0,0,0,0,0});
B53B = subs(B53, {B01,B11,B21,B31,B41,B51,B61,B71},{0,1,0,0,0,0,0,0});
B63B = subs(B63, {B01,B11,B21,B31,B41,B51,B61,B71},{0,1,0,0,0,0,0,0});
B04B = subs(B04, {B01,B11,B21,B31,B41,B51,B61,B71},{0,1,0,0,0,0,0,0});
B14B = subs(B14, {B01,B11,B21,B31,B41,B51,B61,B71},{0,1,0,0,0,0,0,0});
B24B = subs(B24, {B01,B11,B21,B31,B41,B51,B61,B71},{0,1,0,0,0,0,0,0});
B34B = subs(B34, {B01,B11,B21,B31,B41,B51,B61,B71},{0,1,0,0,0,0,0,0});
B44B = subs(B44, {B01,B11,B21,B31,B41,B51,B61,B71},{0,1,0,0,0,0,0,0});
B54B = subs(B54, {B01,B11,B21,B31,B41,B51,B61,B71},{0,1,0,0,0,0,0,0});

%Segment C
B02C = subs(B02, {B01,B11,B21,B31,B41,B51,B61,B71},{0,0,1,0,0,0,0,0});
B12C = subs(B12, {B01,B11,B21,B31,B41,B51,B61,B71},{0,0,1,0,0,0,0,0});
B22C = subs(B22, {B01,B11,B21,B31,B41,B51,B61,B71},{0,0,1,0,0,0,0,0});
B32C = subs(B32, {B01,B11,B21,B31,B41,B51,B61,B71},{0,0,1,0,0,0,0,0});
B42C = subs(B42, {B01,B11,B21,B31,B41,B51,B61,B71},{0,0,1,0,0,0,0,0});
B52C = subs(B52, {B01,B11,B21,B31,B41,B51,B61,B71},{0,0,1,0,0,0,0,0});
B62C = subs(B62, {B01,B11,B21,B31,B41,B51,B61,B71},{0,0,1,0,0,0,0,0});
B72C = subs(B72, {B01,B11,B21,B31,B41,B51,B61,B71},{0,0,1,0,0,0,0,0});
B03C = subs(B03, {B01,B11,B21,B31,B41,B51,B61,B71},{0,0,1,0,0,0,0,0});
B13C = subs(B13, {B01,B11,B21,B31,B41,B51,B61,B71},{0,0,1,0,0,0,0,0});
```

```
B23C = subs(B23, {B01,B11,B21,B31,B41,B51,B61,B71},{0,0,1,0,0,0,0,0});
B33C = subs(B33, {B01,B11,B21,B31,B41,B51,B61,B71},{0,0,1,0,0,0,0,0});
B43C = subs(B43, {B01,B11,B21,B31,B41,B51,B61,B71},{0,0,1,0,0,0,0,0});
B53C = subs(B53, {B01,B11,B21,B31,B41,B51,B61,B71},{0,0,1,0,0,0,0,0});
B63C = subs(B63, {B01,B11,B21,B31,B41,B51,B61,B71},{0,0,1,0,0,0,0,0});
B04C = subs(B04, {B01,B11,B21,B31,B41,B51,B61,B71},{0,0,1,0,0,0,0,0});
B14C = subs(B14, {B01,B11,B21,B31,B41,B51,B61,B71},{0,0,1,0,0,0,0,0});
B24C = subs(B24, {B01,B11,B21,B31,B41,B51,B61,B71},{0,0,1,0,0,0,0,0});
B34C = subs(B34, {B01,B11,B21,B31,B41,B51,B61,B71},{0,0,1,0,0,0,0,0});
B44C = subs(B44, {B01,B11,B21,B31,B41,B51,B61,B71},{0,0,1,0,0,0,0,0});
B54C = subs(B54, {B01,B11,B21,B31,B41,B51,B61,B71},{0,0,1,0,0,0,0,0});

%Segment D
B02D = subs(B02, {B01,B11,B21,B31,B41,B51,B61,B71},{0,0,0,1,0,0,0,0});
B12D = subs(B12, {B01,B11,B21,B31,B41,B51,B61,B71},{0,0,0,1,0,0,0,0});
B22D = subs(B22, {B01,B11,B21,B31,B41,B51,B61,B71},{0,0,0,1,0,0,0,0});
B32D = subs(B32, {B01,B11,B21,B31,B41,B51,B61,B71},{0,0,0,1,0,0,0,0});
B42D = subs(B42, {B01,B11,B21,B31,B41,B51,B61,B71},{0,0,0,1,0,0,0,0});
B52D = subs(B52, {B01,B11,B21,B31,B41,B51,B61,B71},{0,0,0,1,0,0,0,0});
B62D = subs(B62, {B01,B11,B21,B31,B41,B51,B61,B71},{0,0,0,1,0,0,0,0});
B72D = subs(B72, {B01,B11,B21,B31,B41,B51,B61,B71},{0,0,0,1,0,0,0,0});
B03D = subs(B03, {B01,B11,B21,B31,B41,B51,B61,B71},{0,0,0,1,0,0,0,0});
B13D = subs(B13, {B01,B11,B21,B31,B41,B51,B61,B71},{0,0,0,1,0,0,0,0});
B23D = subs(B23, {B01,B11,B21,B31,B41,B51,B61,B71},{0,0,0,1,0,0,0,0});
B33D = subs(B33, {B01,B11,B21,B31,B41,B51,B61,B71},{0,0,0,1,0,0,0,0});
B43D = subs(B43, {B01,B11,B21,B31,B41,B51,B61,B71},{0,0,0,1,0,0,0,0});
B53D = subs(B53, {B01,B11,B21,B31,B41,B51,B61,B71},{0,0,0,1,0,0,0,0});
B63D = subs(B63, {B01,B11,B21,B31,B41,B51,B61,B71},{0,0,0,1,0,0,0,0});
B04D = subs(B04, {B01,B11,B21,B31,B41,B51,B61,B71},{0,0,0,1,0,0,0,0});
B14D = subs(B14, {B01,B11,B21,B31,B41,B51,B61,B71},{0,0,0,1,0,0,0,0});
B24D = subs(B24, {B01,B11,B21,B31,B41,B51,B61,B71},{0,0,0,1,0,0,0,0});
```

```
B34D = subs(B34, {B01,B11,B21,B31,B41,B51,B61,B71},{0,0,0,1,0,0,0,0});
B44D = subs(B44, {B01,B11,B21,B31,B41,B51,B61,B71},{0,0,0,1,0,0,0,0});
B54D = subs(B54, {B01,B11,B21,B31,B41,B51,B61,B71},{0,0,0,1,0,0,0,0});

%Segment E
B02E = subs(B02, {B01,B11,B21,B31,B41,B51,B61,B71},{0,0,0,0,1,0,0,0});
B12E = subs(B12, {B01,B11,B21,B31,B41,B51,B61,B71},{0,0,0,0,1,0,0,0});
B22E = subs(B22, {B01,B11,B21,B31,B41,B51,B61,B71},{0,0,0,0,1,0,0,0});
B32E = subs(B32, {B01,B11,B21,B31,B41,B51,B61,B71},{0,0,0,0,1,0,0,0});
B42E = subs(B42, {B01,B11,B21,B31,B41,B51,B61,B71},{0,0,0,0,1,0,0,0});
B52E = subs(B52, {B01,B11,B21,B31,B41,B51,B61,B71},{0,0,0,0,1,0,0,0});
B62E = subs(B62, {B01,B11,B21,B31,B41,B51,B61,B71},{0,0,0,0,1,0,0,0});
B72E = subs(B72, {B01,B11,B21,B31,B41,B51,B61,B71},{0,0,0,0,1,0,0,0});
B03E = subs(B03, {B01,B11,B21,B31,B41,B51,B61,B71},{0,0,0,0,1,0,0,0});
B13E = subs(B13, {B01,B11,B21,B31,B41,B51,B61,B71},{0,0,0,0,1,0,0,0});
B23E = subs(B23, {B01,B11,B21,B31,B41,B51,B61,B71},{0,0,0,0,1,0,0,0});
B33E = subs(B33, {B01,B11,B21,B31,B41,B51,B61,B71},{0,0,0,0,1,0,0,0});
B43E = subs(B43, {B01,B11,B21,B31,B41,B51,B61,B71},{0,0,0,0,1,0,0,0});
B53E = subs(B53, {B01,B11,B21,B31,B41,B51,B61,B71},{0,0,0,0,1,0,0,0});
B63E = subs(B63, {B01,B11,B21,B31,B41,B51,B61,B71},{0,0,0,0,1,0,0,0});
B04E = subs(B04, {B01,B11,B21,B31,B41,B51,B61,B71},{0,0,0,0,1,0,0,0});
B14E = subs(B14, {B01,B11,B21,B31,B41,B51,B61,B71},{0,0,0,0,1,0,0,0});
B24E = subs(B24, {B01,B11,B21,B31,B41,B51,B61,B71},{0,0,0,0,1,0,0,0});
B34E = subs(B34, {B01,B11,B21,B31,B41,B51,B61,B71},{0,0,0,0,1,0,0,0});
B44E = subs(B44, {B01,B11,B21,B31,B41,B51,B61,B71},{0,0,0,0,1,0,0,0});
B54E = subs(B54, {B01,B11,B21,B31,B41,B51,B61,B71},{0,0,0,0,1,0,0,0});

%Segment F
B02F = subs(B02, {B01,B11,B21,B31,B41,B51,B61,B71},{0,0,0,0,0,1,0,0});
B12F = subs(B12, {B01,B11,B21,B31,B41,B51,B61,B71},{0,0,0,0,0,1,0,0});
B22F = subs(B22, {B01,B11,B21,B31,B41,B51,B61,B71},{0,0,0,0,0,1,0,0});
```

```
B32F = subs(B32, {B01,B11,B21,B31,B41,B51,B61,B71},{0,0,0,0,0,1,0,0});
B42F = subs(B42, {B01,B11,B21,B31,B41,B51,B61,B71},{0,0,0,0,0,1,0,0});
B52F = subs(B52, {B01,B11,B21,B31,B41,B51,B61,B71},{0,0,0,0,0,1,0,0});
B62F = subs(B62, {B01,B11,B21,B31,B41,B51,B61,B71},{0,0,0,0,0,1,0,0});
B72F = subs(B72, {B01,B11,B21,B31,B41,B51,B61,B71},{0,0,0,0,0,1,0,0});
B03F = subs(B03, {B01,B11,B21,B31,B41,B51,B61,B71},{0,0,0,0,0,1,0,0});
B13F = subs(B13, {B01,B11,B21,B31,B41,B51,B61,B71},{0,0,0,0,0,1,0,0});
B23F = subs(B23, {B01,B11,B21,B31,B41,B51,B61,B71},{0,0,0,0,0,1,0,0});
B33F = subs(B33, {B01,B11,B21,B31,B41,B51,B61,B71},{0,0,0,0,0,1,0,0});
B43F = subs(B43, {B01,B11,B21,B31,B41,B51,B61,B71},{0,0,0,0,0,1,0,0});
B53F = subs(B53, {B01,B11,B21,B31,B41,B51,B61,B71},{0,0,0,0,0,1,0,0});
B63F = subs(B63, {B01,B11,B21,B31,B41,B51,B61,B71},{0,0,0,0,0,1,0,0});
B04F = subs(B04, {B01,B11,B21,B31,B41,B51,B61,B71},{0,0,0,0,0,1,0,0});
B14F = subs(B14, {B01,B11,B21,B31,B41,B51,B61,B71},{0,0,0,0,0,1,0,0});
B24F = subs(B24, {B01,B11,B21,B31,B41,B51,B61,B71},{0,0,0,0,0,1,0,0});
B34F = subs(B34, {B01,B11,B21,B31,B41,B51,B61,B71},{0,0,0,0,0,1,0,0});
B44F = subs(B44, {B01,B11,B21,B31,B41,B51,B61,B71},{0,0,0,0,0,1,0,0});
B54F = subs(B54, {B01,B11,B21,B31,B41,B51,B61,B71},{0,0,0,0,0,1,0,0});

%Segment G
B02G = subs(B02, {B01,B11,B21,B31,B41,B51,B61,B71},{0,0,0,0,0,0,1,0});
B12G = subs(B12, {B01,B11,B21,B31,B41,B51,B61,B71},{0,0,0,0,0,0,1,0});
B22G = subs(B22, {B01,B11,B21,B31,B41,B51,B61,B71},{0,0,0,0,0,0,1,0});
B32G = subs(B32, {B01,B11,B21,B31,B41,B51,B61,B71},{0,0,0,0,0,0,1,0});
B42G = subs(B42, {B01,B11,B21,B31,B41,B51,B61,B71},{0,0,0,0,0,0,1,0});
B52G = subs(B52, {B01,B11,B21,B31,B41,B51,B61,B71},{0,0,0,0,0,0,1,0});
B62G = subs(B62, {B01,B11,B21,B31,B41,B51,B61,B71},{0,0,0,0,0,0,1,0});
B72G = subs(B72, {B01,B11,B21,B31,B41,B51,B61,B71},{0,0,0,0,0,0,1,0});
B03G = subs(B03, {B01,B11,B21,B31,B41,B51,B61,B71},{0,0,0,0,0,0,1,0});
B13G = subs(B13, {B01,B11,B21,B31,B41,B51,B61,B71},{0,0,0,0,0,0,1,0});
B23G = subs(B23, {B01,B11,B21,B31,B41,B51,B61,B71},{0,0,0,0,0,0,1,0});
```

```
B33G = subs(B33, {B01,B11,B21,B31,B41,B51,B61,B71},{0,0,0,0,0,0,1,0});
B43G = subs(B43, {B01,B11,B21,B31,B41,B51,B61,B71},{0,0,0,0,0,0,1,0});
B53G = subs(B53, {B01,B11,B21,B31,B41,B51,B61,B71},{0,0,0,0,0,0,1,0});
B63G = subs(B63, {B01,B11,B21,B31,B41,B51,B61,B71},{0,0,0,0,0,0,1,0});
B04G = subs(B04, {B01,B11,B21,B31,B41,B51,B61,B71},{0,0,0,0,0,0,1,0});
B14G = subs(B14, {B01,B11,B21,B31,B41,B51,B61,B71},{0,0,0,0,0,0,1,0});
B24G = subs(B24, {B01,B11,B21,B31,B41,B51,B61,B71},{0,0,0,0,0,0,1,0});
B34G = subs(B34, {B01,B11,B21,B31,B41,B51,B61,B71},{0,0,0,0,0,0,1,0});
B44G = subs(B44, {B01,B11,B21,B31,B41,B51,B61,B71},{0,0,0,0,0,0,1,0});
B54G = subs(B54, {B01,B11,B21,B31,B41,B51,B61,B71},{0,0,0,0,0,0,1,0});

%Segment H
B02H = subs(B02, {B01,B11,B21,B31,B41,B51,B61,B71},{0,0,0,0,0,0,0,1});
B12H = subs(B12, {B01,B11,B21,B31,B41,B51,B61,B71},{0,0,0,0,0,0,0,1});
B22H = subs(B22, {B01,B11,B21,B31,B41,B51,B61,B71},{0,0,0,0,0,0,0,1});
B32H = subs(B32, {B01,B11,B21,B31,B41,B51,B61,B71},{0,0,0,0,0,0,0,1});
B42H = subs(B42, {B01,B11,B21,B31,B41,B51,B61,B71},{0,0,0,0,0,0,0,1});
B52H = subs(B52, {B01,B11,B21,B31,B41,B51,B61,B71},{0,0,0,0,0,0,0,1});
B62H = subs(B62, {B01,B11,B21,B31,B41,B51,B61,B71},{0,0,0,0,0,0,0,1});
B72H = subs(B72, {B01,B11,B21,B31,B41,B51,B61,B71},{0,0,0,0,0,0,0,1});
B03H = subs(B03, {B01,B11,B21,B31,B41,B51,B61,B71},{0,0,0,0,0,0,0,1});
B13H = subs(B13, {B01,B11,B21,B31,B41,B51,B61,B71},{0,0,0,0,0,0,0,1});
B23H = subs(B23, {B01,B11,B21,B31,B41,B51,B61,B71},{0,0,0,0,0,0,0,1});
B33H = subs(B33, {B01,B11,B21,B31,B41,B51,B61,B71},{0,0,0,0,0,0,0,1});
B43H = subs(B43, {B01,B11,B21,B31,B41,B51,B61,B71},{0,0,0,0,0,0,0,1});
B53H = subs(B53, {B01,B11,B21,B31,B41,B51,B61,B71},{0,0,0,0,0,0,0,1});
B63H = subs(B63, {B01,B11,B21,B31,B41,B51,B61,B71},{0,0,0,0,0,0,0,1});
B04H = subs(B04, {B01,B11,B21,B31,B41,B51,B61,B71},{0,0,0,0,0,0,0,1});
B14H = subs(B14, {B01,B11,B21,B31,B41,B51,B61,B71},{0,0,0,0,0,0,0,1});
B24H = subs(B24, {B01,B11,B21,B31,B41,B51,B61,B71},{0,0,0,0,0,0,0,1});
B34H = subs(B34, {B01,B11,B21,B31,B41,B51,B61,B71},{0,0,0,0,0,0,0,1});
```

```
B44H = subs(B44, {B01,B11,B21,B31,B41,B51,B61,B71},{0,0,0,0,0,0,0,1});
B54H = subs(B54, {B01,B11,B21,B31,B41,B51,B61,B71},{0,0,0,0,0,0,0,1});

fprintf('Blending functions :\n');

B04 = [B04A, B04B, B04C, B04D, B04E, B04F, B04G, B04H]; B04 = simplify(B04)
B14 = [B14A, B14B, B14C, B14D, B14E, B14F, B14G, B14H]; B14 = simplify(B14)
B24 = [B24A, B24B, B24C, B24D, B24E, B24F, B24G, B24H]; B24 = simplify(B24)
B34 = [B34A, B34B, B34C, B34D, B34E, B34F, B34G, B34H]; B34 = simplify(B34)
B44 = [B44A, B44B, B44C, B44D, B44E, B44F, B44G, B44H]; B44 = simplify(B44)

fprintf('\n');

fprintf('General Equation of Curve :\n');

P = P0*B04 + P1*B14 + P2*B24 + P3*B34 + P4*B44

fprintf('\n');

fprintf('Actual Equation  :\n');

x = subs(P, ([P0, P1, P2, P3, P4]), ([x0, x1, x2, x3, x4])); x = simplify(x)
y = subs(P, ([P0, P1, P2, P3, P4]), ([y0, y1, y2, y3, y4])); y = simplify(y)

%plotting BFs

tta = linspace(t0, t1);
ttb = linspace(t1, t2);
ttc = linspace(t2, t3);
ttd = linspace(t3, t4);
tte = linspace(t4, t5);
```

```
ttf = linspace(t5, t6);
ttg = linspace(t6, t7);
tth = linspace(t7, t8);

B04aa = subs(B04A, t, tta);
B04bb = subs(B04B, t, ttb);
B04cc = subs(B04C, t, ttc);
B04dd = subs(B04D, t, ttd);
B04ee = subs(B04E, t, tte);
B04ff = subs(B04F, t, ttf);
B04gg = subs(B04G, t, ttg);
B04hh = subs(B04H, t, tth);

B14aa = subs(B14A, t, tta);
B14bb = subs(B14B, t, ttb);
B14cc = subs(B14C, t, ttc);
B14dd = subs(B14D, t, ttd);
B14ee = subs(B14E, t, tte);
B14ff = subs(B14F, t, ttf);
B14gg = subs(B14G, t, ttg);
B14hh = subs(B14H, t, tth);

B24aa = subs(B24A, t, tta);
B24bb = subs(B24B, t, ttb);
B24cc = subs(B24C, t, ttc);
B24dd = subs(B24D, t, ttd);
B24ee = subs(B24E, t, tte);
B24ff = subs(B24F, t, ttf);
B24gg = subs(B24G, t, ttg);
B24hh = subs(B24H, t, tth);
```

```
B34aa = subs(B34A, t, tta);
B34bb = subs(B34B, t, ttb);
B34cc = subs(B34C, t, ttc);
B34dd = subs(B34D, t, ttd);
B34ee = subs(B34E, t, tte);
B34ff = subs(B34F, t, ttf);
B34gg = subs(B34G, t, ttg);
B34hh = subs(B34H, t, tth);

B44aa = subs(B44A, t, tta);
B44bb = subs(B44B, t, ttb);
B44cc = subs(B44C, t, ttc);
B44dd = subs(B44D, t, ttd);
B44ee = subs(B44E, t, tte);
B44ff = subs(B44F, t, ttf);
B44gg = subs(B44G, t, ttg);
B44hh = subs(B44H, t, tth);

figure,
plot(tta, B04aa, 'k-', ttb, B04bb, 'k--', ttc, B04cc, 'k-.', ttd, ...
B04dd, 'b-', tte, B04ee, 'b--', ttf, B04ff, 'b-.', ttg, B04gg, 'r-', tth, B04hh, 'r--');
hold on;
plot(tta, B14aa, 'k-', ttb, B14bb, 'k--', ttc, B14cc, 'k-.', ttd, ...
B14dd, 'b-', tte, B14ee, 'b--', ttf, B14ff, 'b-.', ttg, B14gg, 'r-', tth, B14hh, 'r--');
plot(tta, B24aa, 'k-', ttb, B24bb, 'k--', ttc, B24cc, 'k-.', ttd, ...
B24dd, 'b-', tte, B24ee, 'b--', ttf, B24ff, 'b-.', ttg, B24gg, 'r-', tth, B24hh, 'r--');
plot(tta, B34aa, 'k-', ttb, B34bb, 'k--', ttc, B34cc, 'k-.', ttd, ...
B34dd, 'b-', tte, B34ee, 'b--', ttf, B34ff, 'b-.', ttg, B34gg, 'r-', tth, B34hh, 'r--');
plot(tta, B44aa, 'k-', ttb, B44bb, 'k--', ttc, B44cc, 'k-.', ttd, ...
B44dd, 'b-', tte, B44ee, 'b--', ttf, B44ff, 'b-.', ttg, B44gg, 'r-', tth, B44hh, 'r--');
```

```
xlabel ('t'); ylabel('B'); title('B04 - B14 - B24 - B34 - B44');
legend('A', 'B', 'C', 'D', 'E', 'F', 'G', 'H');
hold off;

%plotting curve

xa = x(1); ya = y(1);
xb = x(2); yb = y(2);
xc = x(3); yc = y(3);
xd = x(4); yd = y(4);
xe = x(5); ye = y(5);
xf = x(6); yf = y(6);
xg = x(7); yg = y(7);
xh = x(8); yh = y(8);

xaa = subs(xa, t, tta); yaa = subs(ya, t, tta);
xbb = subs(xb, t, ttb); ybb = subs(yb, t, ttb);
xcc = subs(xc, t, ttc); ycc = subs(yc, t, ttc);
xdd = subs(xd, t, ttd); ydd = subs(yd, t, ttd);
xee = subs(xe, t, tte); yee = subs(ye, t, tte);
xff = subs(xf, t, ttf); yff = subs(yf, t, ttf);
xgg = subs(xg, t, ttg); ygg = subs(yg, t, ttg);
xhh = subs(xh, t, tth); yhh = subs(yh, t, tth);

X = [x0, x1, x2, x3, x4]; Y = [y0, y1, y2, y3, y4];

figure
subplot(131), plot(tta, xaa, 'k-', ttb, xbb, 'k--', ttc, xcc, 'k-.', ...
ttd, xdd, 'b-', tte, xee, 'b--', ttf, xff, 'b-.', ttg, xgg, 'r-', tth, xhh, 'r--');
xlabel('t'); ylabel('x');  title('t - x'); axis square;
subplot(132), plot(tta, yaa, 'k-', ttb, ybb, 'k--', ttc, ycc, 'k-.', ...
```

```
ttd, ydd, 'b-', tte, yee, 'b--', ttf, yff, 'b-.', ttg, ygg, 'r-', tth, yhh, 'r--');
xlabel('t'); ylabel('y');  title('t - y'); axis square;
subplot(133), plot(xaa, yaa, 'k-', xbb, ybb, 'k--', xcc, ycc, 'k-.', ...
xdd, ydd, 'b-', xee, yee, 'b--', xff, yff, 'b-.', xgg, ygg, 'r-', xhh, yhh, 'r--');
hold on;
scatter(X, Y, 20, 'r', 'filled');
xlabel('x'); ylabel('y'); title('x - y'); axis square; grid;
axis([-1.5 1.5 -1.5 1.5]);
hold off;
```

The concepts for open-uniform and non-uniform splines are similar to those discussed for quadratic splines and the reader is expected to extend the idea for cubic splines.

3.9 CHAPTER SUMMARY

The following points summarize the topics discussed in this chapter:

- Approximating splines in general do not go through any of their CPs.
- *B*-splines are approximating splines proposed to overcome the drawbacks of Bezier splines.
- *B*-splines consist of multiple curve segments with continuity at join points.
- Values of the parametric variable t at the join points are stored in the KV.
- Uniform *B*-splines have uniform gaps in the KV.
- BFs of *B*-splines are calculated using the Cox de Boor algorithm.
- A *B* spline has two defining parameters d related to its degree and n related to the number of CPs.
- Number of CPs can be changed independent of the degree of *B*-splines.
- BFs and equation of *B*-splines consists of multiple parts due to the segments.
- BFs of uniform *B*-splines have the symmetric shapes but shifted from each other.
- Changing the CPs affects only specific segments instead of the entire curve.
- The spatial *B*-spline curve is independent of the KV values.
- Open-uniform *B*-splines have KV with repeated values, called multiplicity.
- Multiplicity forces approximating *B*-splines to behave like hybrid splines.
- Non-uniform *B*-splines have non-uniform gaps in the KV.
- BFs of non-uniform *B*-splines are unsymmetric in shape.

3.10 REVIEW QUESTIONS

1. What are main differences between *B*-splines and Bezier splines?
2. Differentiate between uniform, open-uniform, and non-uniform *B*-splines
3. What is a KV for a *B*-spline curve?
4. How are BFs of *B*-splines calculated, using the Cox de Boor algorithm?
5. Under what conditions can *B*-splines behave like hybrid splines?
6. What is meant by local control property of a *B*-spline?
7. Can the number of CPs be changed independent of the degree of *B*-splines?
8. How does changing the KV affect the spatial *B*-spline curve?

9. Why does a B-spline curve equation have multiple sub-components?

10. What is multiplicity of the KV and how does it affect the BFs and the spline curve?

3.11 PRACTICE PROBLEMS

1. Find the equation of a linear uniform B-spline associated with the CPs (1, 0), (–1, 1), and (1, –1).

2. Derive the BFs of a uniform linear B-spline with $d = 2$ and $n = 3$.

3. Derive the BFs of a uniform quadratic B-spline associated with five CPs.

4. A B-spline has degree 2 and is associated with four CPs. Find an expression for the first BF $B03$ for the first two curve segments A and B, using Cox de Boor algorithm, if the KV is of the form $T = [0, 0.2, 0.5, 0.7, \ldots]$.

5. A uniform B-spline has degree 1 and is associated with three CPs P_0, P_1, and P_2. Derive equation of the second curve segment using Cox de Boor algorithm, if KV is of the form $T = [0, 0.4, 0.5, 0.8, \ldots]$.

6. A non-uniform B-spline has degree 1 and is associated with three CPs P_0, P_1, and P_2. Derive equation of the first and second curve segments if the KV is $T = [0, 4, 5, 8, 9]$.

7. Find the equation of a uniform quadratic B-spline having CPs (2, 5), (4, –1), (5, 8), and (7, –5).

8. Find the equation of a non-uniform quadratic B-spline having CPs (2, 5), (4, –1), (5, 8), and (7, –5) and a KV $T =[0, 4, 5, 8, 9, 13, 15]$.

9. Find the equations of the first two segments of a uniform cubic B-spline with CPs (2, 0), (4, 1), (5, 7), (6, –5), and (8, –1).

10. Find the first two BFs of a non-uniform cubic B-spline with $d = 4$ and $n = 4$, and KV $T = [0, 1, 2, 5, 6, 8, 10, 12, 13]$.

CHAPTER 4

2D Transformations

4.1 INTRODUCTION

Two-dimensional transformations enable us to change the location, orientation and shapes of splines in 2D space. These transformations are translation, rotation, scaling, reflection, and shear (Hearn and Baker, 1996) applied individually or in combination of two or more. Given known coordinates of a point, each of these transformations is represented by a matrix which when multiplied to the original coordinates gives us a new set of coordinates. An entire spline is transformed by transforming all the points of the spline. In the following sections, the transformation matrices will be derived and applied to original points to give new points. Before that, however, we introduce the concept of homogeneous coordinates, which enable us to represent all types of transformation in a homogeneous or uniform manner. To calculate the coordinates and directions, we use a 2D right-handed coordinate system. Although, we have been using coordinates of points from Chapter 1, now we establish a formal definition of a 2D coordinate system as transformation requires a rigorous understanding of how distances and angles are calculated and how they are changed during the course of transformation operations.

The concept of a 2D Cartesian coordinate system is attributed to the 17th century French mathematician Rene Descartes and is widely used to measure location of a point on a 2D plane from a reference point called the origin. The location is represented as a pair of signed distances measured along two mutually perpendicular lines called axes meeting at the origin. Sometimes, the axes are also referred to as the primary axes or principal axes to distinguish them from other lines parallel to them. The first axis is usually depicted as a number line along the horizontal direction and called the X-axis while the second axis as a number line along the vertical direction and called the Y-axis (see Figure 4.1). Distance of a point on the plane measured along the X-axis from the origin is called the x-coordinate or abscissa and distance along the Y-axis is called the y-coordinate or ordinate. The location of a point is, therefore, represented as a pair of ordered numbers (x, y) called its coordinates. Since all distances are measured from the origin, the origin itself has coordinates $(0, 0)$ and denoted by O. In most cases, the origin is visualized as a point in the center of the paper with the x-coordinates being measured positive toward the right and negative toward the left, and the y-coordinates being positive toward the top and negative toward the bottom. However in the context of a computer display, the origin is usually located at the lower-left corner of the screen with the axes toward the right and top, while for some cases the origin is also depicted at the upper left of the screen with the axes toward the right and bottom of the screen. Since the axes are perpendicular to each other the system is often referred to as rectangular coordinate system or orthogonal coordinate system. Numbers measured along the axes can in general be floating point numbers although in some cases like measuring pixel dimensions of images on screen, the numbers are thought to be only integers. The two axes divide the plane into four parts called quadrants such that in the first quadrant (Q1) both x and y are positive, in the second quadrant (Q2), x is negative, in the third quadrant (Q3) both are negative, and in the fourth quadrant (Q4), y is negative. To fix the orientation of the axes with respect to one another a convention called the right-hand rule is often used. If the first finger and the second finger of the right-hand are stretched out at right angles to each other then the direction from the hand to the tip of the finger would represent the positive x and positive y directions, respectively. The angle of rotation around a primary axis is considered positive when in counter-clockwise (CCW) direction when seen from the tip of the axis toward the origin. Another way to determine this is to use the right-handed convention: with the thumb of the right-hand pointing toward the positive end of an axis (away from the origin) the direction of curvature of the fingers indicate the positive direction of rotation around that axis (O'Rourke, 2003).

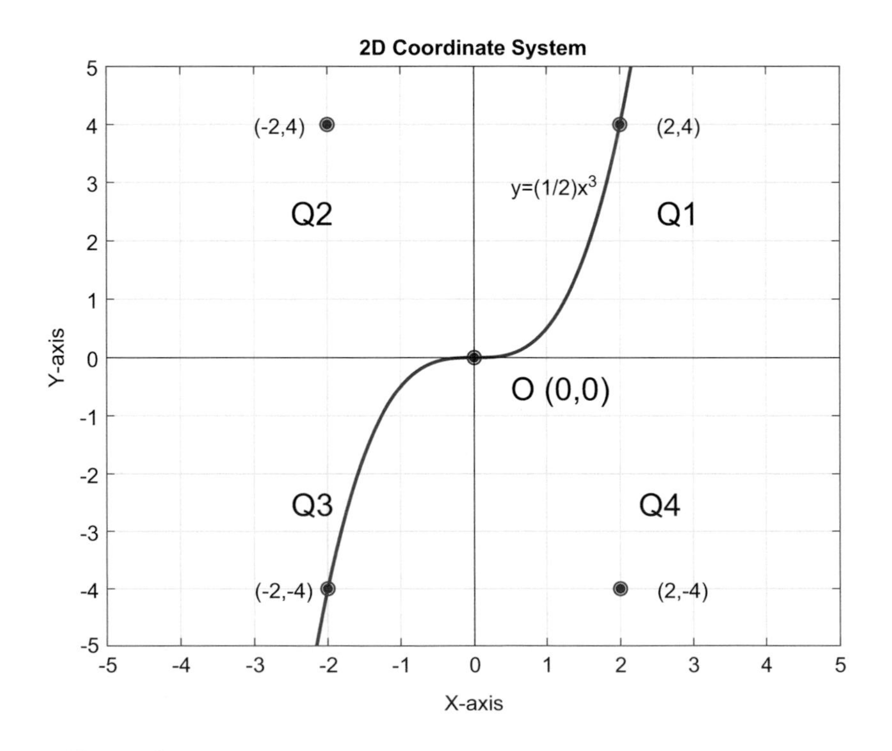

FIGURE 4.1 2D coordinate system.

A Cartesian coordinate system plays one of the most important roles in graphics since it enables representation of spline curves as a series of coordinate points e.g. a spline $y = 0.5x^3$ can be described by the vector containing all points whose coordinates satisfy the equation. Moreover, the number of points and the gap between them can be adjusted according to the resolution of the display system to always produce a smooth curve. This has led to the growth of vector graphics.

4.2 HOMOGENEOUS COORDINATES

A point $P(x_1, y_1)$ when translated by amounts (t_x, t_y) has new coordinates $Q(x_2, y_2)$ given by (Foley et al., 1995):

$$x_2 = x_1 + t_x$$

$$y_2 = y_1 + t_y$$

Written in matrix form this becomes:

$$\begin{bmatrix} x_2 \\ y_2 \end{bmatrix} = \begin{bmatrix} 1 & 0 \\ 0 & 1 \end{bmatrix} \begin{bmatrix} x_1 \\ y_1 \end{bmatrix} + \begin{bmatrix} t_x \\ t_y \end{bmatrix} \tag{4.1}$$

Now for two main reasons, we would prefer a slightly different form of representation of the transformation matrix than that shown above. The first reason is that when multiple transformations are involved one after another we would prefer a multiplicative form of the matrix rather than the additive form. This would enable us to multiply all the transformations together and calculate the final coordinates at the end, rather than calculate intermediate coordinates after each step. The product of two or more transformation matrices is known as the "composite transformation" matrix, and provides the net effect of multiple transformations within a single matrix (Hearn and Baker, 1996). The second reason is that we would prefer a square transformation matrix whose inverse would give us the "inverse transformation." For these reasons, we use the form in Equation (4.2), which is called "homogeneous coordinates" (Hearn and Baker, 1996) in contrast to Equation (4.1) called "Cartesian coordinates." To generate the square matrix, a third row is included, which is typically ignored after the new coordinates are computed.

$$\begin{bmatrix} x_2 \\ y_2 \\ 1 \end{bmatrix} = \begin{bmatrix} 1 & 0 & t_x \\ 0 & 1 & t_y \\ 0 & 0 & 1 \end{bmatrix} \begin{bmatrix} x_1 \\ y_1 \\ 1 \end{bmatrix}$$

Since the matrices have three rows, the homogeneous coordinates are referred to as (x, y, h), which implies Cartesian coordinates of $(x/h, y/h)$. In most cases $h = 1$ so the values are equal, but in some cases when h is not 1, we would need conversion from one system to another. Such examples are also included in this book. The transformation matrices for each type of transformation operation will now be dealt with in more details.

4.3 TRANSLATION

A point $P(x_1, y_1)$ when translated by amounts (t_x, t_y) has new coordinates $Q(x_2, y_2)$ given by:

$$\begin{bmatrix} x_2 \\ y_2 \\ 1 \end{bmatrix} = \begin{bmatrix} 1 & 0 & t_x \\ 0 & 1 & t_y \\ 0 & 0 & 1 \end{bmatrix} \begin{bmatrix} x_1 \\ y_1 \\ 1 \end{bmatrix} \tag{4.2}$$

The inverse transformation is computed by taking the inverse of the matrix as below:

$$\begin{bmatrix} x_1 \\ y_1 \\ 1 \end{bmatrix} = \begin{bmatrix} 1 & 0 & t_x \\ 0 & 1 & t_y \\ 0 & 0 & 1 \end{bmatrix}^{-1} \begin{bmatrix} x_2 \\ y_2 \\ 1 \end{bmatrix} \tag{4.3}$$

It can be verified that the inverse of the matrix is equal to the negative of the arguments.

$$\begin{bmatrix} 1 & 0 & t_x \\ 0 & 1 & t_y \\ 0 & 0 & 1 \end{bmatrix}^{-1} = \begin{bmatrix} 1 & 0 & -t_x \\ 0 & 1 & -t_y \\ 0 & 0 & 1 \end{bmatrix} \tag{4.4}$$

Symbolically, if T denotes the forward translation operation with arguments (t_x, t_y) and T' denotes the reverse translation then the above can be written as:

$$T'(t_x, t_y) = T(-t_x, -t_y)$$

This is the convention followed throughout this book i.e. the operations themselves would be denoted by single letters such as T, S, R, and so on for translation, scaling, and rotation while a specific transformation matrix would be denoted with a letter with a subscript. For example

$$T_1 = T(3, -4) = \begin{bmatrix} 1 & 0 & 3 \\ 0 & 1 & -4 \\ 0 & 0 & 1 \end{bmatrix}$$

Example 4.1

A square having vertices (0, 0), (1, 0), (1, 1), and (0, 1) is translated by amounts (−3, 4). Find its new vertices.

$$\text{Original coordinate matrix: } C = \begin{bmatrix} 0 & 1 & 1 & 0 \\ 0 & 0 & 1 & 1 \\ 1 & 1 & 1 & 1 \end{bmatrix}$$

$$\text{Translation matrix: } T_1 = T(-3, 4) = \begin{bmatrix} 0 & 1 & -3 \\ 0 & 0 & 4 \\ 0 & 0 & 1 \end{bmatrix}$$

$$\text{New coordinate matrix: } D = T_1 * C = \begin{bmatrix} -3 & -2 & -2 & -3 \\ 4 & 4 & 5 & 5 \\ 1 & 1 & 1 & 1 \end{bmatrix}$$

New vertex coordinates are (–3, 4), (–2, 4), (–2, 5), and (–3, 5) (Figure 4.2)

MATLAB® Code 4.1

```
clear all; clc;
X = [0 1 1 0 0];
Y = [0 0 1 1 0];
C = [X; Y; 1 1 1 1 1];
tx = -3; ty = 4;
T1 = [1 0 tx; 0 1 ty; 0 0 1];
D = T1*C;

fprintf('New vertices : \n');
for i=1:4
    fprintf('(%.2f, %.2f) \n',D(1,i), D(2,i));
end
```

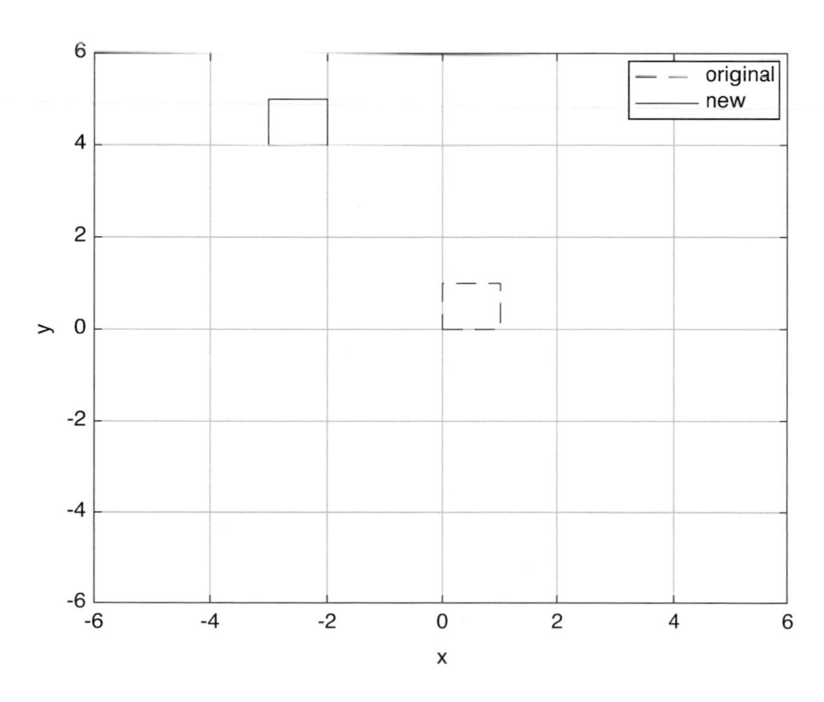

FIGURE 4.2 Plot for Example 4.1.

```
%plotting
plot(C(1,:), C(2,:), 'b--', D(1,:),D(2,:), 'r');
xlabel('x');
ylabel('y');
legend('original', 'new');
axis([-6, 6, -6, 6]);
grid;
```

NOTE

`for`: initiates a for loop for printing out all the vertices

4.4 SCALING

A scaling operation alters the size of graphic objects by multiplying the *X*- and *Y*-coordinates of each point of the object by scaling factors s_x and s_y. If scaling factors are less than 1, they reduce the size of the object; if they are more than 1, they increase the size; and if they are equal to 1, they keep the size unaltered. If the factors are positive, the size increases along the original direction of coordinate axes; if they are negative, the coordinate signs are flipped. If s_x and s_y are equal then scaling is uniform otherwise non-uniform.

A point $P(x_1, y_1)$ when scaled by amounts (s_x, s_y) has new coordinates $Q(x_2, y_2)$ given by:

$$\begin{bmatrix} x_2 \\ y_2 \\ 1 \end{bmatrix} = \begin{bmatrix} s_x & 0 & 0 \\ 0 & s_y & 0 \\ 0 & 0 & 1 \end{bmatrix} \begin{bmatrix} x_1 \\ y_1 \\ 1 \end{bmatrix} \tag{4.5}$$

It can be verified that the inverse of the matrix is equal to the reciprocal of the arguments.

$$\begin{bmatrix} s_x & 0 & 0 \\ 0 & s_y & 0 \\ 0 & 0 & 1 \end{bmatrix}^{-1} = \begin{bmatrix} 1/s_x & 0 & 0 \\ 0 & 1/s_y & 0 \\ 0 & 0 & 1 \end{bmatrix} \tag{4.6}$$

If S denotes the forward scaling operation with arguments (s_x, s_y) and S' denotes the reverse scaling then the above can be written as:

$$S'(s_x, s_y) = S(1/s_x, 1/s_y)$$

Note that the scaling operation pertaining to the above matrix is always with respect to the origin.

Example 4.2

A quadrilateral having vertices (–1, –1), (1, –2), (1, 2), (–1, 1) is scaled by amounts (–2, 3). Find its new vertices.

Original coordinate matrix: $C = \begin{bmatrix} -1 & 1 & 1 & -1 \\ -1 & -2 & 2 & 1 \\ 1 & 1 & 1 & 1 \end{bmatrix}$

Scaling matrix: $S_1 = S(-2, 3) = \begin{bmatrix} -2 & 0 & 0 \\ 0 & 3 & 0 \\ 0 & 0 & 1 \end{bmatrix}$

New coordinate matrix: $D = S_1 \cdot C = \begin{bmatrix} 2 & -2 & -2 & 2 \\ -3 & -6 & 6 & 3 \\ 1 & 1 & 1 & 1 \end{bmatrix}$

New vertex coordinates are (2, –3), (–2, –6), (–2, 6), and (2, 3) (Figure 4.3)

MATLAB Code 4.2

```
clear all; clc;
X = [-1 1 1 -1 -1];
Y = [-1 -2 2 1 -1];
C = [X;Y; 1 1 1 1 1];
sx = -2;
sy = 3;
S1 = [sx 0 0; 0 sy 0; 0 0 1];
D = S1*C;

fprintf('New vertices : \n');
for i=1:4
```

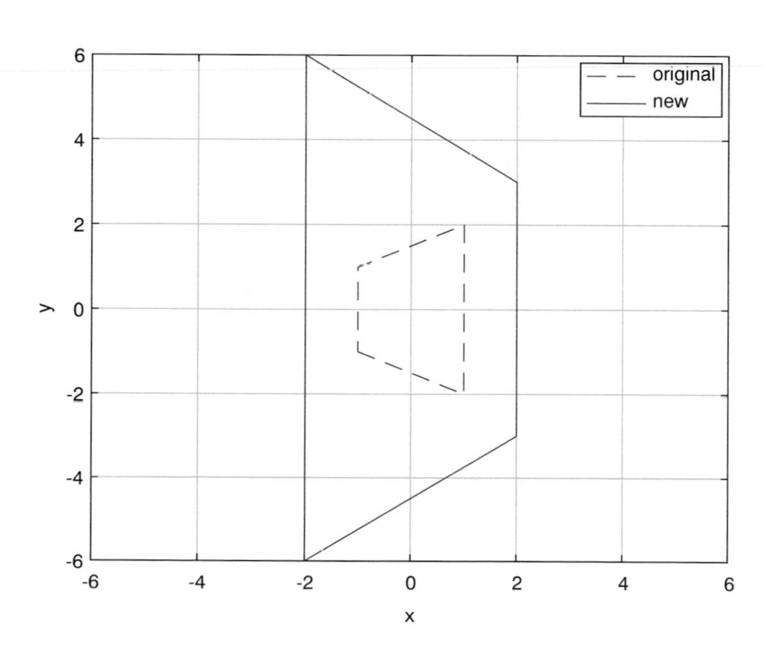

FIGURE 4.3 Plot for Example 4.2.

```
    fprintf('(%.2f, %.2f) \n',D(1,i), D(2,i));
end

%plotting
plot(C(1,:), C(2,:), 'b--', D(1,:),D(2,:), 'r');
xlabel('x');
ylabel('y');
legend('original', 'new');
axis([-6, 6, -6, 6]); grid;
```

Note that here the center of the square is at point (0, 0) so the scaling is uniform with respect to the origin.

4.5 ROTATION

A rotation operation moves a point along the circumference of a circle centered at the origin and radius equal to the distance of the point from the origin. Rotation is considered positive when it is in the CCW direction and negative along the clockwise (CW) direction.

A point $P(x_1, y_1)$ when rotated by angle (θ) has new coordinates $Q(x_2, y_2)$ given by (Figure 4.4):

$$\begin{bmatrix} x_2 \\ y_2 \\ 1 \end{bmatrix} = \begin{bmatrix} \cos\theta & -\sin\theta & 0 \\ \sin\theta & \cos\theta & 0 \\ 0 & 0 & 1 \end{bmatrix} \begin{bmatrix} x_1 \\ y_1 \\ 1 \end{bmatrix} \tag{4.7}$$

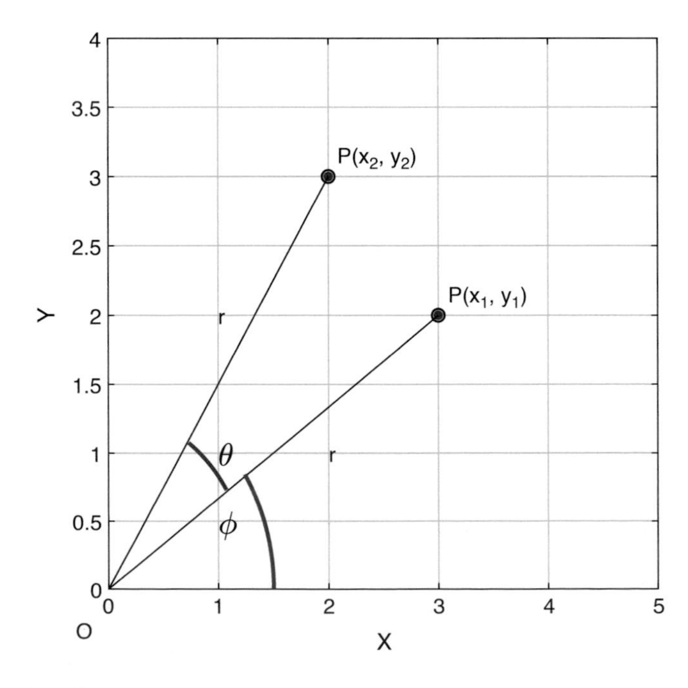

FIGURE 4.4 Deriving the rotation matrix.

To arrive at the above expression let angle between OP and X-axis be φ and r be the length of OP.

Then: $x_1 = r \cdot \cos(\varphi)$, $y_1 = r \cdot \sin(\varphi)$

Now: $x_2 = r \cdot \cos(\varphi + \theta) = r \cdot \cos(\varphi) \cdot \cos(\theta) - r \cdot \sin(\varphi) \cdot \sin(\theta)$

Similarly: $y_2 = r \cdot \sin(\varphi + \theta) = r \cdot \cos(\varphi) \cdot \sin(\theta) + r \cdot \sin(\varphi) \cdot \cos(\theta)$

Simplifying: $x_2 = x_1 \cdot \cos(\theta) - y_1 \cdot \sin(\theta)$

Similarly: $y_2 = x_1 \cdot \sin(\theta) + y_1 \cdot \cos(\theta)$

It can be verified that the inverse of the matrix is equal to the negative of the argument, remembering that $\cos(-\theta) = \cos(\theta)$.

$$\begin{bmatrix} \cos\theta & -\sin\theta & 0 \\ \sin\theta & \cos\theta & 0 \\ 0 & 0 & 1 \end{bmatrix}^{-1} = \begin{bmatrix} \cos\theta & \sin\theta & 0 \\ -\sin\theta & \cos\theta & 0 \\ 0 & 0 & 1 \end{bmatrix} \tag{4.8}$$

If R denotes the forward rotation operation with arguments (θ) and R' denotes the reverse rotation then the above can be written as:

$$R'(\theta) = R(-\theta)$$

Note that the rotation operation pertaining to the above matrix is always with respect to the origin.

Example 4.3

A square having vertices (–1, –1), (1, –1), (1, 1), and (–1, 1) is rotated by angle 30° around the origin along CCW direction. Find its new vertices.

Original coordinate matrix: $C = \begin{bmatrix} -1 & 1 & 1 & -1 \\ -1 & -1 & 1 & 1 \\ 1 & 1 & 1 & 1 \end{bmatrix}$

Rotation matrix: $R_1 = R(30) = \begin{bmatrix} \cos 30 & -\sin 30 & 0 \\ \sin 30 & \cos 30 & 0 \\ 0 & 0 & 1 \end{bmatrix} = \begin{bmatrix} 0.87 & -0.5 & 0 \\ 0.5 & 0.87 & 0 \\ 0 & 0 & 1 \end{bmatrix}$

New coordinate matrix: $D = R_1 \cdot C = \begin{bmatrix} -0.37 & 1.37 & -0.37 & 1.37 \\ -1.37 & -0.37 & 1.37 & 0.37 \\ 1 & 1 & 1 & 1 \end{bmatrix}$

New vertex coordinates are (–0.37, –1.37), (1.37, –0.37), (0.37, 1.37), and (–1.37, 0.37) (Figure 4.5)

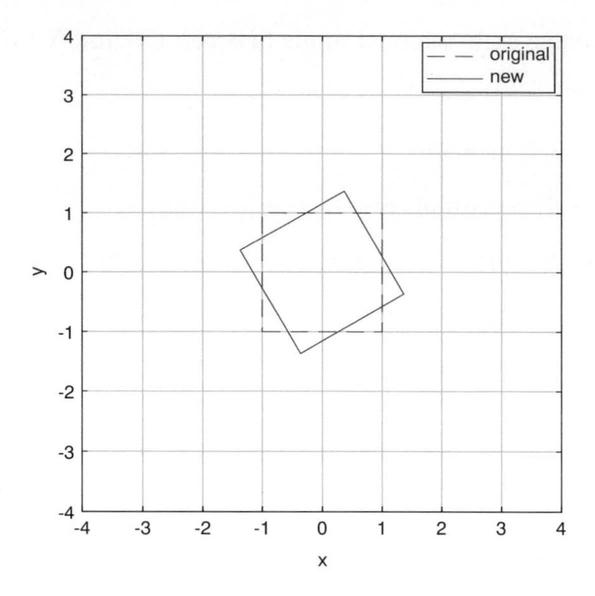

FIGURE 4.5 Plot for Example 4.3.

MATLAB Code 4.3

```
clear all; clc;
X = [-1 1 1 -1 -1];
Y = [-1 -1 1 1 -1];
C = [X;Y; 1 1 1 1 1];
A = deg2rad(30);
R1 =[cos(A) -sin(A) 0; sin(A) cos(A) 0; 0 0 1];
D = R1*C;

fprintf('New vertices : \n');
for i=1:4
    fprintf('(%.2f, %.2f) \n',D(1,i), D(2,i));
end

%plotting
plot(C(1,:), C(2,:), 'b--', D(1,:),D(2,:), 'r');
xlabel('x');
ylabel('y');
legend('original', 'new');
axis([-4, 4, -4, 4]); grid;
axis square;
```

NOTE

Here, the center of the square is at point (0, 0) so the rotation is with respect to the origin.

`cos`: calculates cosine of an angle in radians
`deg2rad`: converts degree to radian values

4.6 FIXED-POINT SCALING

For a general fixed-point scaling with respect to a fixed point $\left(x_f, y_f\right)$, the following steps are taken:

1. Translate object so that fixed-point moves to origin: $T_1 = T\left(-x_f, -y_f\right)$
2. Scale object about origin: $S_1 = S\left(s_x, s_y\right)$
3. Reverse translate the object to original location: $T_2 = T\left(x_f, y_f\right)$.

Composite transformation: $M = T_2 \cdot S_1 \cdot T_1$

Example 4.4

A square having vertices (–1, –1), (1, –1), (1, 1), and (–1, 1) is scaled by amounts (–2, 3) about one of its vertices (–1, –1). Find its new vertices.

Original coordinate matrix: $C = \begin{bmatrix} -1 & 1 & 1 & -1 \\ -1 & -1 & 1 & 1 \\ 1 & 1 & 1 & 1 \end{bmatrix}$

Forward translation matrix: $T_1 = T(1, 1) = \begin{bmatrix} 0 & 1 & 1 \\ 0 & 0 & 1 \\ 0 & 0 & 1 \end{bmatrix}$

Scaling matrix: $S_1 - S(-2, 3) = \begin{bmatrix} 2 & 0 & 0 \\ 0 & 3 & 0 \\ 0 & 0 & 1 \end{bmatrix}$

Reverse translation matrix: $T_2 = T(-1, -1) = \begin{bmatrix} 0 & 1 & -1 \\ 0 & 0 & -1 \\ 0 & 0 & 1 \end{bmatrix}$

Composite transformation matrix: $M = T_2 \cdot S_1 \cdot T_1 = \begin{bmatrix} -2 & 0 & -3 \\ 0 & 3 & 2 \\ 0 & 0 & 1 \end{bmatrix}$

New coordinate matrix: $D = M \cdot C = \begin{bmatrix} -1 & -5 & -5 & -1 \\ -1 & -1 & 5 & 5 \\ 1 & 1 & 1 & 1 \end{bmatrix}$

New vertex coordinates are (–1, –1), (–5, –1), (–5, 5), and (–1, 5) (Figure 4.6)

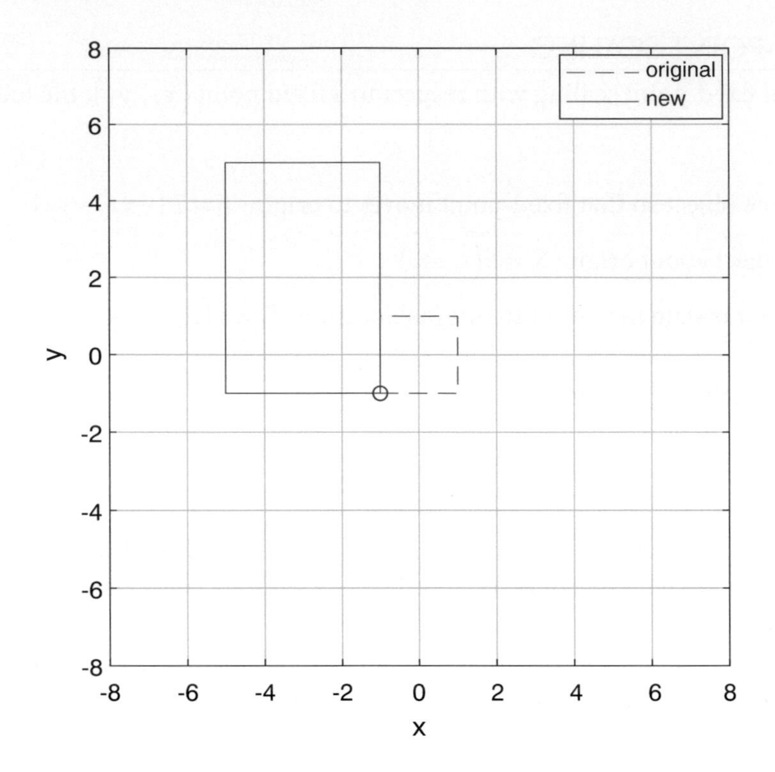

FIGURE 4.6 Plot for Example 4.4.

MATLAB Code 4.4

```
clear all; clc;
X = [-1 1 1 -1 -1];
Y = [-1 -1 1 1 -1];
C = [X;Y; 1 1 1 1 1];
sx = -2; sy = 3;
xf = -1; yf = -1;
T1 = [1, 0, -xf ; 0, 1, -yf ; 0, 0, 1];
S1 = [sx, 0, 0 ; 0, sy, 0 ; 0, 0, 1];
T2 = [1, 0, xf ; 0, 1, yf ; 0, 0, 1];
M = T2 * S1 * T1 ;
D = M*C;

fprintf('New vertices : \n');
for i=1:4
    fprintf('(%.2f, %.2f) \n',D(1,i), D(2,i));
end

%plotting
plot(C(1,:), C(2,:), 'b--', D(1,:),D(2,:), 'r', xf, yf, 'ro');
xlabel('x');
ylabel('y');
```

```
legend('original', 'new');
axis([-8, 8, -8, 8]);
axis square; grid;
```

4.7 FIXED-POINT ROTATION

For a general fixed-point rotation with respect to a fixed point (x_f, y_f), the following steps are taken (Foley et al., 1995):

1. Translate object so that fixed-point moves to origin: $T_1 = T(-x_f, -y_f)$
2. Rotate object about origin: $R_1 = R(\theta)$
3. Reverse translate the object to original location: $T_2 = T(x_f, y_f)$
4. Calculate composite transformation matrix: $M = T_2 \cdot R_1 \cdot T_1$.

Example 4.5

A square having vertices (–1, –1), (1, –1), (1, 1), and (–1, 1) is rotated by angle 30° along CCW direction about one of its vertices (–1, –1). Find its new vertices.

$$\text{Original coordinate matrix: } C = \begin{bmatrix} -1 & 1 & 1 & -1 \\ -1 & -1 & 1 & 1 \\ 1 & 1 & 1 & 1 \end{bmatrix}$$

$$\text{Forward translation matrix: } T_1 = T(1, 1) = \begin{bmatrix} 0 & 1 & 1 \\ 0 & 0 & 1 \\ 0 & 0 & 1 \end{bmatrix}$$

$$\text{Rotation matrix: } R_1 = R(30) = \begin{bmatrix} \cos 30 & -\sin 30 & 0 \\ \sin 30 & \cos 30 & 0 \\ 0 & 0 & 1 \end{bmatrix} = \begin{bmatrix} 0.87 & -0.5 & 0 \\ 0.5 & 0.87 & 0 \\ 0 & 0 & 1 \end{bmatrix}$$

$$\text{Reverse translation matrix: } T_2 = T(-1, -1) = \begin{bmatrix} 0 & 1 & -1 \\ 0 & 0 & -1 \\ 0 & 0 & 1 \end{bmatrix}$$

$$\text{Composite transformation matrix: } M = T_2 \cdot R_1 \cdot T_1 = \begin{bmatrix} 0.87 & -0.5 & -0.63 \\ 0.5 & 0.87 & 0.37 \\ 0 & 0 & 1 \end{bmatrix}$$

$$\text{New coordinate matrix: } D = M \cdot C = \begin{bmatrix} -1 & 0.73 & -0.27 & -2 \\ -1 & 0 & 1.73 & 0.73 \\ 1 & 1 & 1 & 1 \end{bmatrix}$$

New vertex coordinates are (–1, –1), (0.73, 0), (–0.27, 1.73), and (–2, 0.73) (Figure 4.7)

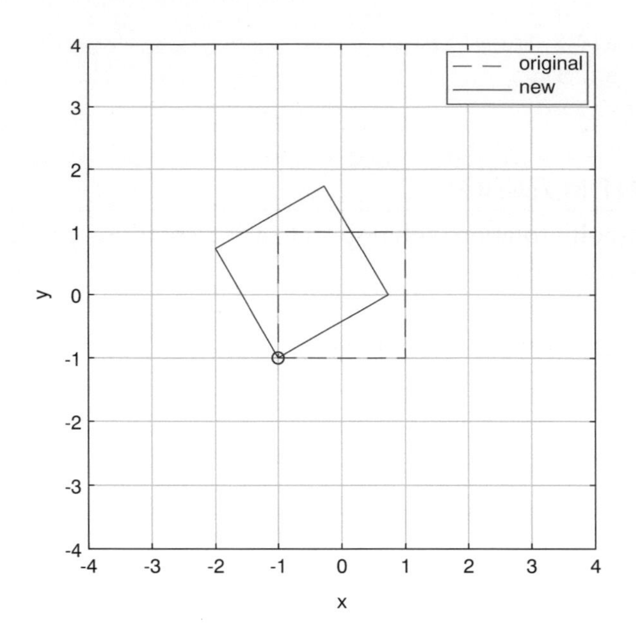

FIGURE 4.7 Plot for Example 4.5.

MATLAB Code 4.5

```
clear all; clc;
X = [-1 1 1 -1 -1];
Y = [-1 -1 1 1 -1];
C = [X ; Y ; 1 1 1 1 1];
xf =-1;
yf =-1;
A = 30;
T1 = [1, 0, -xf ; 0, 1, -yf ; 0, 0, 1];
R1 = [cosd(A), -sind(A), 0 ; sind(A), cosd(A), 0 ; 0, 0, 1];
T2 = [1, 0, xf ; 0, 1, yf ; 0, 0, 1];
M = T2 * R1 * T1 ;
D = M * C;

fprintf('New vertices : \n');
for i=1:4
    fprintf('(%.2f, %.2f) \n',D(1,i), D(2,i));
end

%plotting
plot(C(1,:), C(2,:), 'b--', D(1,:),D(2,:), 'r', xf, yf, 'ro');
xlabel('x');
ylabel('y');
legend('original', 'new');
axis([-4, 4, -4, 4]);
axis square; grid;
```

NOTE

`cosd`: calculates cosine of an angle in degrees
`sind`: calculates sine of an angle in degrees

4.8 REFLECTION

A reflection operation about an axis reverses the sign of the coordinate perpendicular to the axis. Thus, for example a reflection about the X-axis reverses the y-coordinate of a point. The reflection matrices, therefore, are given by the following, where the subscript denotes the axis about which the reflection takes place:

$$F_x = \begin{bmatrix} 1 & 0 & 0 \\ 0 & -1 & 0 \\ 0 & 0 & 1 \end{bmatrix} \tag{4.9}$$

$$F_y = \begin{bmatrix} -1 & 0 & 0 \\ 0 & 1 & 0 \\ 0 & 0 & 1 \end{bmatrix} \tag{4.10}$$

A reflection can also take place about the origin in which case both the x- and y-coordinates gets reversed. This is indicated by the subscript "o."

$$F_o = \begin{bmatrix} -1 & 0 & 0 \\ 0 & -1 & 0 \\ 0 & 0 & 1 \end{bmatrix} \tag{4.11}$$

Example 4.6

A triangle having vertices at (1, 2), (3, 2), and (3, 4) is reflected about the X-axis, Y-axis, and the origin. Find its new coordinates.

Original coordinate matrix: $C = \begin{bmatrix} 1 & 3 & 3 \\ 2 & 2 & 4 \\ 0 & 0 & 1 \end{bmatrix}$

For reflection about X-axis: $F_x = \begin{bmatrix} 1 & 0 & 0 \\ 0 & -1 & 0 \\ 0 & 0 & 1 \end{bmatrix}$

New coordinate matrix: $D_x = F_x \cdot C = \begin{bmatrix} 1 & 3 & 3 \\ -2 & -2 & -4 \\ 0 & 0 & 1 \end{bmatrix}$

For reflection about Y-axis: $F_y = \begin{bmatrix} -1 & 0 & 0 \\ 0 & 1 & 0 \\ 0 & 0 & 1 \end{bmatrix}$

New coordinate matrix: $D_y = F_y \cdot C = \begin{bmatrix} -1 & -3 & -3 \\ 2 & 2 & 4 \\ 0 & 0 & 1 \end{bmatrix}$

For reflection about origin: $F_o = \begin{bmatrix} -1 & 0 & 0 \\ 0 & -1 & 0 \\ 0 & 0 & 1 \end{bmatrix}$

New coordinate matrix: $D_o = F_o \cdot C = \begin{bmatrix} -1 & -3 & -3 \\ -2 & -2 & -4 \\ 0 & 0 & 1 \end{bmatrix}$ (Figure 4.8)

MATLAB Code 4.6

```
clear all; clc;
X = [1, 3, 3, 1];
Y = [2, 2, 4, 2];
C = [X; Y; 1 1 1 1];
```

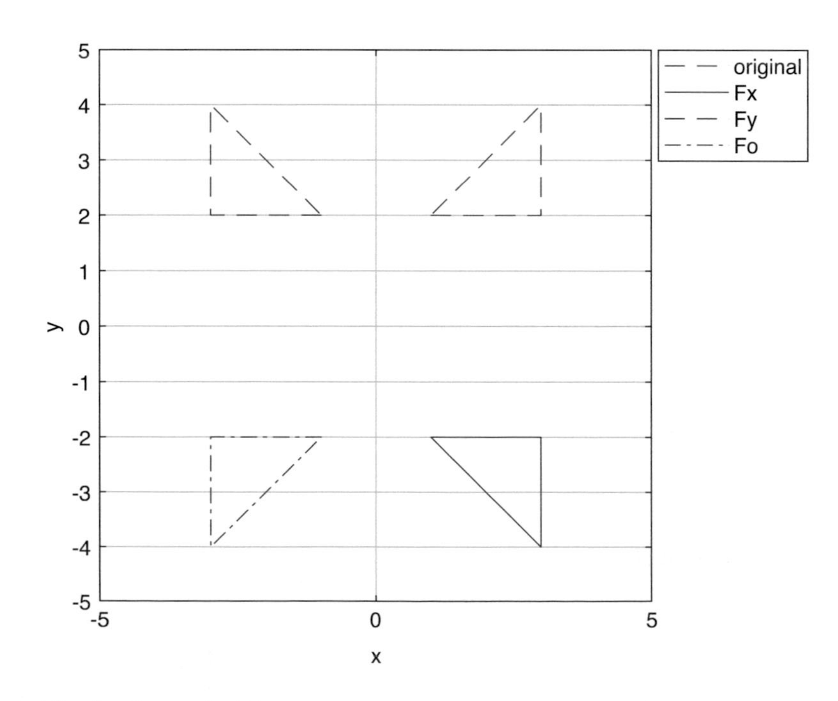

FIGURE 4.8 Plot for Example 4.6.

```
Fx = [1 0 0; 0 -1 0; 0 0 1];
Fy = [-1 0 0; 0 1 0; 0 0 1];
Fo = [-1 0 0; 0 -1 0; 0 0 1];
Dx = Fx*C;
Dy = Fy*C;
Do = Fo*C;

fprintf('New vertices X-axis : \n');
for i=1:3
    fprintf('(%.2f, %.2f) \n',Dx(1,i), Dx(2,i));
end

fprintf('New vertices Y-axis : \n');
for i=1:3
    fprintf('(%.2f, %.2f) \n',Dy(1,i), Dy(2,i));
end

fprintf('New vertices origin: \n');
for i=1:3
    fprintf('(%.2f, %.2f) \n',Do(1,i), Do(2,i));
end

%plotting
plot(C(1,:), C(2,:), 'b--', Dx(1,:),Dx(2,:), 'r', ...
    Dy(1,:), Dy(2,:), 'r--', Do(1,:), Do(2,:), 'r-.');
xlabel('x'); ylabel('y');
axis([-5, 5, -5, 5]);
legend('original', 'Fx', 'Fy', 'Fo');
axis square; grid;
```

4.9 FIXED-LINE REFLECTION

Transformation matrix for reflection about the fixed line L: $y = mx + c$ is obtained using following steps (Chakraborty, 2010):

1. Intercept point of L with Y-axis is $(0, c)$. Translate point to origin: $T_1 = T(0, -c)$
2. Rotate L about origin by $-\theta$, where $\theta = \arctan(m)$: $R_1 = R(\theta)$
3. Apply reflection about X-axis: F_x
4. Reverse rotate around X-axis: $R_2 = R(-\theta)$
5. Reverse translate to original location: $T_2 = T(0, c)$
6. Calculate composite transformation: $M = T_2 \cdot R_2 \cdot F_x \cdot R_1 \cdot T_1$ (Figure 4.9).

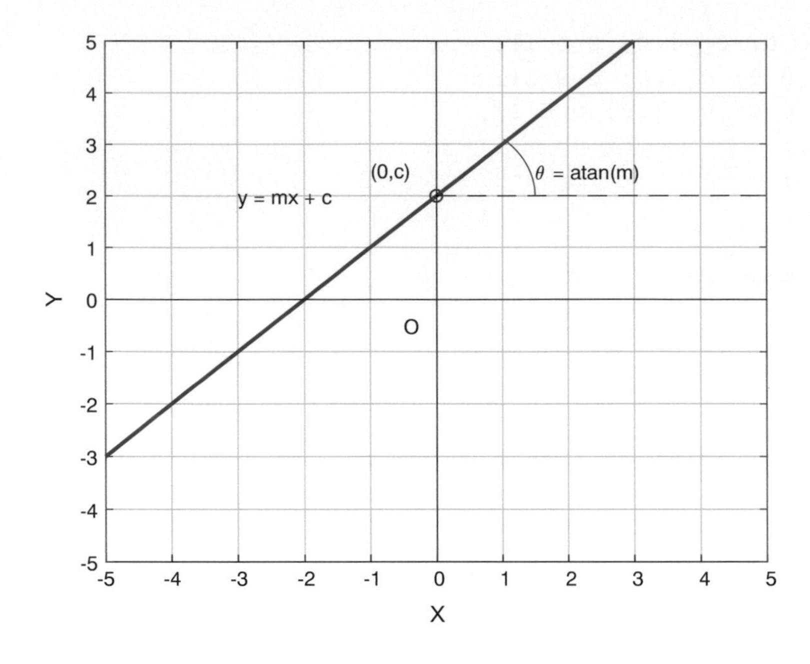

FIGURE 4.9 Fixed-line reflection.

Example 4.7

A triangle having vertices (2, 4), (4, 7), and (5, 6) is reflected about the line $y = 0.5x + 2$. Find its new coordinates.

Original coordinate matrix: $C = \begin{bmatrix} 2 & 4 & 5 \\ 4 & 7 & 6 \\ 0 & 0 & 1 \end{bmatrix}$

For given line L, we have $m = 0.5$, $c = 2$, $k = \arctan(m) = 0.463$ radians $= 26.56°$

Forward translation matrix: $T_1 = T(0, -c) = \begin{bmatrix} 1 & 0 & 0 \\ 0 & 1 & -2 \\ 0 & 0 & 1 \end{bmatrix}$

Forward rotation matrix: $R_1 = R(k) = \begin{bmatrix} \cos k & -\sin k & 0 \\ \sin k & \cos k & 0 \\ 0 & 0 & 1 \end{bmatrix}$

$$= \begin{bmatrix} 0.89 & 0.45 & 0 \\ -0.45 & 0.89 & 0 \\ 0 & 0 & 1 \end{bmatrix}$$

Reflection about X-axis: $F_x = \begin{bmatrix} 1 & 0 & 0 \\ 0 & -1 & 0 \\ 0 & 0 & 1 \end{bmatrix}$

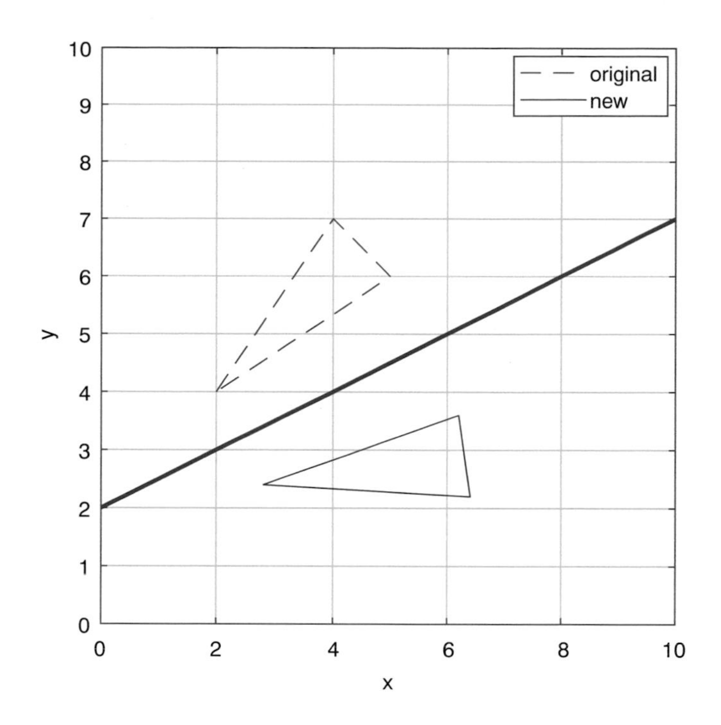

FIGURE 4.10 Plot for Example 4.7.

Reverse rotation matrix: $R_2 = R(-k) = \begin{bmatrix} \cos k & \sin k & 0 \\ -\sin k & \cos k & 0 \\ 0 & 0 & 1 \end{bmatrix}$

$$= \begin{bmatrix} 0.89 & -0.45 & 0 \\ 0.45 & 0.89 & 0 \\ 0 & 0 & 1 \end{bmatrix}$$

Reverse translation matrix: $T_2 = T(0, c) = \begin{bmatrix} 1 & 0 & 0 \\ 0 & 1 & 2 \\ 0 & 0 & 1 \end{bmatrix}$

Composite transformation matrix: $M = T_2 \cdot R_2 \cdot F_x \cdot R_1 \cdot T_1 = \begin{bmatrix} 0.6 & 0.8 & -1.6 \\ 0.8 & -0.6 & 3.2 \\ 0 & 0 & 1 \end{bmatrix}$

New coordinate matrix: $D = M \cdot C = \begin{bmatrix} 2.8 & 6.4 & 6.2 \\ 2.4 & 2.2 & 3.6 \\ 1 & 1 & 1 \end{bmatrix}$

New vertex coordinates are (2.8, 2.4), (6.4, 2.2), and (6.2, 3.6) (Figure 4.10).

MATLAB Code 4.7

```
clear all; clc;
m = 0.5; c = 2;
T1 = [1 0 0 ; 0 1 -c ; 0 0 1];
k = atan(m);
R1 = [cos(k), sin(k), 0 ; -sin(k), cos(k), 0 ; 0 0 1];
Fx = [1 0 0; 0 -1 0; 0 0 1];
R2 = inv(R1);
T2 = inv(T1);
M = T2*R2*Fx*R1*T1;
C = [2 4 5 2; 4 7 6 4; 1 1 1 1];
D = M*C;

fprintf('New vertices : \n');
for i=1:3
    fprintf('(%.2f, %.2f) \n',D(1,i), D(2,i));
end

%plotting
xx = linspace(0,10);
yy = m*xx + c;
plot(C(1,:),C(2,:), 'b--', D(1,:), D(2,:), 'r');
hold on;
plot(xx, yy, 'b-', 'LineWidth', 1.5);
legend('original', 'new'); axis([0, 10, 0, 10]);
grid; axis square;
xlabel('x'); ylabel('y');
hold off;
```

NOTE

`atan`: computes inverse tangent in radian values

4.10 SHEAR

A shear operation distorts a graphic object by changing a set of coordinate values while keeping other values constant (Chakraborty, 2010). There can be two types of shear: one along the *X*-direction and the other along the *Y*-direction. For an *X*-direction, shear the *x*-coordinates of points are shifted in value by an amount proportional to their *y*-coordinates, while the *y*-coordinates themselves remain constant (see Figure 4.11a).

It is evident here that: $x_2 = x_1 + h \cdot y_1$, where h is a constant of proportionality. This means that larger is the *y*-coordinate of a point more is its shift along the *x*-coordinate. Since the *y*-coordinates remain unchanged, $y_2 = y_1$. In matrix form:

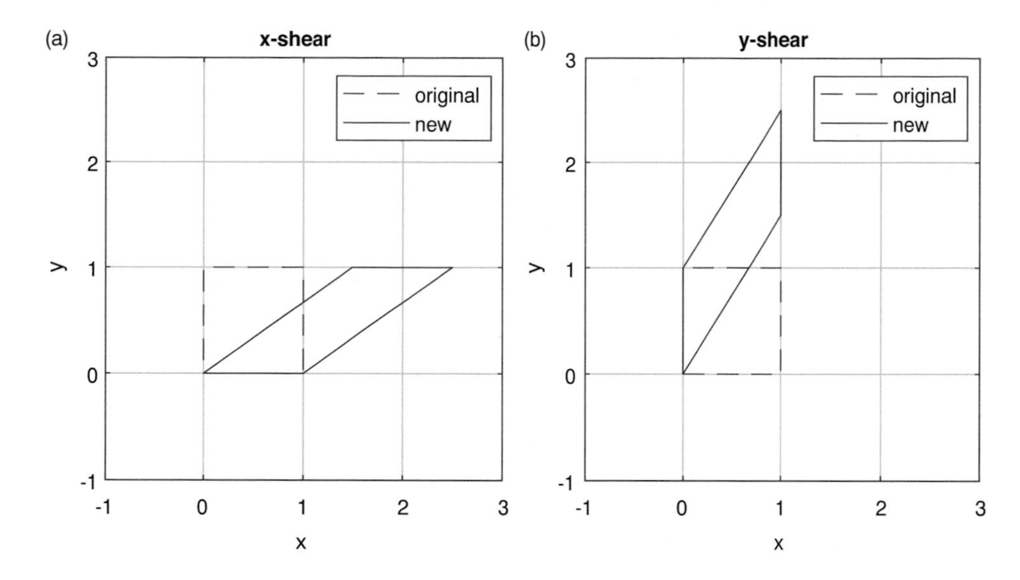

FIGURE 4.11 (a and b) Shear along x- and y-directions.

$$H_x = \begin{bmatrix} 1 & h & 0 \\ 0 & 1 & 0 \\ 0 & 0 & 1 \end{bmatrix} \tag{4.12}$$

Similarly a y-direction shear (Figure 4.11b) would be represented as:

$$H_y = \begin{bmatrix} 1 & 0 & 0 \\ h & 1 & 0 \\ 0 & 0 & 1 \end{bmatrix} \tag{4.13}$$

Example 4.8

A square having vertices at (–1, –1), (1, –1), (1, 1), and (–1, 1) is subjected to a shear of 1.5 along the X-axis and then a shear of 2 along the y-axis. Find its new vertices.

Original coordinate matrix: $C = \begin{bmatrix} -1 & 1 & 1 & -1 \\ -1 & -1 & 1 & 1 \\ 1 & 1 & 1 & 1 \end{bmatrix}$

Shear along X-direction: $H_1 = H_x(1.5) = \begin{bmatrix} 1 & 1.5 & 0 \\ 0 & 1 & 0 \\ 0 & 0 & 1 \end{bmatrix}$

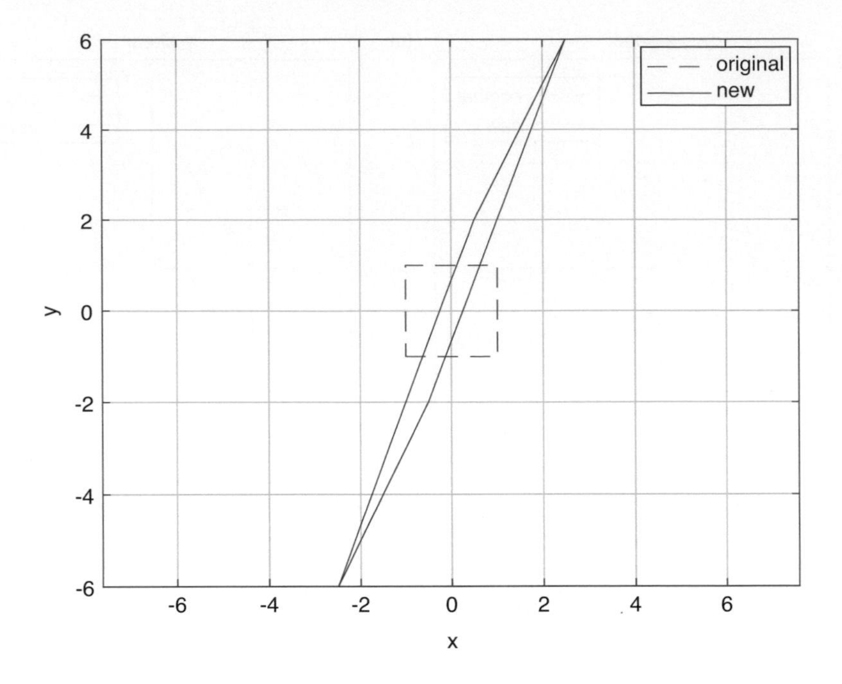

FIGURE 4.12 Plot for Example 4.8.

$$\text{Shear along } Y\text{-direction: } H_2 = H_y(2) = \begin{bmatrix} 1 & 0 & 0 \\ 2 & 1 & 0 \\ 0 & 0 & 1 \end{bmatrix}$$

New coordinate matrix: $D = H_2 \cdot H_1 \cdot C$
New vertex coordinates are (–2.5, –6), (–0.5, –2), (2.5, 6), and (0.5, 2) (Figure 4.12)

MATLAB Code 4.8

```
clear all; clc;
X=[-1 1 1 -1 -1]; Y=[-1 -1 1 1 -1];
C = [X;Y; 1 1 1 1 1];
hx = 1.5; hy = 2;
Hx = [1, hx, 0 ; 0, 1, 0 ; 0, 0, 1];
Hy = [1, 0, 0 ; hy, 1, 0 ; 0, 0, 1];
D = Hy*Hx*C;

fprintf('New vertices : \n');
for i=1:4
    fprintf('(%.2f, %.2f) \n',D(1,i), D(2,i));
end

plot(C(1,:), C(2,:), 'b--', D(1,:),D(2,:), 'r');
xlabel('x');
```

```
ylabel('y');
legend('original', 'new');
axis equal; grid;
```

4.11 AFFINE TRANSFORMATIONS

A composite transformation involving a combination of translation, rotation, scaling, and shear is referred to as affine transformation (Hearn and Baker, 1996; Shirley, 2002). An affine transformation preserves the following property: if P was an arbitrary point located on the line segment joining endpoints A and B before the transformation, then after the transformation the transformed point P' will still be located on the line segment joining the transformed endpoints A' and B'. In general, a rectangle under affine transformation will be converted to a parallelogram (Rovenski, 2010).

Recalling that,

$$T = \begin{bmatrix} 1 & 0 & t_x \\ 0 & 1 & t_y \\ 0 & 0 & 1 \end{bmatrix}$$

$$S = \begin{bmatrix} s_x & 0 & 0 \\ 0 & s_y & 0 \\ 0 & 0 & 1 \end{bmatrix}$$

$$R = \begin{bmatrix} \cos\theta & -\sin\theta & 0 \\ \sin\theta & \cos\theta & 0 \\ 0 & 0 & 1 \end{bmatrix}$$

$$H = \begin{bmatrix} 1 & h_x & 0 \\ h_y & 1 & 0 \\ 0 & 0 & 1 \end{bmatrix}$$

Composite transformation can be calculated as:

$$M = T \cdot S \cdot R \cdot H = \begin{bmatrix} s_x \cdot \cos\theta - h_y \cdot s_x \cdot \sin\theta & h_x \cdot s_x \cdot \cos\theta - s_x \cdot \sin\theta & t_x \\ s_y \cdot \sin\theta + h_y \cdot s_y \cdot \cos\theta & s_y \cdot \cos\theta + h_x \cdot s_y \cdot \sin\theta & t_y \\ 0 & 0 & 1 \end{bmatrix}$$

$$= \begin{bmatrix} a & b & c \\ d & e & f \\ 0 & 0 & 1 \end{bmatrix} \tag{4.14}$$

Thus under affine transformation, the transformed coordinates x', y' of a point are related to the original coordinates x, y by the general relation of the form:

$$x' = ax + by + c$$
$$y' = dx + ey + f \tag{4.15}$$

Reflection can also be added to the set of transformations, which will simply change the sign of some of the coefficients depending on the type of reflection.

Example 4.9

Find the new coordinates of a square with vertices (1, 1), (1, −1), (−1, −1), and (−1, 1) under the composite transformations involving the following: T(2, −4), S(3, −1), R(π/2), and H(1, −2).

Here,

$$T = \begin{bmatrix} 1 & 0 & 2 \\ 0 & 1 & -4 \\ 0 & 0 & 1 \end{bmatrix}$$

$$S = \begin{bmatrix} 3 & 0 & 0 \\ 0 & -1 & 0 \\ 0 & 0 & 1 \end{bmatrix}$$

$$R = \begin{bmatrix} \cos\pi/2 & -\sin\pi/2 & 0 \\ \sin\pi/2 & \cos\pi/2 & 0 \\ 0 & 0 & 1 \end{bmatrix} = \begin{bmatrix} 0 & -1 & 0 \\ 1 & 0 & 0 \\ 0 & 0 & 1 \end{bmatrix}$$

$$H = \begin{bmatrix} 1 & 1 & 0 \\ -2 & 1 & 0 \\ 0 & 0 & 1 \end{bmatrix}$$

$$\text{Composite transformation matrix: } M = \begin{bmatrix} 6 & -3 & 2 \\ -1 & -1 & -4 \\ 0 & 0 & 1 \end{bmatrix}$$

$$\text{Original coordinate matrix: } C = \begin{bmatrix} -1 & 1 & 1 & -1 \\ -1 & -1 & 1 & 1 \\ 1 & 1 & 1 & 1 \end{bmatrix}$$

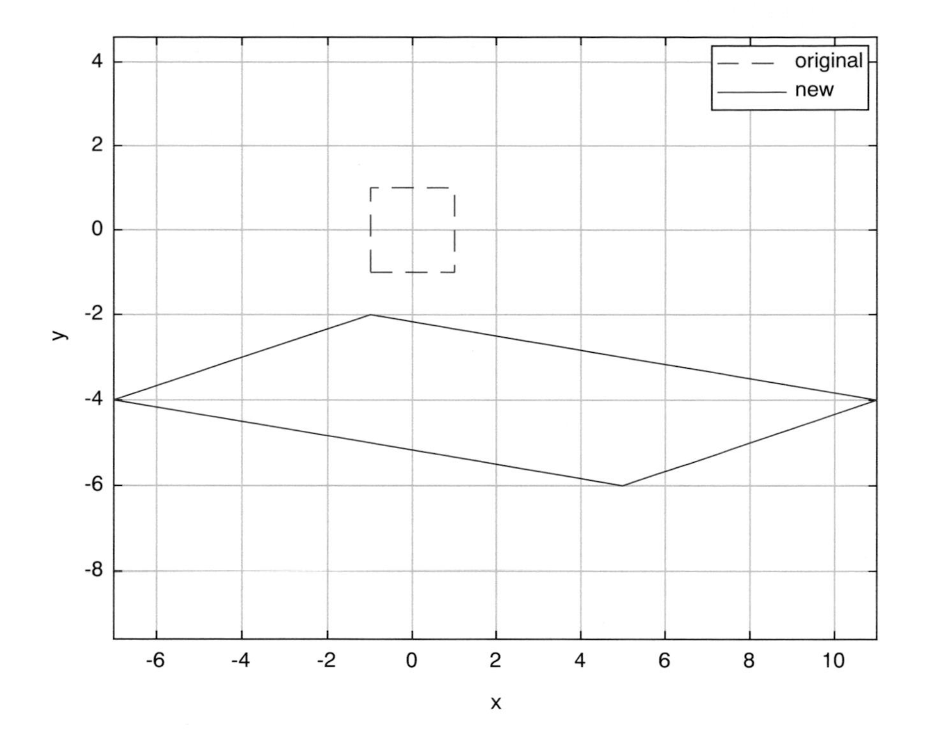

FIGURE 4.13 Plot for Example 4.9.

New coordinate matrix: $D = M \cdot C = \begin{bmatrix} -1 & 11 & 5 & -7 \\ -2 & -4 & -6 & -4 \\ 1 & 1 & 1 & 1 \end{bmatrix}$

New vertex coordinates are (−1, −2), (11, −4), (5, −6), and (−7, −4) (Figure 4.13).

MATLAB Code 4.9

```
clear all; clc;
syms tx ty sx sy A hx hy ;
X = [-1 1 1 -1 -1];
Y = [-1 -1 1 1 -1];
C = [X ; Y ; 1 1 1 1 1];

T = [1 0 tx; 0 1 ty; 0 0 1];
S = [sx 0 0; 0 sy 0; 0 0 1];
R = [cos(A) -sin(A) 0; sin(A) cos(A) 0; 0 0 1];
H = [1 hx 0; hy 1 0; 0 0 1];

M = T*S*R*H;
M1 = subs(M, [tx, ty, sx, sy, A, hx, hy], [2, -4, 3, -1, pi/2, 1, -2]);
D = M1 * C;

fprintf('New vertices : \n');
for i=1:4
    fprintf('(%.2f, %.2f) \n',D(1,i), D(2,i));
end

%plotting
plot(C(1,:), C(2,:), 'b--', D(1,:),D(2,:), 'r');
xlabel('x');ylabel('y');
legend('original', 'new');
axis equal; grid;
```

4.12 PERSPECTIVE TRANSFORMATIONS

We have seen above that the affine transformation matrix described by Equation (4.14) converts a rectangle into a parallelogram and is described as below, where (x', y') are the transformed coordinates of the point (x, y).

$$\begin{bmatrix} x' \\ y' \\ 1 \end{bmatrix} = \begin{bmatrix} a & b & c \\ d & e & f \\ 0 & 0 & 1 \end{bmatrix} \begin{bmatrix} x \\ y \\ 1 \end{bmatrix} \tag{4.16}$$

The mapping relations between the old and new coordinates are given by Equation (4.15). During this transformation, a set of constraints are applied, which is responsible for the parallelogram shape. If a rectangle has a new set of vertices with coordinates $(x_0, y_0), (x_1, y_1), (x_2, y_2)$, and (x_3, y_3) then the following constraints are applied: $(x_1 - x_0) = (x_2 - x_3)$ and $(y_1 - y_0) = (y_2 - y_3)$. These constraints simply state the lengths of the opposite sides should be equal and this forces the new figure to be a parallelogram. However, if the constraints are not applied then a rectangle will be converted to an arbitrary quadrilateral and the corresponding transformation does not remain affine any more (Rovenski, 2010). The new transformation is called a perspective (or projective) transformation and is described by the transformation matrix shown below:

$$\begin{bmatrix} x' \\ y' \\ w \end{bmatrix} = \begin{bmatrix} a & b & c \\ d & e & f \\ g & h & 1 \end{bmatrix} \begin{bmatrix} x \\ y \\ 1 \end{bmatrix} \tag{4.17}$$

where x' and y' are in homogeneous coordinates. The Cartesian coordinates are $X = x'/w$ and $Y = y'/w$. From Equation (4.17), we can derive the mapping relations between the old and new coordinates:

$$\begin{aligned} X &= \frac{x'}{w} = \frac{ax + by + c}{gx + hy + 1} \\ Y &= \frac{y'}{w} = \frac{dx + ey + f}{gx + hy + 1} \end{aligned} \tag{4.18}$$

Details about the mapping process are given in Chapter 8.

Example 4.10

Find the new coordinates of a square with vertices (0,0), (1,0), (1,1), and (0,1) under the following transformations: (a) $\begin{bmatrix} 5 & 2 & 5 \\ 2 & 5 & 5 \\ 0 & 0 & 1 \end{bmatrix}$ ***and (b)*** $\begin{bmatrix} 5 & 2 & 5 \\ 2 & 5 & 5 \\ 5 & 2 & 1 \end{bmatrix}$. ***Specify the type of transformation in each case.***

(a)

$$\text{Original coordinate matrix: } C = \begin{bmatrix} 0 & 1 & 1 & 0 \\ 0 & 0 & 1 & 1 \\ 1 & 1 & 1 & 1 \end{bmatrix}$$

$$\text{Transformation matrix: } M = \begin{bmatrix} 5 & 2 & 5 \\ 2 & 5 & 5 \\ 0 & 0 & 1 \end{bmatrix}$$

$$\text{New coordinate matrix: } D = M \cdot C = \begin{bmatrix} 5 & 10 & 12 & 7 \\ 5 & 7 & 12 & 10 \\ 1 & 1 & 1 & 1 \end{bmatrix}$$

New vertex coordinates are (5.00, 5.00), (10.00, 7.00), (12.00, 12.00), and (7.00, 10.00)
Let $x_0 = 5$, $y_0 = 5$, $x_1 = 10$, $y_1 = 7$, $x_2 = 12$, $y_2 = 12$, $x_3 = 7$, $y_3 = 10$
Now $d_1 = x_1 - x_0 = 5$, $d_2 = x_2 - x_3 = 5$, $d_3 = y_1 - y_0 = 2$, $d_4 = y_2 - y_3 = 2$
Since $d_1 = d_2$ and $d_3 = d_4$, the transformation is affine in nature (Figure 4.14a)

(b)

$$\text{Transformation matrix: } M = \begin{bmatrix} 5 & 2 & 5 \\ 2 & 5 & 5 \\ 5 & 2 & 1 \end{bmatrix}$$

New coordinate matrix: $D_h = M \cdot C = \begin{bmatrix} 5 & 10 & 12 & 7 \\ 5 & 7 & 12 & 10 \\ 1 & 6 & 8 & 3 \end{bmatrix}$ (in homogeneous coordinates)

New coordinate matrix: $D = \begin{bmatrix} 5 & 1.67 & 1.5 & 2.33 \\ 5 & 1.17 & 1.5 & 3.33 \\ 1 & 1 & 1 & 1 \end{bmatrix}$ (in Cartesian coordinates)

New vertex coordinates are (5.00, 5.00), (1.67, 1.17), (1.50, 1.50), and (2.33, 3.33)
Let $x_0 = 5$, $y_0 = 5$, $x_1 = 1.67$, $y_1 = 1.17$, $x_2 = 1.5$, $y_2 = 1.5$, $x_3 = 2.33$, $y_3 = 3.33$
Now $d_1 = x_1 - x_0 = -3.33$, $d_2 = x_2 - x_3 = -0.83$, $d_3 = y_1 - y_0 = -3.83$, $d_4 = y_2 - y_3 = -1.83$
Since $d_1 \neq d_2$ and $d_3 \neq d_4$, the transformation is perspective in nature (Figure 4.14b).

MATLAB Code 4.10

```
clear all; clc; format compact;

% (a)

C = [0 1 1 0 0 ; 0 0 1 1 0 ; 1 1 1 1 1]
M = [5 2 5 ; 2 5 5 ; 0 0 1]
D = M*C
```

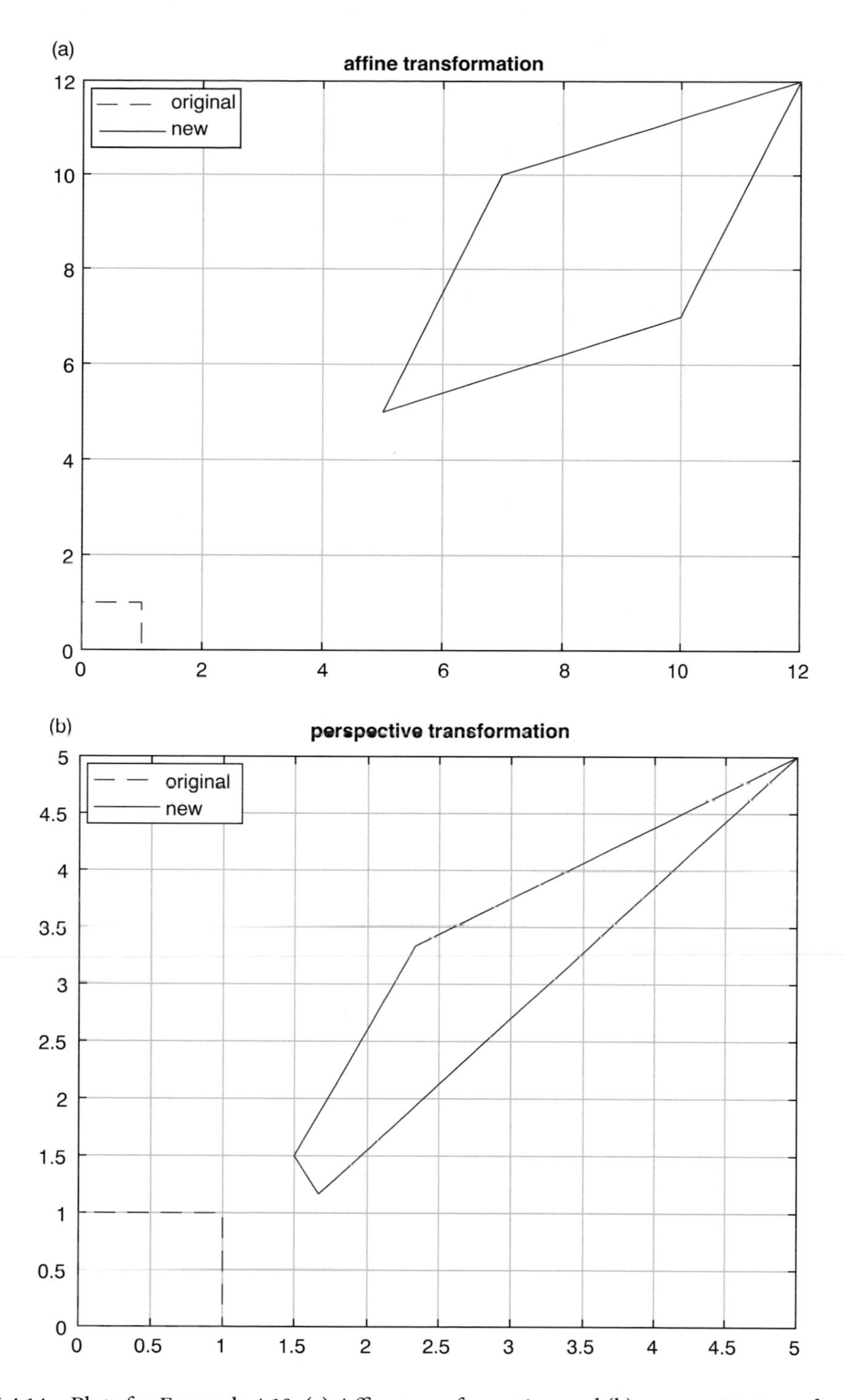

FIGURE 4.14 Plots for Example 4.10. (a) Affine transformation and (b) perspective transformation.

```
figure,
plot (C(1,:), C(2,:), 'b--', D(1,:), D(2,:), 'r-'); grid;
legend('original', 'new');
title('affine transformation');

fprintf('New vertices : \n');
for i=1:4
    fprintf('(%.2f, %.2f) \n', D(1,i), D(2,i));
end

x0 = D(1,1); y0 = D(2,1);
x1 = D(1,2); y1 = D(2,2);
x2 = D(1,3); y2 = D(2,3);
x3 = D(1,4); y3 = D(2,4);

d1 = x1 - x0, d2 = x2 - x3,
d3 = y1 - y0, d4 = y2 - y3,

if d1 == d2 && d3 == d4
    fprintf('Transformation is affine\n');
else
    fprintf('Transformation is perspective\n');
end

fprintf('\n\n');
% (b)

M = [5 2 5 ; 2 5 5 ; 5 2 1]
Dh = M*C;

for i=1:length(Dh)
    D(:,i) = Dh(:,i)/Dh(3,i);
end

fprintf('New vertices : \n');
for i=1:4
    fprintf('(%.2f, %.2f) \n', D(1,i), D(2,i));
end

figure
plot (C(1,:), C(2,:), 'b--', D(1,:), D(2,:), 'r-'); grid;
legend('original', 'new');
title('perspective transformation');

x0 = D(1,1); y0 = D(2,1);
x1 = D(1,2); y1 = D(2,2);
```

```
x2 = D(1,3); y2 = D(2,3);
x3 = D(1,4); y3 = D(2,4);

d1 = x1 - x0, d2 = x2 - x3,
d3 = y1 - y0, d4 = y2 - y3,

if d1 == d2 && d3 == d4
    fprintf('Transformation is affine\n');
else
    fprintf('Transformation is perspective\n');
end
```

4.13 VIEWING TRANSFORMATIONS

Viewing transformations are associated with displaying rendered graphics output on a display device. It has already been mentioned that 2D graphics objects like splines and polygons are created on a plane and stored using a coordinate system. The coordinate system is used to measure location of points and store them in vectors. The creation plane can extend on all sides indefinitely, limited only by the hardware resources. When a portion of this plane is to be displayed on the output devices, the graphics display system needs two additional functional components to make this possible: the window and the viewport. The window is defined by the four corners of a rectangular area using the coordinate system of the graphics application software, to select a particular portion of the creation plane that needs to be displayed. The viewport is a mapped version of the window to the coordinates of the output device like a monitor, to enable the hardware to display the stored graphics data on screen. The viewport is also defined by the four corners of a rectangular area but uses the coordinate system of the display device. The mapping between the window and the viewport is collectively referred to as "viewing transformations" (Hearn and Baker, 1996; Shirley, 2002) and involve both translation and scaling (see Figure 4.15).

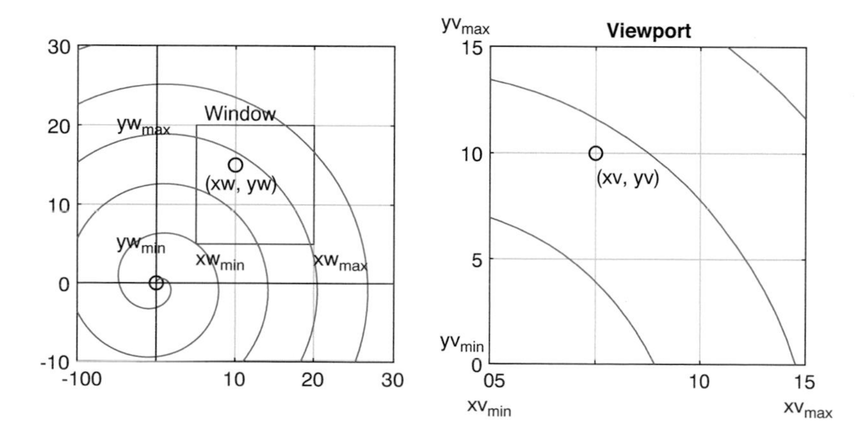

FIGURE 4.15 Window and viewport.

The left-hand figure shows a spline generated using application specified coordinates, and a rectangular window on it with its vertices defined as

$$A\left(xw_{\min}, yw_{\min}\right), B\left(xw_{\max}, yw_{\min}\right), C\left(xw_{\max}, yw_{\max}\right), D\left(xw_{\min}, yw_{\max}\right)$$

The contents inside the window is mapped onto a device viewport showed on the right side with its vertices defined as

$$A'\left(xv_{\min}, yv_{\min}\right), B'\left(xv_{\max}, yv_{\min}\right), C'\left(xv_{\max}, yv_{\max}\right), D'\left(xv_{\min}, yv_{\max}\right)$$

The mapping from the window to the viewport involve the following steps (Foley et al., 1995):

1. Translate A to A': $T_1 = T\left(xv_{\min} - xw_{\min}, yv_{\min} - yw_{\min}\right)$
2. Translate P to origin: $T_2 = T\left(-xv_{\min}, -yv_{\min}\right)$
3. Scale about origin: $S_1 = S\left(s_x, s_y\right)$
4. Reverse translate back to A': $T_3 = T\left(xv_{\min}, yv_{\min}\right)$
5. Composite transformation: $M = T_3 \cdot S_1 \cdot T_2 \cdot T_1$

The scaling factors involve changing the dimension of the window to the viewport. The width of the viewport is $\left(xv_{\max} - xv_{\min}\right)$ and the width of the window is $\left(xw_{\max} - xw_{\min}\right)$. The horizontal scaling factor s_x will be a ratio of these i.e.

$$s_x = \frac{xv_{\max} - xv_{\min}}{xw_{\max} - xw_{\min}} \tag{4.19}$$

The height of the viewport is $\left(yv_{\max} - yv_{\min}\right)$ and the height of the window is $\left(yw_{\max} - yw_{\min}\right)$. The vertical scaling factor s_y will be a ratio of these i.e.

$$s_y = \frac{yv_{\max} - yv_{\min}}{yw_{\max} - yw_{\min}} \tag{4.20}$$

Example 4.11

Obtain a transformation that maps a window whose lower-left corner is at A(1, 1) and upper-right corner at C(3, 5) onto a viewport that has a lower-left corner at P(0, 0) and upper-right corner at Q(0.5, 0.5).

Translate A to P: $T_1 = T(-1, -1)$

Apply scaling about P, since P is already at the origin: $S_1 = S\left(s_x, s_y\right)$

From Equation (4.19),

$$s_x = \frac{xv_{\max} - xv_{\min}}{xw_{\max} - xw_{\min}} = \frac{0.5}{2} = 0.25$$

From Equation (4.20),

$$s_y = \frac{yv_{\max} - yv_{\min}}{yw_{\max} - yw_{\min}} = \frac{0.5}{4} = 0.125$$

$$\text{Composite transformation } M = S_1 \cdot T_1 = \begin{bmatrix} 0.25 & 0 & -0.25 \\ 0 & 0.125 & -0.125 \\ 0 & 0 & 1 \end{bmatrix}$$

MATLAB Code 4.11

```
clear all; clc;

xwmin=1;
ywmin=1;
xwmax=3;
ywmax=5;
xvmin=0;
yvmin=0;
xvmax=0.5;
yvmax=0.5;

tx = xvmin-xwmin;
ty = yvmin-ywmin;
T1 = [1 0 tx; 0 1 ty;  0 0 1];

sx = (xvmax - xvmin)/(xwmax - xwmin);
sy = (yvmax - yvmin)/(ywmax - ywmin);
S1 = [sx 0 0; 0 sy 0 ; 0 0 1];

M = S1*T1
```

Another point of interest in a viewing transformation operation is to find the new coordinates of a specific point in a window, after it is mapped to the viewport. Let a point (xw, yw) in a designated window be mapped to viewport coordinates (xv, yv). To keep relative placements same we require the following to hold true:

$$\frac{xv - xv_{\min}}{xw - xw_{\min}} = \frac{xv_{\max} - xv_{\min}}{xw_{\max} - xw_{\min}} \tag{4.21}$$

$$\frac{yv - yv_{\min}}{yw - yw_{\min}} = \frac{yv_{\max} - yv_{\min}}{yw_{\max} - yw_{\min}} \tag{4.22}$$

Solving for (xv, yv) we get:

$$\begin{aligned} xv &= xv_{\min} + s_x \cdot (xw - xw_{\min}) \\ yv &= yv_{\min} + s_y \cdot (yw - yw_{\min}) \end{aligned} \tag{4.23}$$

Example 4.12

A user works on a coordinate system, which is a square window with corner coordinates P(–2, –5) and Q(8, 5). The user's area is mapped to a square viewport whose corner coordinates are P′(400, 500) and Q′(600, 800). (a) Find the window to viewport transformation matrix and (b) find out the origin of the user's coordinate system after mapping to the viewport.

(a)

Here, $xw_{\min} = -2$, $yw_{\min} = -5$, $xw_{\max} = 8$, $yw_{\max} = 5$, $xv_{\min} = 400$, $yv_{\min} = 500$, $xv_{\max} = 600$, $yv_{\max} = 800$

$sx = (xv_{\max} - xv_{\min})/(xw_{\max} - xw_{\min}) = 20$

$sy = (yv_{\max} - yv_{\min})/(yw_{\max} - yw_{\min}) = 30$

Forward translation $T_1 = T(xv_{\min} - xw_{\min}, yv_{\min} - yw_{\min})$

Translation to origin $T_2 = T(-xv_{\min}, -yv_{\min})$

Scale about origin $S_1 = S(sx, sy)$

Reverse translation $T_3 = T(xv_{\min}, yv_{\min})$

Composite transformation: $M = T_3 * S_1 * T_2 * T_1 = \begin{bmatrix} 20 & 0 & 440 \\ 0 & 30 & 650 \\ 0 & 0 & 1 \end{bmatrix}$

Verification: $M*P = P'$, $M*Q = Q'$

(b)

Here, $xw = 0$, $yw = 0$, $xv = ?$, $yv = ?$

From Equation (4.18)

$xv = xv_{\min} + (xw - xw_{\min})*sx = 440$

$yv = yv_{\min} + (yw - yw_{\min})*sy = 650$

Thus, point (0, 0) of the window maps to point (440, 650) of the viewport.

MATLAB Code 4.12

```
clear all; clc;

xwmin = -2;
ywmin = -5;
xwmax = 8;
ywmax = 5;
```

```
xvmin = 400;
yvmin = 500;
xvmax = 600;
yvmax = 800;
sx = (xvmax - xvmin)/(xwmax - xwmin);
sy = (yvmax - yvmin)/(ywmax - ywmin);
T1 = [1, 0, xvmin - xwmin ; 0, 1, yvmin - ywmin ; 0, 0, 1];
T2 = [1, 0, 0 - xvmin ; 0, 1, 0 - yvmin ; 0, 0, 1];
S1 = [sx, 0, 0 ; 0, sy, 0 ; 0, 0, 1];
T3 = [1, 0, xvmin ; 0, 1, yvmin ; 0, 0, 1];
M = T3 * S1 * T2 * T1

xw = 0; yw = 0;
xv = xvmin + (xw - xwmin)*sx; xv = round(xv)
yv = yvmin + (yw - ywmin)*sy; yv = round(yv)
```

4.14 COORDINATE SYSTEM TRANSFORMATIONS

Consider two Cartesian systems with origins at (0, 0) and (x_0, y_0) and angle θ between the x- and X-axes. To transform points from x–y system to X–Y system following steps are followed (Hearn and Baker, 1996; Shirley, 2002):

1. Translate origin (x_0, y_0) to (0, 0) point: $T_1 = T(-x_0, -y_0)$
2. Rotate X-axis onto the x-axis: $R_1 = R(-\theta)$
3. Composite transformation: $M = R_1 \cdot T_1$ (Figure 4.16)

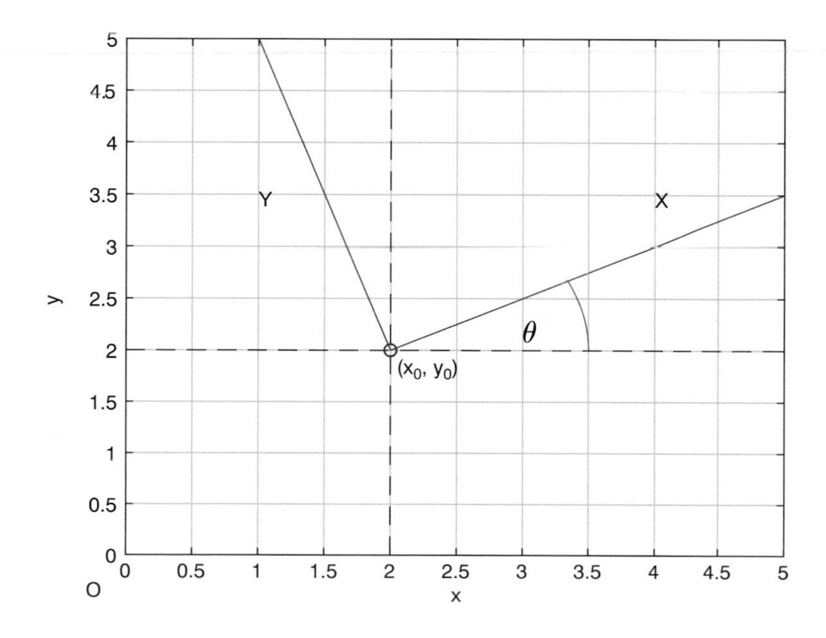

FIGURE 4.16 Coordinate system transformation.

Example 4.13

Find the equation of a straight line $y = 2x + 3$ in the X–Y coordinate system, which results from a 90° rotation of the x–y coordinate system.

Original equation: $y = mx + c$

Two points on the original line $A(0, c)$ and $B(-c/m, 0)$

To transform from x–y to X–Y system, the latter has to be rotated by −90° to coincide with x–y system

Hence, rotation matrix is $R_1 = R(-90)$

New coordinates: $P = R_1 {*} A = (c, 0)$

New coordinates: $Q = R_1 {*} B = (0, c/m)$

Equation of line joining P and Q: $(Y - 0)/(X - c) = (c/m - 0)/(0 - c)$

Simplifying: $Y = (-1/m)X + (1/m)c$

Substituting values of m and c we get:

Original equation: $y = 2x + 3$

New equation: $Y = (-1/2)X + (3/2)$

MATLAB Code 4.13

```
clear all; clc;
m = 2; c = 3;
syms x y X Y;
T = -90;
R = [cosd(T), -sind(T), 0 ; sind(T), cosd(T), 0 ; 0, 0, 1];
A = [0 ; c ; 1];
B = [-c/m ; 0 ; 1];
P = R*A;
Q = R*B;
fprintf('Original equation : \n'); y = m*x + c
fprintf('New equation : \n');
X1 = P(1); Y1 = P(2); X2 = Q(1); Y2 = Q(2);
M = (Y2-Y1)/(X2-X1);
Y = M*(X - X1) + Y1;
Y = eval(Y)
```

4.15 CHAPTER SUMMARY

The following points summarize the topics discussed in this chapter:

- Transformations change the location, dimension, orientation, and shape of splines.
- These transformations are translation, scaling, rotation, reflection, and shear.
- Homogeneous coordinates are used to represent all transformations in a uniform manner.
- Transformation matrices are multiplied with the original coordinates to get new coordinates.

- Translation changes location of splines by adding increments to their coordinates.
- Scaling changes the dimensions of splines by multiplying scaling factors to their coordinates.
- Rotation changes the orientation of splines by moving them along a circular path.
- Reflection creates a mirror image of splines by flipping some of their coordinates.
- Shear distorts a spline by changing some of their coordinates while keeping others constant.
- Affine transformations convert a rectangle to a parallelogram.
- Perspective transformations convert a rectangle to an arbitrary quadrilateral.
- Transformations are by default with respect to the origin of the coordinate system.
- Viewing transformation maps data from an application window to a device viewport.
- Coordinate system transformation involves translation of origin and rotation of axes.

4.16 REVIEW QUESTIONS

1. What is meant by a right-handed coordinate system?
2. What are the advantages of homogeneous coordinates over Cartesian coordinates?
3. How does the original coordinates of a point change during translation and scaling operations?
4. What are the effects of a negative and a fractional scaling factor?
5. What is the difference between a general rotation and a fixed-point rotation operation?
6. What is reflection about an axis and reflection about the origin?
7. What is the difference between affine and perspective transformations?
8. What is the difference between x- and y-direction shears?
9. What is mapping between window and viewport?
10. How can coordinates be transformed from one coordinate system to another?

4.17 PRACTICE PROBLEMS

1. Show that a reflection about the line $y=x$ is equivalent to a reflection relative to the x-axis followed by a CCW rotation of 90°.
2. Obtain a transformation that reduces rectangle $ABCD$ formed from points $A(0, 0)$, $B(5, 0)$, $C(5, 4)$, and $D(0, 4)$ to half its size keeping the point D fixed.
3. Reflect the triangle (2, 5), (3, 7), and (4, 6) about the line $x=2y-1$ and find its new coordinates.

4. A square is placed within an animation sequence with its center at (5, 5). It is subsequently shrunk by 1/50th of its size in each successive frame. Determine the corresponding transformation matrix for each frame.

5. What transformation maps triangle ABC: $A(0, 0)$, $B(5, 0)$, and $C(5, 4)$ onto a single point C?

6. Two symbols consisting of a triangle (−1, 0), (0, 1), and (1, 0) and a square (0, 0), (0, 1), (1, 1), and (1, 0) is combined to create a design with the triangle at (1, 1), (1.5, 2), and (2, 1) and square at (1, 1), (1, 2), (2, 2), and (2, 1). Find the corresponding transformations.

7. Find the equation of a straight line $Y = mX + c$ in the x–y coordinate system if the X–Y coordinate system results from a 90° rotation of the x–y coordinate system.

8. What is the composite transformation if one shear $x' = ax + by$ is followed by another shear $y' = bx + ay$, where a and b are constants.

9. Find the transformation, which uses the rectangle $A(1, 1)$, $B(5, 3)$, $C(4, 5)$, and $D(0, 3)$ as a window and a viewport with coordinates (0, 0), (1, 0), (1, 1), and (0, 1).

10. Find the new coordinates of a square with vertices (0, 0), (1, 0), (1, 1), and (0, 1) under the following transformations: (a) $\begin{bmatrix} -9 & -8 & 7 \\ 6 & -5 & -4 \\ 0 & 0 & 1 \end{bmatrix}$ and (b) $\begin{bmatrix} -9 & -8 & 7 \\ 6 & -5 & -4 \\ 3 & 2 & 1 \end{bmatrix}$.

CHAPTER 5

Spline Properties

5.1 INTRODUCTION

This chapter discusses few common properties of splines and how these can be calculated from spline equations (Mathews, 2004). First, it discusses the critical points namely minimum and maximum of spline curves. Additionally, for splines of degree 3 or more the point of inflection (POI) is of interest, which is where the curvature changes from concave to convex or vice versa. Next, it discusses how the tangent and normal to a spline curve can be calculated. The tangent to a curve is the derivative of the curve equation, while the normal is the line perpendicular to the tangent. The tangent and the normal can be represented by line equations or as specific vectors if they are calculated for a specific point on the curve. Computations of the tangent and normal are discussed both for a parametric curve as well as an implicit curve. The third property is calculation of length of a spline curve between any two given points. This can be calculated both from a spatial curve and a parametric curve. The length of the curve segment is approximated by summing over a number of small line segments. The fourth property is to calculate the area under a curve. This is discussed both for curves of the form $y = f(x)$ and also $x = f(y)$. The area is calculated by considering a very thin rectangular area under the curve and then summing over all such rectangular areas. An extension to this is calculation of area bounded by two curves. This is the area under the upper curve minus the area under the lower curve. The fifth property is the centroid of an area under a curve, which is computed using moments. The next section deals with various ways of interpolation and curve fitting. Interpolation is done when we are interested in some intermediate value between some given data points. Interpolation can either be linear by considering straight lines connecting adjacent data points, or non-linear by considering higher degree curves connecting the data points. This section also mentions built-in MATLAB® functions for the purpose. Curve fitting is trying to fit a polynomial curve to given data points and estimating the coefficients of the polynomial. This enables representation of arbitrary data by polynomial functions. This chapter ends by reviewing few 2D plotting functions included in MATLAB. These functions are broadly categorized into two types: one using symbolic variables and the other using a collection of values. These functions can also be either explicit or implicit or parametric. Arguments to some of these functions can be used to specify color and transparency values.

5.2 CRITICAL POINTS

Critical points denote the maximum and minimum points of a spline and Point of Inflection POI for splines of degree 3 or above (see Figure 5.1). For both minimum and maximum points, the slope of the curve is zero as the line is horizontal. Both points of the curve $y = f(x)$ can be found by computing roots of $f'(x) = 0$.

Let r be a root of the equation $f'(x) = 0$. To determine whether the root corresponds to a minimum or maximum, consider a small displacement δ to the left and right of r. For a minimum point, left-hand side (LHS) slope should be negative and right-hand size (RHS) slope should be positive. This implies:

$$f'(r - \delta) < 0 \text{ and } f'(r + \delta) > 0 \tag{5.1}$$

Alternatively, since the slope changes from negative to positive, rate of change of slope is positive i.e.

$$f''(r) > 0 \tag{5.2}$$

For a maximum point, LHS slope should be positive and RHS slope should be negative. This implies:

$$f'(r - \delta) > 0 \text{ and } f'(r + \delta) < 0 \tag{5.3}$$

Alternatively, since the slope changes from positive to negative, rate of change of slope is negative i.e.

$$f''(r) < 0 \tag{5.4}$$

The coordinates of the minimum or maximum point is given by $[r, f(r)]$.

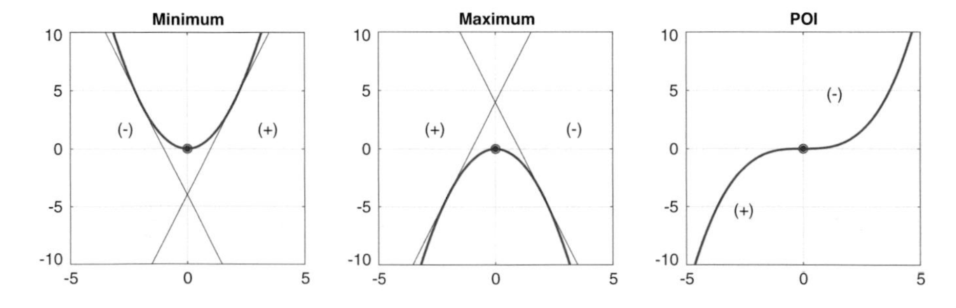

FIGURE 5.1 Critical points.

A POI is where curvature changes from positive (convex) to negative (concave) or vice versa. The necessary conditions to be satisfied are that the curvature should be zero at the POI and the sign of curvature on either side should be opposite. Let r be a root of the equation $f''(x)=0$. Consider a small displacement δ to the left and right of r. Conditions necessary for a POI to exist are:

$$\begin{gathered} f''(x)=0 \\ \text{sign}\{f''(r-\delta)\} \neq \text{sign}\{f''(r+\delta)\} \end{gathered} \tag{5.5}$$

Example 5.1

Find the critical points for the cubic curve: $y=2+13x-31x^2+18x^3$.

From the given equation:

$$f(x)=2+13x-31x^2+18x^3$$

$$f'(x)=13-62x+54x^2$$

$$f''(x)=-62+108x$$

Setting $f'(x)=13-62x+54x^2=0$ roots are $r_1=0.87$, $r_2=0.28$. Let $\delta=0.1$

Now, $f'(r_1-\delta)=f'(0.87-0.1)-13-62(0.77)+54(0.77)^2=-2.72<0$

Also, $f'(r_1+\delta)=f'(0.87+0.1)=13-62(0.97)+54(0.97)^2=+3.67>0$

Thus, there is a minimum point at $x=0.87$.

Since $f(0.87)=1.7$, coordinates of minimum point is (0.87, 1.7)

Verification: $f''(0.87)=-62+108(0.87)=-62+93.96=31.96>0$

Also, $f'(r_2-\delta)=f'(0.28-0.1)=13-62(0.18)+54(0.18)^2=+3.59>0$

And, $f'(r_2+\delta)=f'(0.28+0.1)=13-62(0.38)+54(0.38)^2=-2.76<0$

Thus, there is a maximum point at $x=0.28$

Since $f(0.28)=3.6$, coordinates of maximum point is (0.28, 3.6)

Verification: $f''(0.28)=-62+108(0.28)=-62+30.24=-31.76<0$

For POI, setting $f''(x)=-62+108x=0$, root $r=0.574$

Now $\text{sign}\{f''(r-\delta)\}=\text{sign}\{f''(0.574-0.1)\}=\text{sign}\{-62+108(0.474)\}=$ $\text{sign}(-10.8)=-1$

And $\text{sign}\{f''(r+\delta)\}=\text{sign}\{f''(0.574+0.1)\}=\text{sign}\{-62+108(0.674)\}=$ $\text{sign}(10.8)=+1$

Since signs are opposite, there is a POI at $r=0.574$

Coordinates of the POI: $[0.574, f(0.574)]$ i.e. (0.574, 2.662) (Figure 5.2)

MATLAB Code 5.1

```
clear all; clc;
%ax^3 + bx^2 + cx + d
p = [18, -31, 13, 2];
dp = polyder(p);
d2p = polyder(dp);
r = roots(dp);

if polyval(d2p, r(1)) > 0, m1 = 0; else m1 = 1; end;
b1 = polyval(p, r(1));
if m1 == 0, fprintf('minimum point : (%.2f, %.2f)\n', r(1), b1); end;
if m1 == 1, fprintf('maximum point : (%.2f, %.2f)\n', r(1), b1); end;

if polyval(d2p, r(2)) > 0, m2 = 0; else m2 = 1; end;
b2 = polyval(p, r(2));
if m2 == 0, fprintf('minimum point : (%.2f, %.2f)\n', r(2), b2); end;
if m2 == 1, fprintf('maximum point : (%.2f, %.2f)\n', r(2), b2); end;

s = roots(d2p);
s1 =  polyval(d2p, s - 0.1);
s2 =  polyval(d2p, s + 0.1);
b3 = polyval(p, s);
if (sign(s1) == sign(s2)), fprintf('no poi');
else fprintf('poi : (%.2f, %.2f)\n', s, b3); end;

%plotting
syms x;
y = p(1)*x^3 + p(2)*x^2 + p(3)*x + p(4);
X = [r(1), r(2), s]; Y = [b1, b2, b3];
xx = linspace(min([r(1), r(2), s]), max([r(1), r(2), s]));
```

```
yy = subs(y, x, xx);
plot(xx, yy, 'b-', 'LineWidth', 1.5); hold on;
scatter(X, Y, 20, 'r', 'filled');
plot(X, Y, 'ko');
grid; xlabel('x'); ylabel('y');

%labeling
if m1 == 0, text(r(1), b1+0.1, 'Minimum');
else text(r(1), b1+0.1, 'Maximum');
end;
if m2 == 0, text(r(2), b2+0.1, 'Minimum'):
else text(r(2), b2+0.1, 'Maximum');
end;
text(s, b3+0.1, 'POI');
hold off;
```

NOTE

`polyder`: differentiate polynomial
`roots`: find roots of polynomial equation
`polyval`: evaluate polynomial at the specified value
`sign`: returns sign of the argument +1, 0, or –1

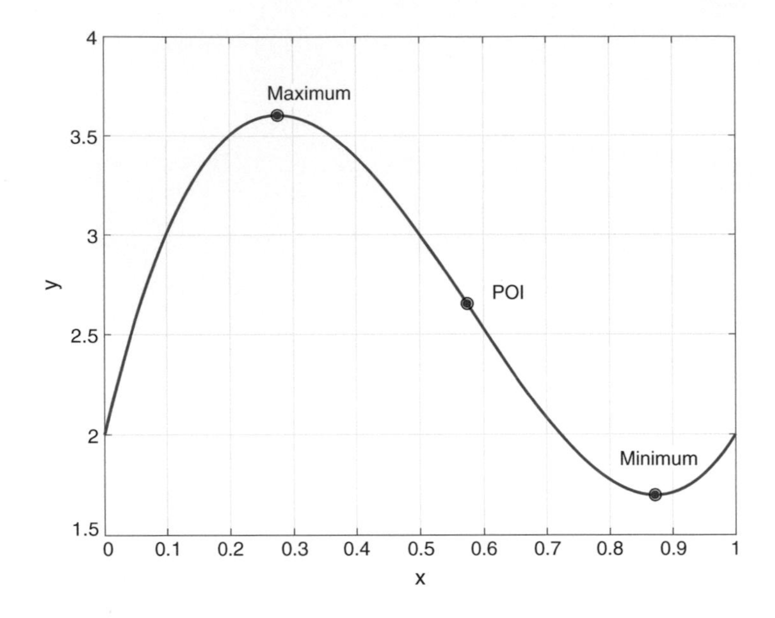

FIGURE 5.2 Plot for Example 5.1.

5.3 TANGENT AND NORMAL

A parametric curve can be expressed as an ordered pair i.e.

$$C(t) = \{x(t),\ y(t)\} \tag{5.6}$$

Tangent vector is obtained from the first derivative:

$$T(t) = C'(t) = \{x'(t),\ y'(t)\} \tag{5.7}$$

Unit tangent vector is obtained by dividing by its magnitude:

$$T(t)/|T(t)| = \{x'(t),\ y'(t)\}/|x'(t),\ y'(t)| \tag{5.8}$$

Normal vector is obtained by rotating tangent vector by 90 degrees:

$$N(t) = R(90) \cdot T(t) \tag{5.9}$$

Expanding:

$$N(t) = \begin{bmatrix} 0 & -1 & 0 \\ 1 & 0 & 0 \\ 0 & 0 & 1 \end{bmatrix} \cdot \begin{bmatrix} x'(t) \\ y'(t) \\ 1 \end{bmatrix} = \begin{bmatrix} -y'(t) \\ x'(t) \\ 1 \end{bmatrix} \tag{5.10}$$

Writing as an ordered pair:

$$N(t)=\{-y'(t),\ x'(t)\} \tag{5.11}$$

Unit normal vector is obtained by dividing by its magnitude:

$$N(t)/|N(t)|=\{-y'(t),\ x'(t)\}/|-y'(t),\ x'(t)| \tag{5.12}$$

Equation of the tangent line through a point (x_1, y_1) on the curve:

$$\frac{y-y_1}{x-x_1}=\frac{y'(t)}{x'(t)} \tag{5.13}$$

Equation of the normal line through a point (x_1, y_1) on the curve:

$$\frac{y-y_1}{x-x_1}=\frac{-x'(t)}{y'(t)} \tag{5.14}$$

Example 5.2

For the circle C(t) = {cos(t), sin(t)}, find the unit tangent vector, unit normal vector, and tangent line and normal line at point P(1/√2, 1/√2).

Given curve: $C(t)=\{\cos(t), \sin(t)\}$

Tangent vector: $T(t)=C'(t)=\{x'(t),\ y'(t)\}=\{-\sin(t), \cos(t)\}=$ unit tangent vector

Normal vector: $R(90)*C'(t)=\{-y'(t), x'(t)\}=\{-\cos(t), -\sin(t)\}=$ unit normal vector

Now at point $P(1/\sqrt{2},\ 1/\sqrt{2})$, solving for $\cos(t)=1/\sqrt{2}$ we must have $t=\pi/4$

Unit tangent vector at P: $\{-\sin(\pi/4), \cos(\pi/4)\}=\{-1/\sqrt{2}, 1/\sqrt{2}\}$

Unit normal vector at P: $\{-\cos(\pi/4), -\sin(\pi/4)\}=\{-1/\sqrt{2}, -1/\sqrt{2}\}$

Tangent line at P: $\dfrac{y-y_1}{x-x_1}=\dfrac{y'(t)}{x'(t)}$

Substituting values: $\dfrac{y-1/\sqrt{2}}{x-1/\sqrt{2}}=\dfrac{\cos(\pi/4)}{-\sin(\pi/4)}=-1$

Simplifying: $x+y-\sqrt{2}=0$

Normal line at P: $\dfrac{y-y_1}{x-x_1}=\dfrac{-x'(t)}{y'(t)}$

Substituting values: $\dfrac{y-1/\sqrt{2}}{x-1/\sqrt{2}}=\dfrac{\sin(\pi/4)}{\cos(\pi/4)}=1$

Simplifying: $x-y=0$ (Figure 5.3)

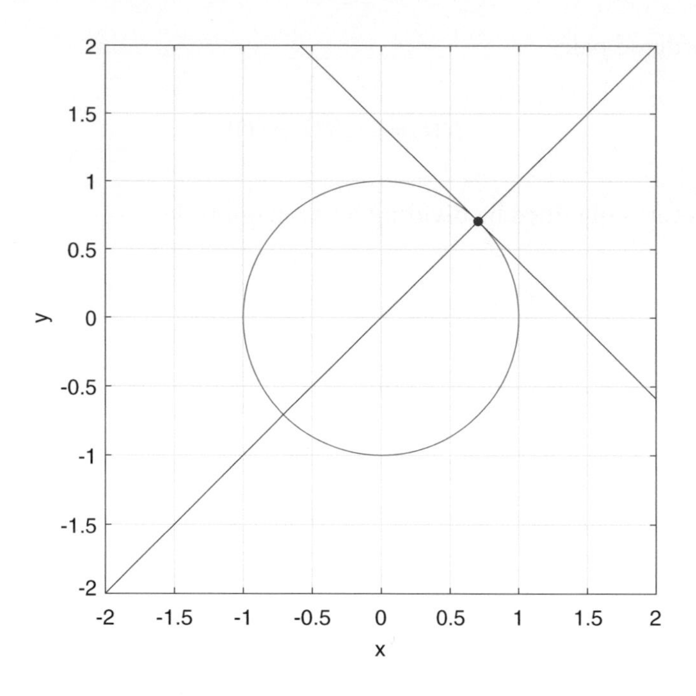

FIGURE 5.3 Plot for Example 5.2.

MATLAB Code 5.2

```
clear all; clc; format compact;
P = [1/sqrt(2), 1/sqrt(2)];
syms t;
x = cos(t); y = sin(t);
C = [x, y];
fprintf('Tangent vector \n');
T = [diff(x), diff(y)]
fprintf('Normal vector \n');
N = [-diff(y), diff(x)]
R = solve(sin(t)==1/sqrt(2), cos(t)==1/sqrt(2));
fprintf('Tangent vector at P \n');
TP = subs(T, 't', R)
fprintf('Normal vector at P \n');
NP = subs(N, 't', R)

syms X Y;
fprintf('Tangent Line at P\n');
Y1 = (TP(2)/TP(1))*(X - P(1))+P(2)
fprintf('Normal Line at P\n');
Y2 = (-TP(1)/TP(2))*(X - P(1))+P(2)
```

```
%plotting
ezplot(x, y);hold on;
axis([-2 2 -2 2]);
xx = linspace(-2,2);
yy1 = subs(Y1, X, xx);
yy2 = subs(Y2, X, xx);
plot(xx, yy1, 'r-', xx, yy2, 'r-');
scatter(P(1), P(2), 20, 'r', 'filled');
grid; hold off;
```

NOTE

`diff`: calculates derivatives and partial derivatives
`solve`: generates solution of equations

If the equation of the curve is expressed in implicit form $f(x, y)=0$, then the equation of the tangent line to the curve at $P(x_0, y_0)$ is derived by first computing the partial derivative of the curve equation at the given point, which gives us the normal vector at that point.

$$N(x, y)=\left(\frac{\partial f}{\partial x}, \frac{\partial f}{\partial y}\right) \tag{5.15}$$

So normal at point P is given by:

$$N_p=\left(\frac{\partial f_P}{\partial x}, \frac{\partial f_P}{\partial y}\right) \tag{5.16}$$

Equation of the tangent line at P is computed as:

$$T_p: \frac{\partial f_P}{\partial x}\cdot(x-x_0)+\frac{\partial f_P}{\partial y}\cdot(y-y_0) \tag{5.17}$$

Example 5.3

Find the normal vector and tangent line to the curve $x^3+2xy+y^2=9$ at point P(1, 2).

Here, $f=x^3+2xy+y^2-9$

Partial derivatives: $f_x(x, y)=\frac{\partial f}{\partial x}=3x^2+2y$ and $f_y(x, y)=\frac{\partial f}{\partial y}=2x+2y$

Normal vector: $N(x,y)=\left[f_x(x,y), f_y(x,y)\right]=\left[3x^2+2y, 2x+2y\right]$

Normal vector at point P: $N(1,2)=[7,6]$

Equation of tangent line at point P: $7(x-1)+6(y-2)=0$ i.e. $7x+6y-19=0$ (Figure 5.4)

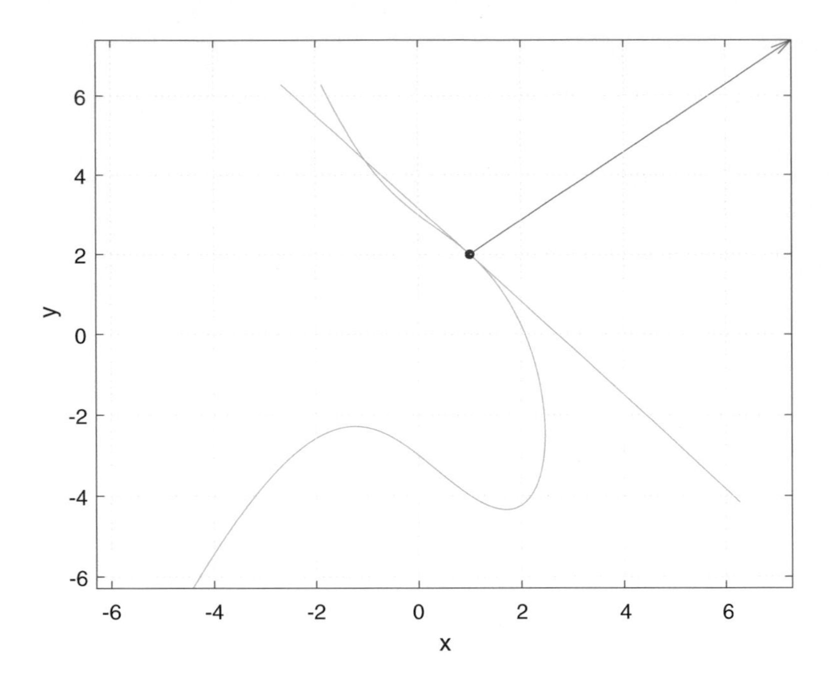

FIGURE 5.4 Plot for Example 5.3.

MATLAB Code 5.3

```
clear all; clc;
syms x y;
f = x^3 + 2*x*y + y^2 - 9;
df = [diff(f, x), diff(f, y)];
p = [1, 2];
fprintf('Normal vector : \n');
n = subs(df, [x, y], [p(1), p(2)])
fprintf('Tangent line : \n');
t = dot(n, [x-1, y-2])

% plotting
ezplot(f); hold on; grid;
ezplot(t);
quiver(p(1), p(2), n(1), n(2));
scatter(p(1), p(2), 20, 'r', 'filled');
hold off;
```

NOTE

`dot`: calculates vector dot product
`quiver`: depicts vectors as arrows with direction and magnitude

5.4 LENGTH OF A CURVE

Length of curve $y = f(x)$ in spatial domain between points $x = a$ and $x = b$ is given by:

$$L = \int_a^b \sqrt{1 + \left(\frac{dy}{dx}\right)^2}\, dx \tag{5.18}$$

The derivation of this expression assumes that a curve can be approximated by a number of small line segments (see Figure 5.5). For such a small line segment if δx be the horizontal distance and δy be the vertical distance then the length of the segment can be approximated by:

$$\delta r = \sqrt{(\delta x)^2 + (\delta y)^2} \tag{5.19}$$

The length of the entire curve is the sum of these small segment distances between $x = a$ and $x = b$. Integrating:

$$L = \int_a^b \sqrt{(dx)^2 + (dy)^2} \tag{5.20}$$

Simplifying:

$$L = \int_a^b \sqrt{\left(\frac{dx}{dx}\right)^2 + \left(\frac{dy}{dx}\right)^2}\, dx = \int_a^b \sqrt{1 + \left(\frac{dy}{dx}\right)^2}\, dx \tag{5.21}$$

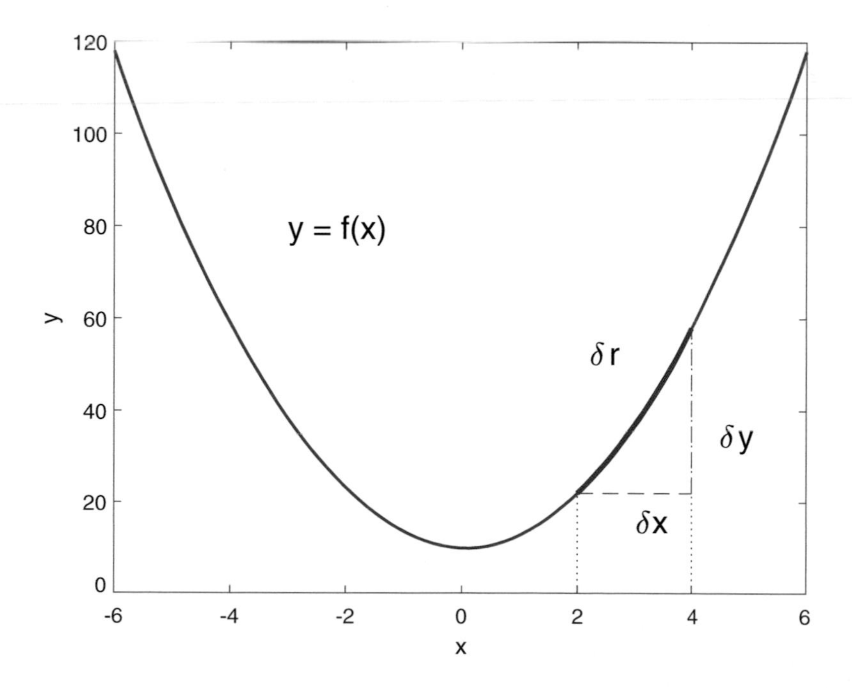

FIGURE 5.5 Derivation of curve length.

Example 5.4

Find the length of the curve $y = 3x + 2$ between $x = 1$ and $x = 5$.

Here, $\frac{dy}{dx} = 3$

From Equation (5.18) we get:

$$L = \int_a^b \sqrt{1 + \left(\frac{dy}{dx}\right)^2}\, dx = \int_1^5 \sqrt{1 + (3)^2}\, dx = \int_1^5 \sqrt{10}\, dx = 4\sqrt{10} = 12.65$$

Verification: $y(1) = 5, y(5) = 17, P_1 = (1,5), P_2 = (5,17), \text{distance}(P_1, P_2) = \sqrt{16 + 144} = 12.65$

MATLAB Code 5.4

```
clear all; clc;
syms x;
y = 3*x+2;
d = diff(y);
e = sqrt(1 + d^2);
f = int(e, 1, 5);
fprintf('Length of curve : %f\n', eval(f));
```

NOTE

`int`: integrate symbolic expression

If the equation of the curve is expressed in parametric form i.e. $x(t)$ and $y(t)$, then from Equation (5.19):

$$dr = \sqrt{(dx)^2 + (dy)^2}$$

Differentiating with respect to t:

$$\frac{dr}{dt} = \sqrt{\left(\frac{dx}{dt}\right)^2 + \left(\frac{dy}{dt}\right)^2}$$

Integrating:

$$L = \int_a^b \sqrt{\left(\frac{dx}{dt}\right)^2 + \left(\frac{dy}{dt}\right)^2}\, dt \tag{5.22}$$

Example 5.5

Determine the length of the parametric curve $x = 3 \cdot \sin t$, $y = 3 \cdot \cos t$, for $0 \le t \le 2\pi$

Here, $\frac{dx}{dt} = 3 \cdot \cos(t)$ and $\frac{dy}{dt} = -3 \cdot \sin(t)$

From Equation (5.22):

$$L=\int_a^b \sqrt{\left(\frac{dx}{dt}\right)^2+\left(\frac{dy}{dt}\right)^2}\,dt=\int_0^{2\pi}\sqrt{(3\cos t)^2+(-3\sin t)^2}\,dt=\int_0^{2\pi}3dt=6\pi$$

Verification: Since this represents a circle with radius 3, length of circumference $= 2\pi(3) = 6\pi$

MATLAB Code 5.5

```
clear all; clc;
syms t;
x = 3*sin(t);
y = 3*cos(t);
dx = diff(x);
dy = diff(y);
r = sqrt((dx)^2 + (dy)^2);
f = int(r, 0, 2*pi);
fprintf('Length : %f\n', eval(f));
```

5.5 AREA UNDER A CURVE

Area below a curve $y=f(x)$ between $x=a$ and $x=b$ is computed by considering a very small rectangular area of width dx and height $y=f(x)$ (see Figure 5.6). The area of this rectangular area is $f(x)\cdot dx$.

To find the entire area, we integrate the small rectangular area over the specified limits.

$$A=\int_a^b f(x)dx \tag{5.23}$$

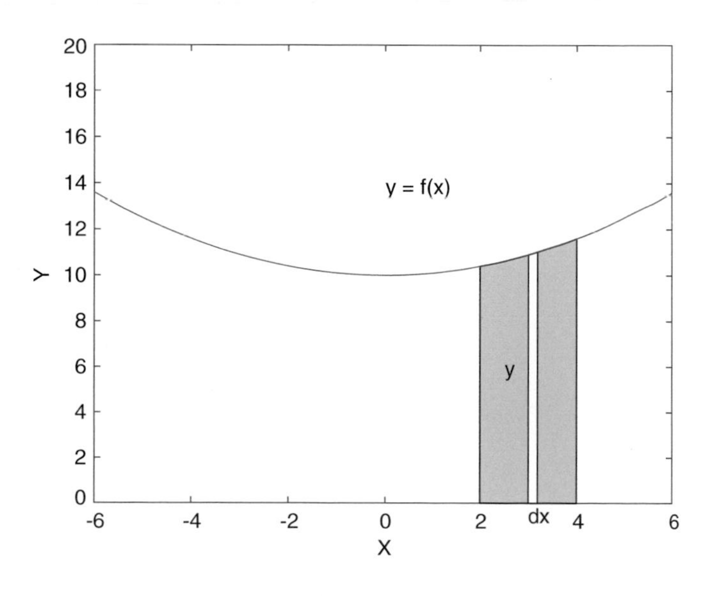

FIGURE 5.6 Area under curve.

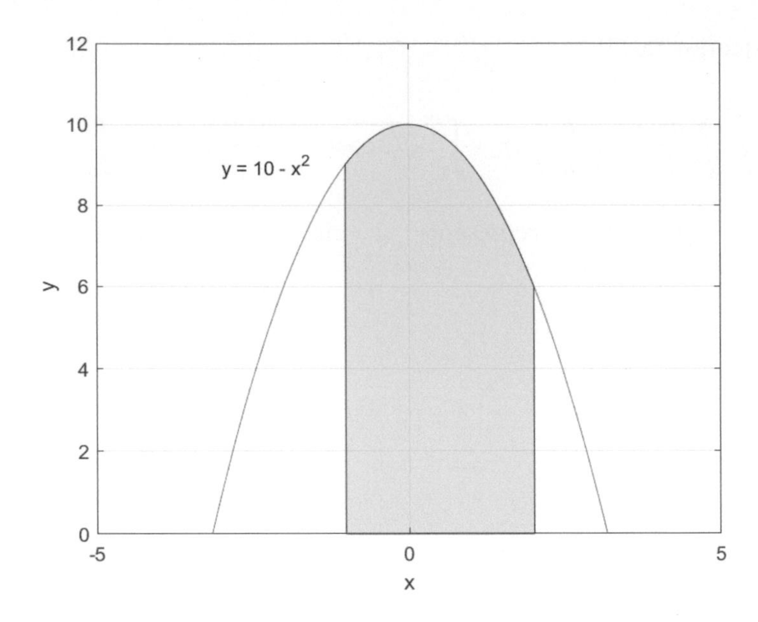

FIGURE 5.7 Plot for Example 5.6.

Example 5.6

Find the area under the curve $y = 10 - x^2$ between the values $x = -1$ and $x = 2$.

From Equation (5.23) we get (Figure 5.7):

$$A = \int_a^b f(x)dx = \int_{-1}^{2} \left(10 - x^2\right)dx = 27$$

MATLAB Code 5.6

```
clear; clc;
syms x;
y = 10 - x^2;
f = int(y, -1, 2);
eval(f);
fprintf('Area : %f\n', eval(f));

%plotting
xx = linspace(-4, 4);
yy = subs(y, x, xx);
plot(xx, yy);
xlabel('x'); ylabel('y');
grid; hold on;
axis([-5 5, 0 12]);
text(-3, 9, 'y = 10 - x^2');

%filling
x = linspace(-1, 2);
```

```
y1 = 10 - x.^2;
y2 = zeros(1,100);
X = [x,fliplr(x)];
Y = [y1,fliplr(y2)];
fill(X,Y,'g');
alpha(0.25);
hold off;
```

NOTE

`zeros`: generates a matrix filled with zeros
`fliplr`: flip array in left–right direction
`fill`: fills a polygon with color
`alpha`: sets transparency values

The above area computed is actually the area between the curve $f(x)$ and the x-axis. However, if the curve itself lies below the x-axis then the area computed by the above formula comes out as negative. In such cases, we need to take the absolute value of the area. It is, therefore, always best to sketch the curve before finding areas under curves.

Example 5.7

Find the area under the curve $y = x^3$ between the values $x = -1$ and $x = 2$

From Equation (5.23) we get (Figure 5.8):

$$A = \int_a^b f(x)dx = \left|\int_{-1}^{0} x^3\,dx\right| + \int_0^2 x^3\,dx = \left|\frac{x^4}{4}\right|_{-1}^{0} + \left|\frac{x^4}{4}\right|_0^2 = \frac{1}{4} + \frac{16}{4} = \frac{17}{4} = 4.25$$

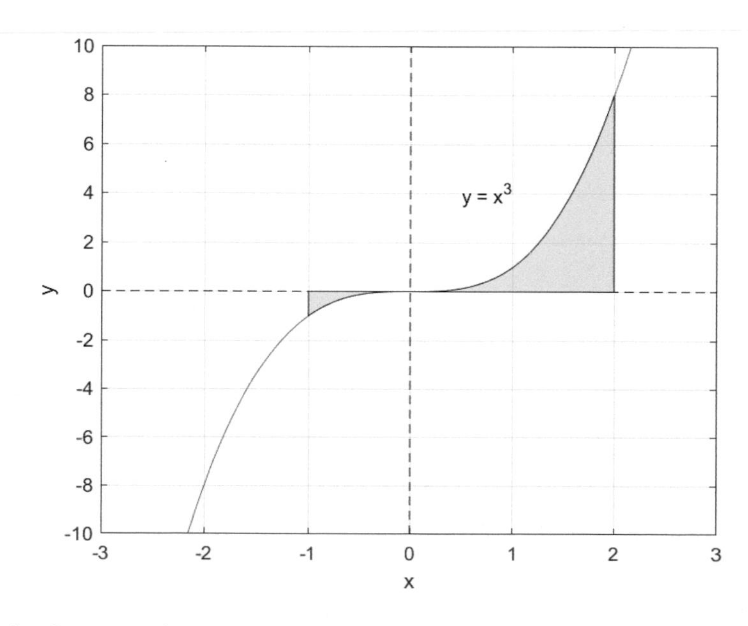

FIGURE 5.8 Plot for Example 5.7.

Alternatively, take absolute value of the function before integration:

$$A = \int_{-1}^{2} |x^3| dx = \left| \frac{x^4 * \text{sign}(x)}{4} \right|_{-1}^{2} = \frac{16}{4} - \frac{1(-1)}{4} = \frac{17}{4}$$

Compare with the incorrect result if we directly take the integration without considering the sign

$$A = \int_{-1}^{2} x^3 dx = \left| \frac{x^4}{4} \right|_{-1}^{2} = \frac{16}{4} - \frac{1}{4} = \frac{15}{4}$$

MATLAB Code 5.7

```
clear; clc;
syms x;
y = x^3;
f1 = abs(int(y, -1, 0));
f2 = int(y, 0, 2);
f = f1 + f2;
fprintf('Area : %f\n', eval(f));
%alternative
f3 = int(abs(y), -1, 2);
fprintf('Alternatively : %f\n', eval(f3));
%plotting
xx = linspace(-3, 3);
yy = subs(y, x, xx);
plot(xx, yy);
xlabel('x'); ylabel('y');
hold on;
X = [-3 3]; Y = [0, 0]; plot(X, Y, 'k--');
X = [0 0]; Y = [-10, 10]; plot(X, Y, 'k--');
axis([-3 3, -10 10]); grid;
text(0.5, 4, 'y = x^3');
x = linspace(-1, 2);
y1 = x.^3;
y2 = zeros(1,100);
X = [x,fliplr(x)];
Y = [y1,fliplr(y2)];
fill(X,Y,'g');  alpha(0.25);
hold off;
```

If the area is required between a curve and the y-axis then by an extension of earlier ideas we first express the curve in the form $x = f(y)$ and then by drawing a series of small rectangle of width dy and height $x = f(y)$, we integrate over the specified limits c to d. In this case:

$$A = \int_{c}^{d} f(y) dy \tag{5.24}$$

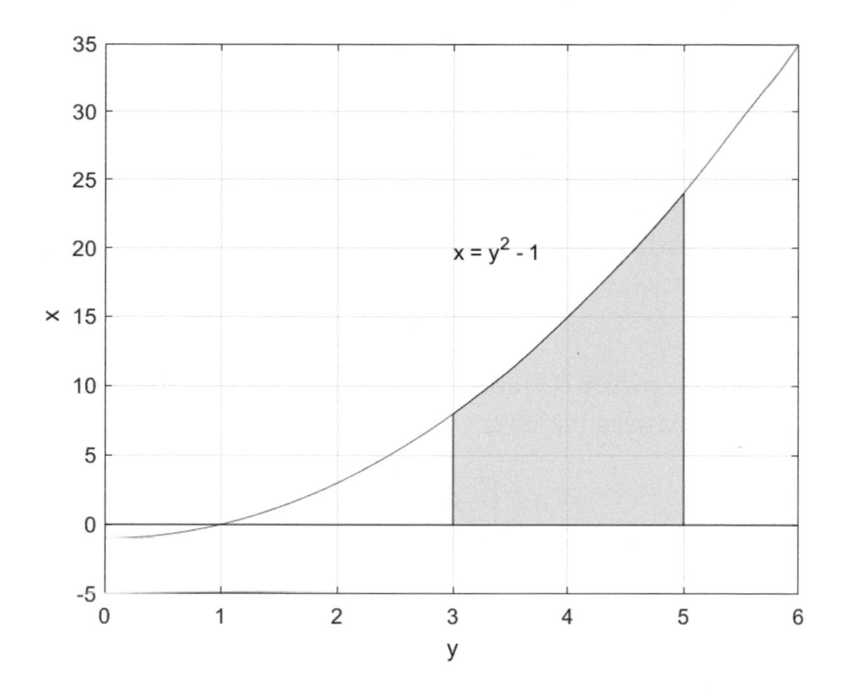

FIGURE 5.9 Plot for Example 5.8.

Example 5.8

Find the area under the curve $y = \sqrt{x+1}$ between the values $y = 3$ and $y = 5$.

Rewriting: $x = f(y) = y^2 - 1$

From Equation (5.24) we get (Figure 5.9):

$$A = \int_c^d f(y)dy = \int_3^5 \left(y^2 - 1\right) dy = 30.67$$

MATLAB Code 5.8

```
clear all; clc;
syms y;
x = y^2 - 1;
f =  int(x, 3, 5);
fprintf('Area : %f\n', eval(f));

%plotting
yy = linspace(0, 6);
xx = subs(x, y, yy);
plot(yy, xx);
xlabel('y'); ylabel('x');
hold on;
X = [3, 3]; Y = [0, subs(x, y, 3)]; plot(X, Y, 'r');
X = [5, 5]; Y = [0, subs(x, y, 5)]; plot(X, Y, 'r');
```

```
plot([0, 6], [0, 0], 'k-'); grid;
x = linspace(3, 5);
y1 = x.^2 - 1;
y2 = zeros(1,100);
X = [x,fliplr(x)];
Y = [y1,fliplr(y2)];
fill(X,Y,'g');  alpha(0.25);
text(3, 20, 'x = y^2 - 1');
hold off;
```

Area bounded by two curves $f(x)$ and $g(x)$ is the area between the upper curve and the x-axis minus the area between the lower curve and x-axis:

$$A = \int_a^b |f(x) - g(x)| dx \tag{5.25}$$

Example 5.9

Find the area bounded by the curves $y = x^3$ and $y = x$ between the values $x = -1$ and $x = 1$

From Equation (5.25) we get (Figure 5.10):

$$A = \int_a^b |f(x) - g(x)| dx = \int_{-1}^{1} |x^3 - x| dx = 0.5$$

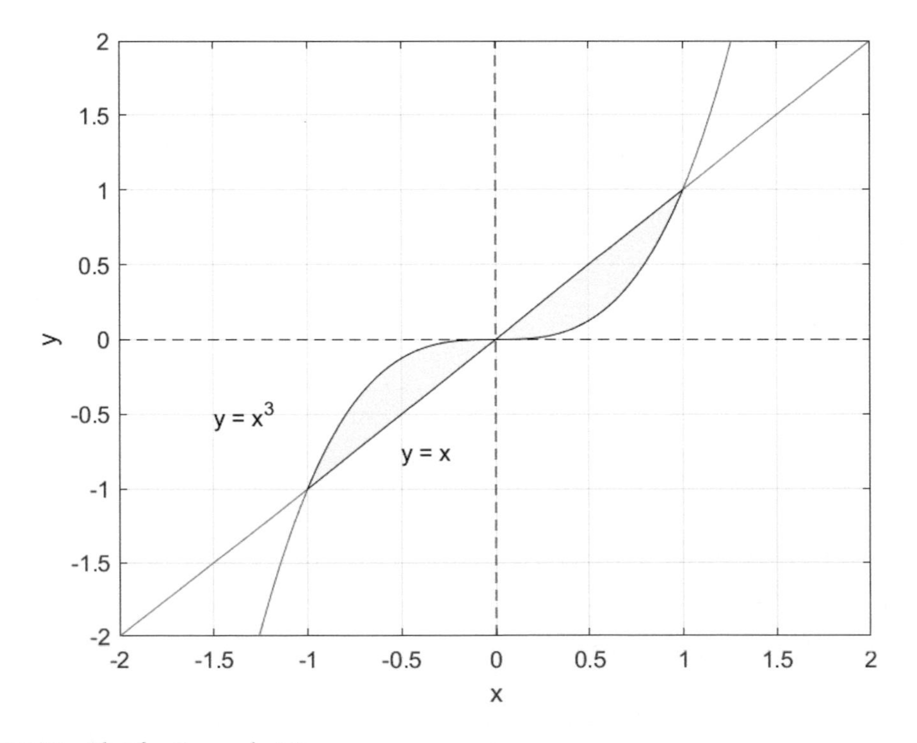

FIGURE 5.10 Plot for Example 5.9.

MATLAB Code 5.9

```
clear; clc;
syms x;
y1 = x;
y2 = x^3;
f =  int(abs(y1 - y2), -1, 1);
fprintf('Area : %f\n', eval(f));

%plotting
xx = linspace(-2,2);
yy1 = subs(y1, x, xx);
yy2 = subs(y2, x, xx);
plot(xx, yy1, xx, yy2);
xlabel('x'); ylabel('y');
axis([-2 2 -2 2]); hold on; grid;
plot([-2, 2], [0, 0], 'k--');
plot([0, 0], [-2, 2], 'k--');
text(-1.5, -0.5, 'y = x^3');
text(-0.5, -0.75, 'y = x');
x = linspace(-1, 1);
y1 = x;
y2 = x.^3;
X = [x,fliplr(x)];
Y = [y1,fliplr(y2)];
fill(X,Y,'y');  alpha(0.25);
hold off;
```

5.6 CENTROID

The earth exerts gravitational force on each particle of a solid object. If all these forces are replaced by a single equivalent force then this force will act through a single point called center of gravity. Assuming the density of the object is uniform, the center of gravity will coincide with the centroid of mass. If the object of uniform density is a thin plate then the center of mass will coincide with the centroid of the area. The centroid of the plate is, therefore, a single point through which the entire weight of the plate can be balanced. It is obvious that for rectangular or circular shapes, the centroid is exactly at the center. Before finding the centroid of an arbitrary shape, let us first consider a polygon of some random dimensions as shown in Figure 5.11.

To find the centroid of the polygon, it is divided into two rectangles and the centroid of each rectangle is at its center. The left rectangle is of width 4 and height 4 so its area is $A_1 = 4 \times 4 = 16$ and its center is at $(x_1, y_1) = (-3, 1)$. The right rectangle is of width 5 and height 6 so its area is $A_2 = 5 \times 6 = 30$ and its center is at $(x_2, y_2) = (1.5, 2)$. To find the centroid of the polygon, we compute moments of the component rectangles from both the *X*- and *Y*-axes and equate their sum with the moment of the entire area. Let $\bar{x}$ and $\bar{y}$ denote the coordinates of the centroid and *A* be its entire area. Equating moments with respect to the *X*- and *Y*-axes we get:

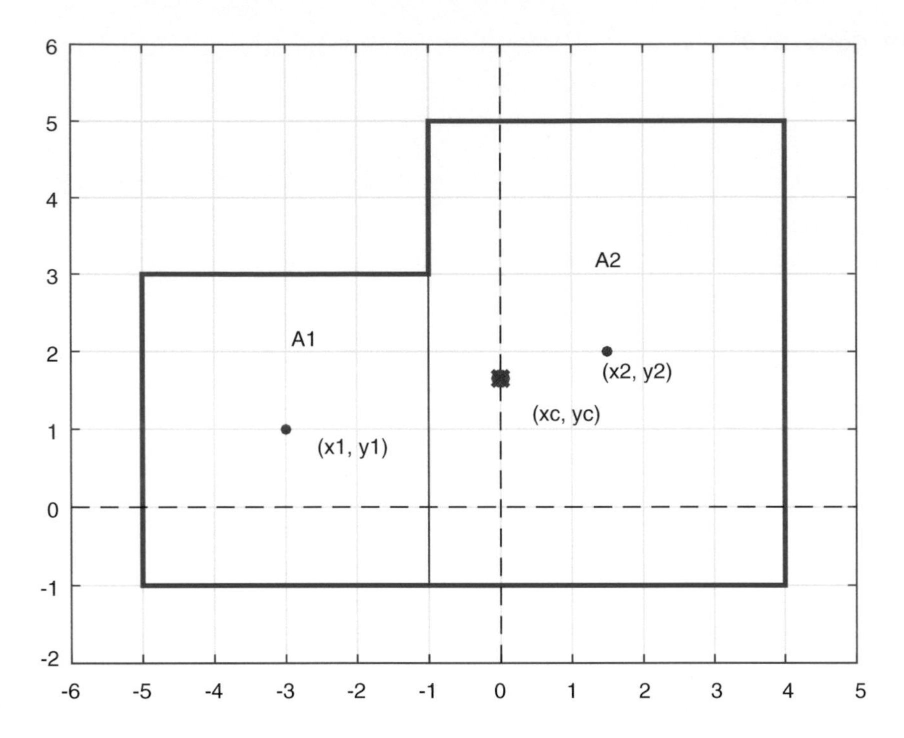

FIGURE 5.11 Deriving centroid of a polygon.

$$A_1 \cdot x_1 + A_2 \cdot x_2 = A \cdot \overline{x}$$

$$A_1 \cdot y_1 + A_2 \cdot y_2 = A \cdot \overline{y}$$

Substituting values, the centroid of the entire area is given by:

$$\overline{x} = \frac{A_1 \cdot x_1 + A_2 \cdot x_2}{A} = \frac{16(-3)+30(1.5)}{16+30} = -0.065$$

$$\overline{y} = \frac{A_1 \cdot y_1 + A_2 \cdot y_2}{A} = \frac{16(1)+30(2)}{16+30} = 1.652$$

From the above, we can formulate a general rule that centroid of an area is equal to the sum of the moments of its component parts divided by the whole area.

Figure 5.12 depicts the graph of a function $y = f(x)$ and it is required to find the centroid of an area below the curve bounded by the lines $x = a$ and $x = b$. Consider a very thin strip of width dx at a distance of x from the Y-axis and of height $y = f(x)$ so that the area of the strip is $y{\cdot}dx$. Moment of the strip along x-direction is thus $(y{\cdot}dx){\cdot}x$ and moment of the strip along y-direction is $(y \cdot dx) \cdot \left(\frac{y}{2}\right)$ since the center of the strip is exactly at the middle

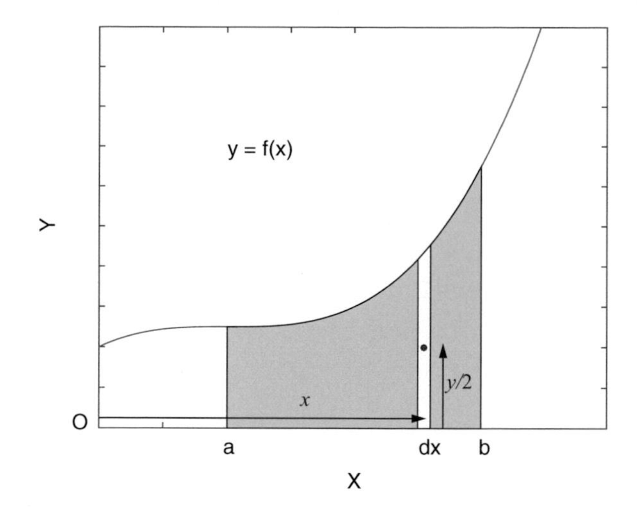

FIGURE 5.12 Deriving centroid of an area under curve.

along the height of the strip. So equating the moments of all such strips within the area and dividing by the area itself we obtain the centroid of the area as given below:

$$\bar{x} = \frac{\int_a^b x \cdot f(x) \cdot dx}{\int_a^b f(x) \cdot dx}$$
$$\bar{y} = \frac{\int_a^b \{f(x)\}^2 \cdot dx}{2\int_a^b f(x) \cdot dx} \tag{5.26}$$

Example 5.10

Find the centroid of the area bounded by the curves $y = x^3$, $x = 0$, and $x = 2$.

From Equation (5.23), area: $\int_a^b f(x) \cdot dx = \int_0^2 x^3 \cdot dx = 4$

From Equation (5.26), centroid coordinates:

$$\bar{x} = \frac{\int_0^2 x \cdot x^3 \cdot dx}{4} = 1.6$$

$$\bar{y} = \frac{\int_0^2 x^6 \cdot dx}{2 \times 4} = 2.29$$

MATLAB Code 5.10

```
clear all; clc;
syms x y;
y = x^3;

a = int(y, 0, 2);
eval(a);
fprintf('Area : %f\n', eval(a));

m1 = int(x*y, 0, 2);
m2 = int(0.5*y^2, 0, 2);

xc = m1/a;
yc = m2/a;

fprintf('Centroid : (%.2f, %.2f) \n', eval(xc), eval(yc));
```

If the region is bounded by two curves $f(x)$ and $g(x)$ on the interval [a, b] then the centroid of the bounded region is given as follows:

$$\overline{x} = \frac{\int_a^b x \cdot \{f(x) - g(x)\} \cdot dx}{\int_a^b \{f(x) - g(x)\} \cdot d}$$

$$\overline{y} = \frac{\int_a^b \left[\{f(x)\}^2 - \{g(x)\}^2 \right] \cdot dx}{2\int_a^b \{f(x) - g(x)\} \cdot dx} \tag{5.27}$$

5.7 INTERPOLATION AND CURVE FITTING

To end the topic on splines, we finally take a look at few functions provided by MATLAB to perform two important tasks related to splines namely interpolation and curve fitting. These tasks are specifically done using programming tools only so there will be no associated numerical problem that can be computed manually.

Interpolation is done when there is some coarse data points and we are interested in finding some intermediate value between the points, which are not directly provided (Mathews, 2004). So the given data can be interpolated to find these values. The first function MATLAB provides is "interp1," which stands for 1D interpolation which by default uses a linear interpolation between the given data. It also has a number of other options as illustrated in Figure 5.13 using two types of data pattern. The first and third columns show a step function, the second and fourth columns show a sine function. The first row indicates interpolation with the "linear" and the "previous neighbor" options, the second row indicates interpolation with the "nearest neighbor" and the "next neighbor" options.

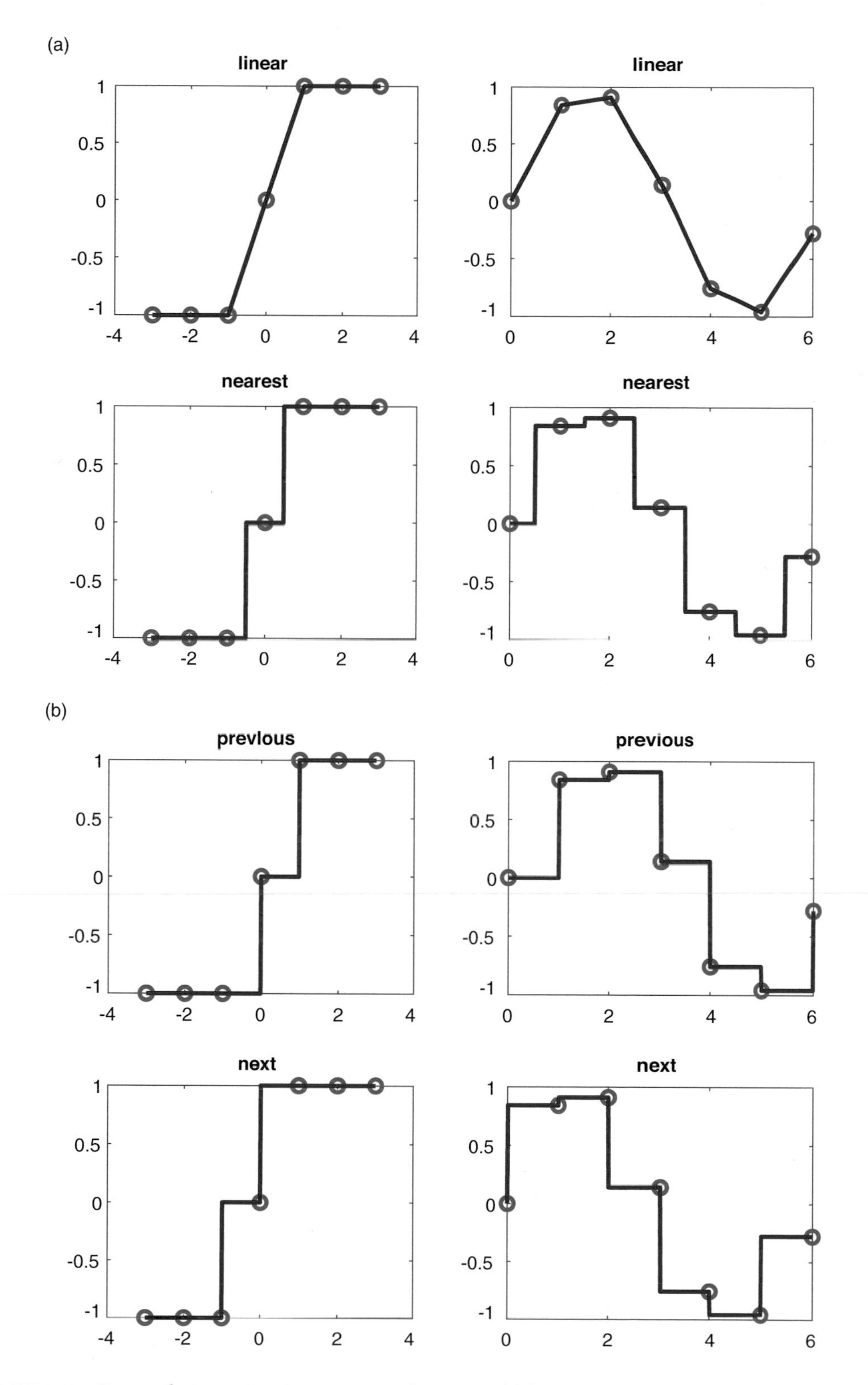

FIGURE 5.13 Interpolation using `interp1` with options (a) linear and nearest neighbor (b) previous neighbor and next neighbor

The corresponding code for generating the plots is given below:

MATLAB Code 5.11

```
clear; clc;
x = -3:3;
y = [-1 -1 -1 0 1 1 1];
t = -3:.01:3;

subplot (221)
plot(x,y,'o',t,interp1(x,y,t, 'linear'), 'r-', 'LineWidth', 2); title('linear');
subplot (223)
plot(x,y,'o',t,interp1(x,y,t, 'nearest'), 'r-', 'LineWidth', 2); title('nearest');

x = 0:2*pi;
y = sin(x);
t = 0:.01:2*pi;

subplot (222)
plot(x,y,'o',t,interp1(x,y,t, 'linear'), 'r-', 'LineWidth', 2); title('linear');
subplot (224)
plot(x,y,'o',t,interp1(x,y,t, 'nearest'), 'r-', 'LineWidth', 2); title('nearest');

figure
x = -3:3;
y = [-1 -1 -1 0 1 1 1];
t = -3:.01:3;
```

```
subplot (221)
plot(x,y,'o',t,interp1(x,y,t, 'previous'), 'r-', 'LineWidth', 2); title('previous');
subplot (223)
plot(x,y,'o',t,interp1(x,y,t, 'next'), 'r-', 'LineWidth', 2);title('next');

x = 0:2*pi;
y = sin(x);
t = 0:.01:2*pi;

subplot (222)
plot(x,y,'o',t,interp1(x,y,t, 'previous'), 'r-', 'LineWidth', 2); title('previous');
subplot (224)
plot(x,y,'o',t,interp1(x,y,t, 'next'), 'r-', 'LineWidth', 2);title('next');
```

NOTE

interp1: performs 1-D interpolation

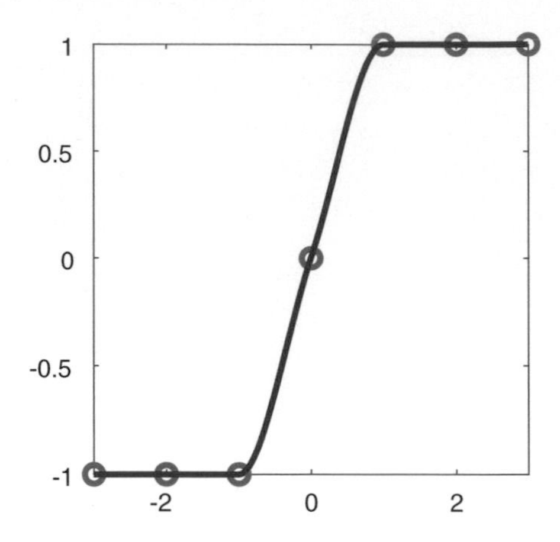

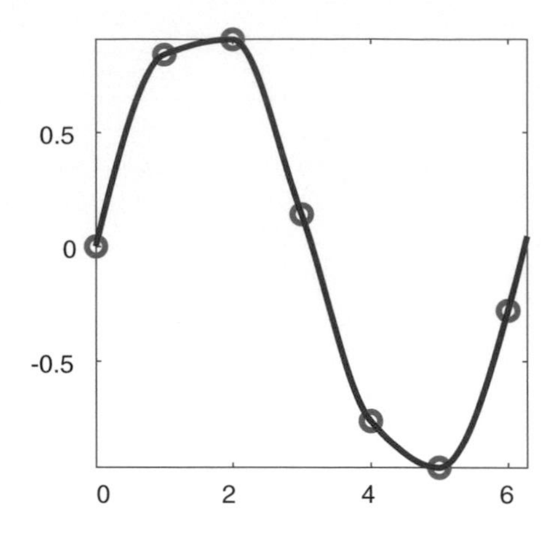

FIGURE 5.14 Interpolation using `pchip`.

The second function is "pchip," which stands for "Piecewise Cubic Hermite Interpolating Polynomial." It provides a piecewise polynomial form of a shape-preserving piecewise cubic Hermite interpolate to the values. For each sub-interval, it interpolates between the end points and also maintains that the slopes at the endpoints be continuous. This is illustrated in Figure 5.14 for the step function on the left and the sine function on the right.

The corresponding code for generating the plots is given below:

MATLAB Code 5.12

```
clear; clc;
x = -3:3; y = [-1 -1 -1 0 1 1 1]; t = -3:.01:3;

subplot (121)
plot(x,y,'o',t, pchip(x,y,t), 'r-', 'LineWidth', 2);
axis tight; axis square;

x = 0:2*pi; y = sin(x); t = 0:.01:2*pi;

subplot (122)
plot(x,y,'o',t, pchip(x,y,t), 'r-', 'LineWidth', 2);
axis tight; axis square;
```

The third function is "spline," which stands for "Piecewise Cubic Spline Interpolating Polynomial." It provides a piecewise polynomial form of a cubic spline over the data values. It is illustrated in Figure 5.15 for the step function and sine function.

The corresponding code for generating the plots is given below:

MATLAB Code 5.13

```
clear; clc;
x = -3:3; y = [-1 -1 -1 0 1 1 1]; t = -3:.01:3;
```

```
subplot (121)
plot(x,y,'o',t, spline(x,y,t), 'r-', 'LineWidth', 2);
axis tight; axis square;

x = 0:2*pi; y = sin(x); t = 0:.01:2*pi;

subplot (122)
plot(x,y,'o',t,spline(x,y,t), 'r-', 'LineWidth', 2);
axis tight; axis square;
```

A comparison between the three is provided in Figure 5.16. The "interp1" function joins data points by straight lines, the smoothest curve is provided by "spline" function while the "pchip" function produces reduced oscillations at the end points.

For fitting a curve to a set of given data points, MATLAB provides a function called "polyfit" that creates a fitting polynomial of a specified degree whose sum of square errors from the data points is minimum, and returns the coefficients of the polynomial. The code below illustrates the process. Two sets of data points have been generated and curves of various degrees have been used to fit the data. The results are plotted by varying the value of d. The coefficients of the polynomials are returned by the function. Figure 5.17 shows a step function in the upper row and a sine function in the lower row. The columns indicate the degree of the polynomials used for curve fitting i.e. 1, 3, and 9.

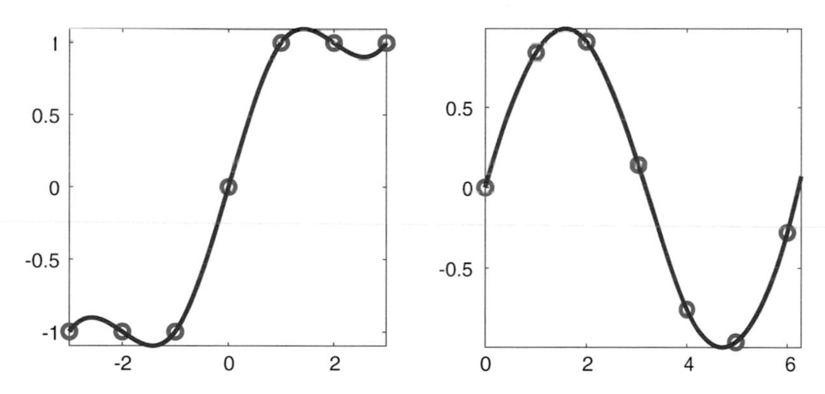

FIGURE 5.15 Interpolation using `spline`.

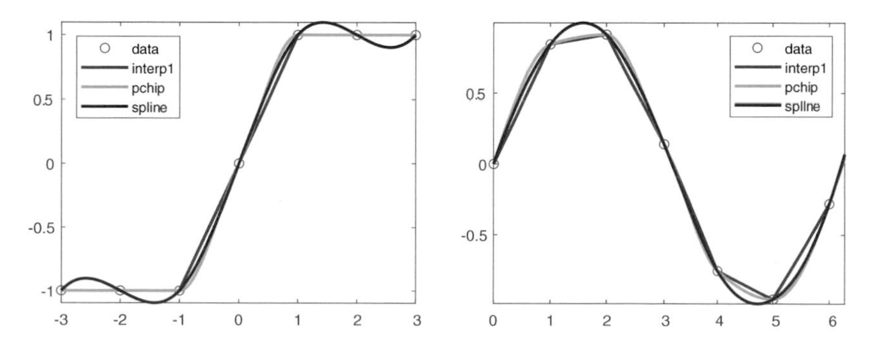

FIGURE 5.16 Comparison between the interpolation methods.

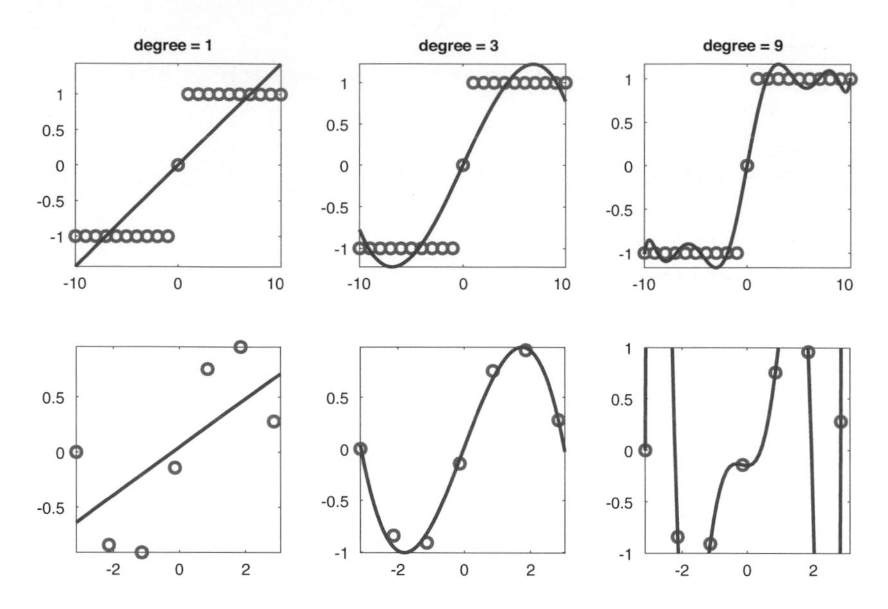

FIGURE 5.17 Curve fitting using `polyfit`.

The corresponding code for generating the plots is given below:

```
clear; clc;
x = -10:10;
y = [-1 -1 -1 -1 -1 -1 -1 -1 -1 -1 0 1 1 1 1 1 1 1 1 1 1];
t = -10:.01:10;

subplot (231)
d = 1; pf = polyfit(x, y, d);
pv = polyval(pf, t);
plot(x,y,'o',t, pv, 'LineWidth', 2);
axis tight; axis square;
title('degree = 1');

subplot (232)
d = 3; pf = polyfit(x, y, d);
pv = polyval(pf, t);
plot(x,y,'o',t, pv, 'LineWidth', 2);
axis tight; axis square;
title('degree = 3');

subplot (233)
d = 9; pf = polyfit(x, y, d);
pv = polyval(pf, t);
plot(x,y,'o',t, pv, 'LineWidth', 2);
axis tight; axis square;
title('degree = 9');
```

```
clear x y t pf pv;
x = -pi:pi;
y = sin(x);
t = -pi:.1:pi;

subplot (234)
d = 1; pf = polyfit(x, y, d);
pv = polyval(pf, t);
plot(x,y,'o',t, pv, 'LineWidth', 2);
axis tight; axis square;

subplot (235)
d = 3; pf = polyfit(x, y, d);
pv = polyval(pf, t);
plot(x,y,'o',t, pv, 'LineWidth', 2);
axis tight; axis square;

subplot (236)
d = 9; pf = polyfit(x, y, d);
pv = polyval(pf, t);
plot(x,y,'o',t, pv, 'LineWidth', 2);
axis tight; axis square;
axis([-pi pi -1 1]);
```

NOTE

`polyfit`: generates a polynomial to fit a given data

5.8 NOTES ON 2D PLOTTING FUNCTIONS

Before we shift our primary focus from 2D to 3D domain, this section summarizes the MATLAB 2D plotting functions used and some additional ones (Marchand, 2002). The reader is encouraged to explore further details about these functions from MATLAB documentations.

(a) ezplot: This function can be used to plot using symbolic variables:

(i) one variable (Figure 5.18):

```
ezplot('1 - 2.25*t + 1.25*t^2')
```

(ii) two variables (Figure 5.19):

```
ezplot('x^4 + y^3 = 2*x*y')
```

(iii) parametric variable (Figure 5.20):

```
ezplot('cos(t)', 'sin(t)')
```

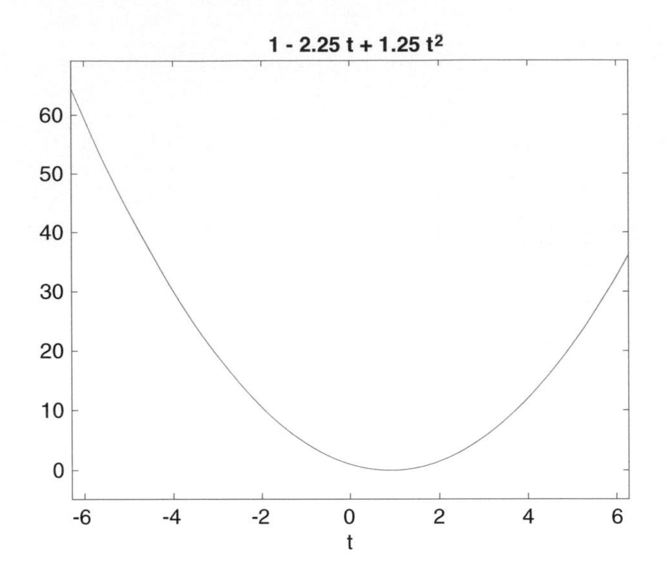

FIGURE 5.18 Plotting with one variable using `ezplot`.

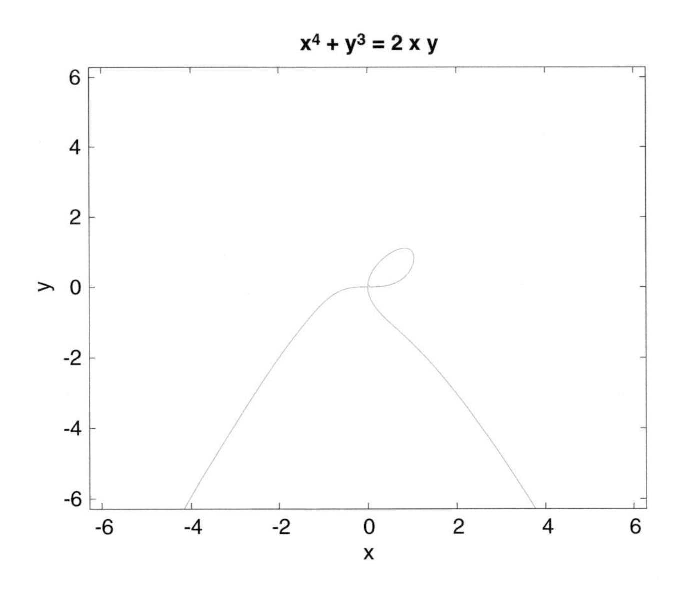

FIGURE 5.19 Plotting with two variables using `ezplot`.

(b) plot: This function can be used to plot using a vector of values:

(i) parametric equations (Figure 5.21):

```
t = 0:pi/50:10*pi; plot(t.*sin(t),t.*cos(t));
```

(ii) explicit equations (Figure 5.22):

```
x = -pi:.1:pi; y = tan(sin(x)) - sin(tan(x)); plot(x,y);
```

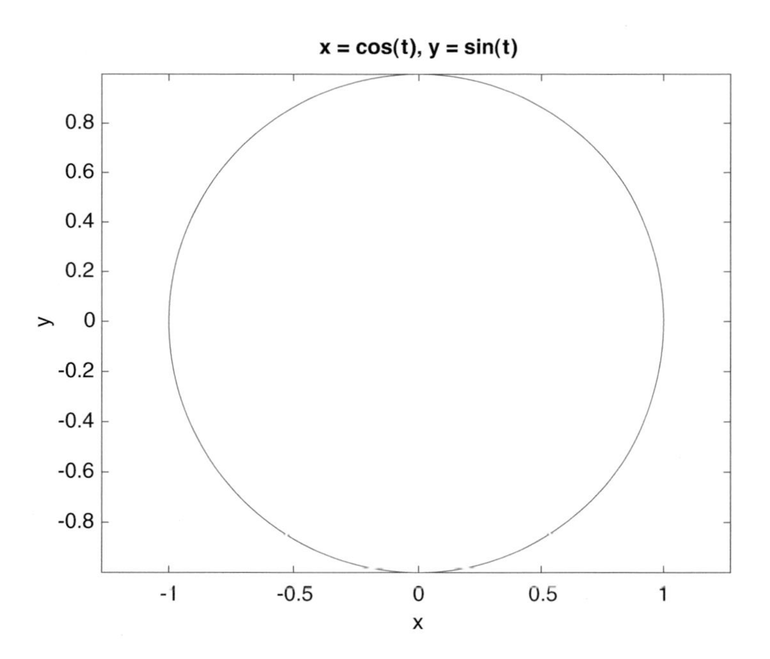

FIGURE 5.20 Plotting with parametric variable using `ezplot`.

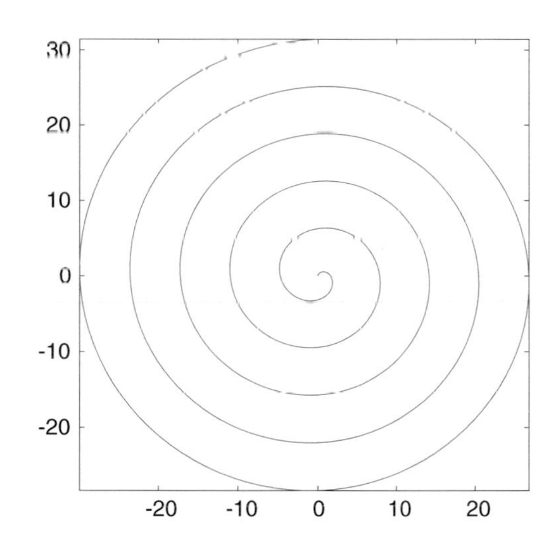

FIGURE 5.21 Plotting with one variable using `plot`.

(c) ezcontour: Contour plots, one version with only the edges and the other being a filled version (Figure 5.23)

```
ezcontour ('x*exp(-x^2 - y^2)')
ezcontourf ('x*exp(-x^2 - y^2)')
```

(d) fimplicit: Plots implicit functions (introduced from MATLAB version 2016). The version of an installed MATLAB package can be checked by typing ver at the command line (Figure 5.24).

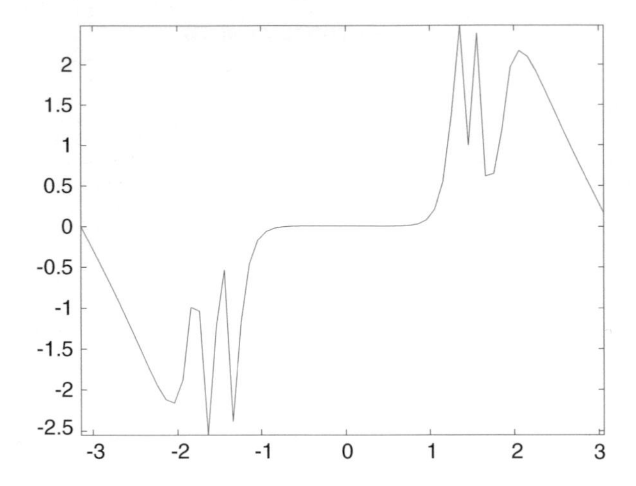

FIGURE 5.22 Plotting explicit equations using `plot`.

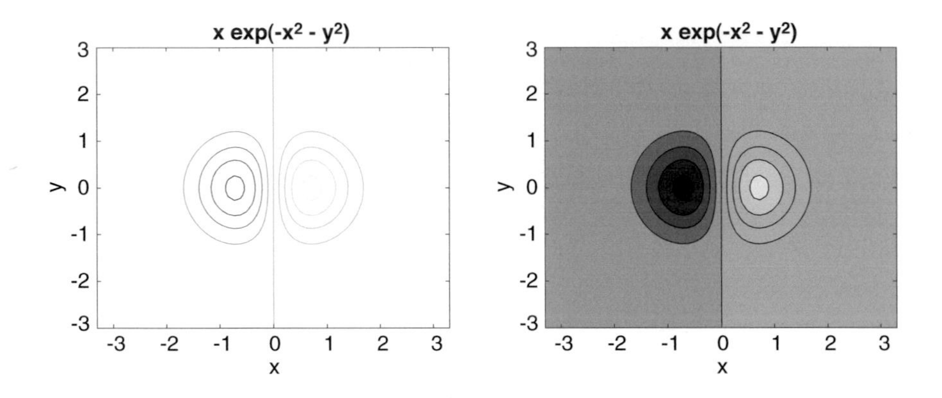

FIGURE 5.23 Contour plots using `ezcontour`.

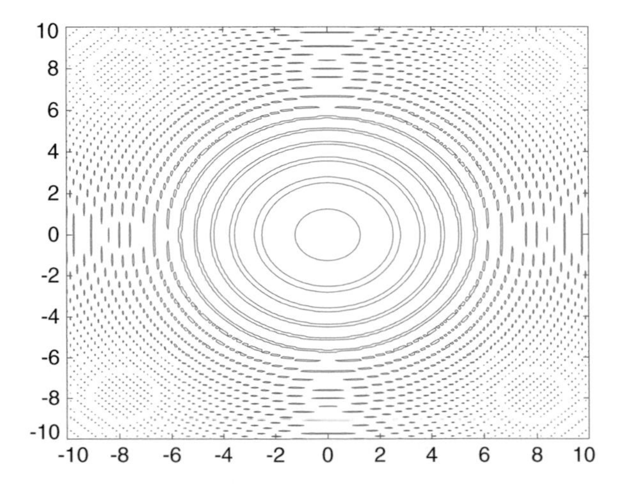

FIGURE 5.24 Implicit function plots using `fimplicit`.

```
fimplicit(@(x,y) sin(x.^2 + y.^2) + cos(x.^2 + y.^2) - 1, [-10 10])
```

(e) patch: Creates filled polygons given the vertices and colors (Figure 5.25)

```
x = [4 6 11 9];
y = [2 7 9 4];
c = [0 4 6 8];
colormap(jet);
patch(x,y,c);
colorbar;
hold;
v1 = [2 4; 2 10; 8 4];
patch('Vertices', v1, 'FaceColor', 'red', 'FaceAlpha', 0.3, 'EdgeColor', 'red');
axis([0 12 0 12]);
```

(f) fplot: plots a continuous or piecewise functions (Figure 5.26)

```
fplot(@(x) sin(x),[-2*pi pi],'b');
hold on;
fplot(@(x) cos(x),[-pi 2*pi],'r');
hold off;
```

NOTE

`colormap`: specifies a color scheme using predefined color look-up tables
`colorbar`: creates a color bar by appending colors in the colormap

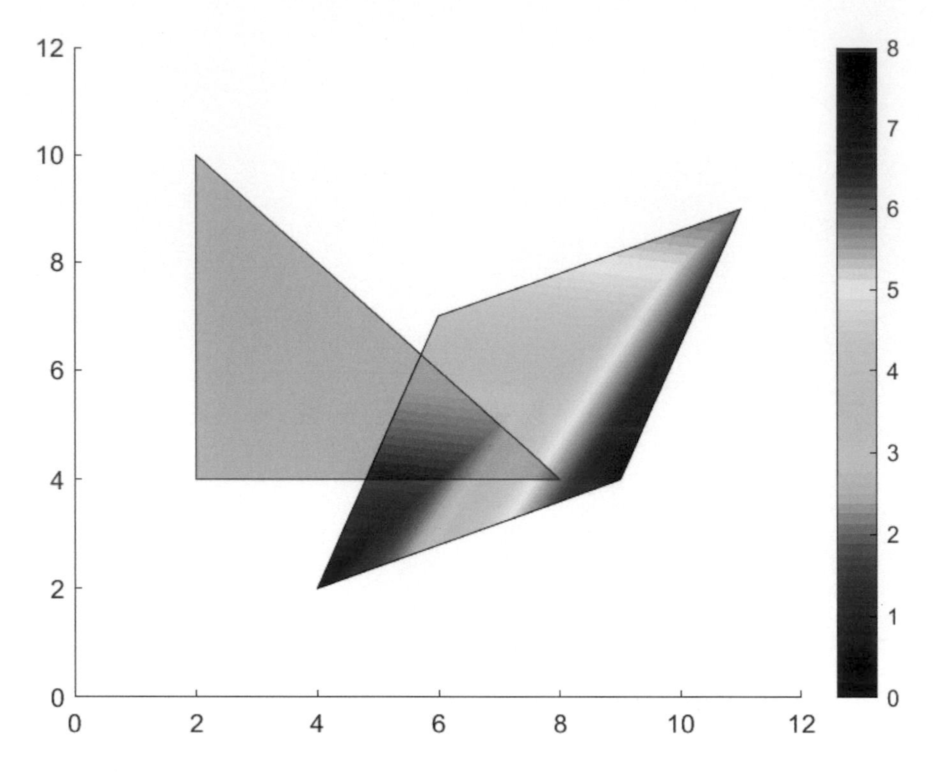

FIGURE 5.25 Plotting with `patch`.

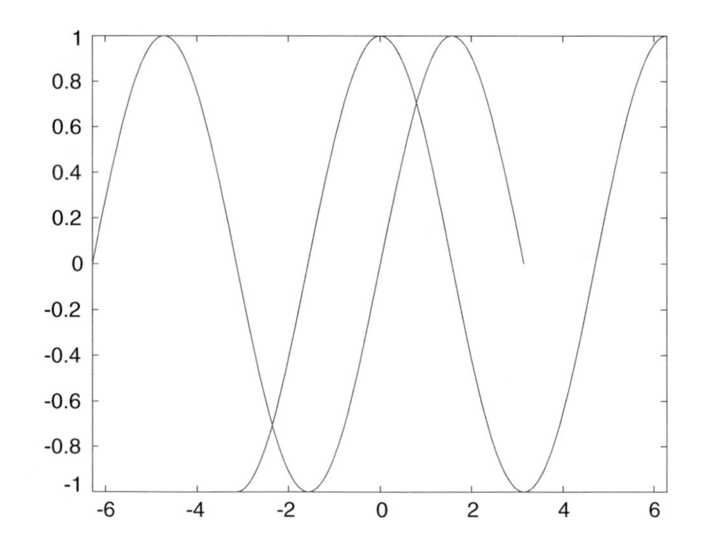

FIGURE 5.26 Plotting with `fplot`.

5.9 CHAPTER SUMMARY

The following points summarize the topics discussed in this chapter:

- Critical points of splines are the maximum, minimum and POI.
- For a minimum point the rate of change of slope is positive.
- For a maximum point the rate of change of slope is negative.
- For a POI, the curvature should be zero and should have opposite signs on either side.
- The tangent of a curve is obtained by the first derivative of the curve equation.
- The normal of a curve is obtained by rotating the tangent by 90°.
- The length of a curve segment is obtained by summing up the lengths of very small segments.
- The area under a curve is obtained by summing up areas of very small rectangular segments.
- Area is positive above the x-axis and negative below the x-axis.
- Area bounded by two curves is the absolute difference of the areas under the two curves.
- Interpolation between two data points can be done linearly using a number of options.
- Cubic spline interpolation can provide smoother interpolation curves.
- Curve fitting to a set of data points can be done using a specific degree of polynomial.
- MATLAB provides a number of plotting functions for implicit, explicit, and parametric expressions.

5.10 REVIEW QUESTIONS

1. What is meant by critical points of a spline?
2. How are the minimum and maximum points differentiated by the gradients of a curve?
3. How is the POI of a curve determined? Why does a cubic curve always have a single POI?
4. Representing a parametric curve by an ordered pair, how is the tangent and normal computed?
5. How is the expression for length of a curve between two given points determined?
6. How is the area between a curve and the x-axis bounded by two vertical lines determined?
7. What convention is followed for computing area above and below the x-axis?

8. How is the area bounded between two given curves determined?
9. What is the observed difference between a linear interpolation and a cubic spline interpolation?
10. For obtaining the best fitting curve to a set of data points what criteria needs to be fulfilled?

5.11 PRACTICE PROBLEMS

1. Find the length of $y = x^{1.5}$ between $x = 0$ and $x = 5$.
2. Determine the length of the curve $y = \ln(\sec x)$ for $0 \leq x \leq \pi/4$.
3. Find the area between $y = x^2$ and $x = y^2$ from $x = 0$ to $x = 1$.
4. Find the area fully enclosed between the curves $x = y + 1$ and $x = 0.5y^2 - 3$.
5. Find the minimum, maximum and inflection points for the curve: $y = -2.67x + 0.67x^3$.
6. Find the tangent vector, the normal vector, and equation of tangent line of the curve (t, t^2) at point (1, 1).
7. Find the unit tangent vector and unit normal vector to the parametric curve $x = t^3$, $y = t^2$ at (−8, 4).
8. For the implicit curve $x^3 + y^2 - x = 4$, find the tangent line and normal vector at point (−1, 2).
9. For the cycloid $(t + \sin t, 1 - \cos t)$, find the tangent vector and normal vector at $t = \pi/2$.
10. Find the centroid of the area bounded by the curves $y = \sqrt{x}$ and $y = x^3$.

CHAPTER 6

Vectors

6.1 INTRODUCTION

Vectors involve both magnitude and direction. Two vectors are equal if they have the same magnitude and direction. Vectors a and b are not equal even if they have same magnitude. Vector $-c$ is defined as having same magnitude but reverse direction as c. Multiplying a vector by a scalar changes its magnitude but keeps direction same e.g. $2a$. By default vectors are "free" i.e. parallel shifting does not change their magnitude or direction. Vectors can also be "bound" i.e. cannot be shifted e.g. position vector of point P with respect to origin O (Olive, 2003) (see Figure 6.1).

Vector addition implies finding the resultant of two vectors. There are two methods to do this, both of which are essentially equivalent. The triangle rule states that, if p and q represents two sides of a triangle, then $p+q$ is given by the third side. The parallelogram rule states that if P and Q be two adjacent sides of a parallelogram, $P+Q$ is given by its diagonal (Shirley, 2002). If there are more than two vectors, then we use the polygon rule, which says that addition of any number of vectors is obtained by arranging them end to end and closing the final side of resulting polygon i.e. $r=a+b+c+d$ (see Figure 6.2). For 3D vectors, they are to be joined end to end in 3D space.

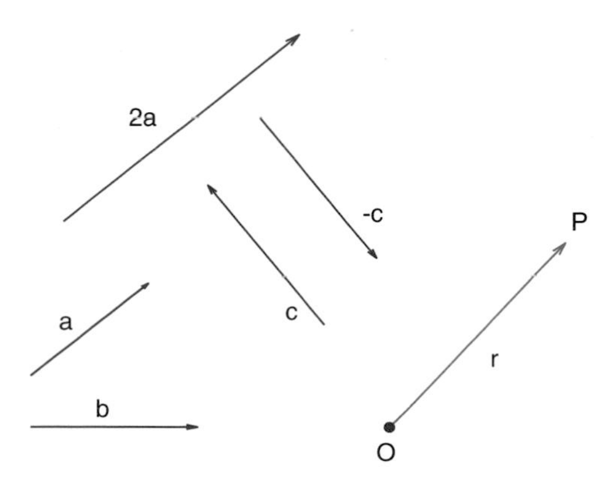

FIGURE 6.1 Examples of vectors.

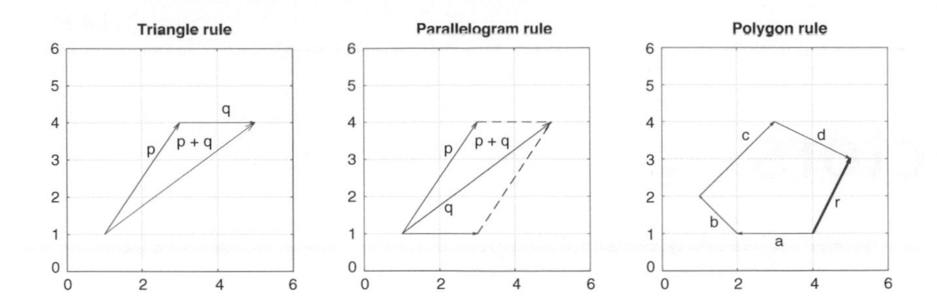

FIGURE 6.2 Vector addition.

Vectors can be represented in terms of some chosen reference components. In practice reference, vectors are chosen to be orthogonal (perpendicular) and of unit length. The standard notations for the unit vectors are i along X-axis, j along Y-axis, and k along Z-axis. Position vector measured from the origin of the coordinate system is written as $P = ai + bj$ (2-D coordinate system) and $P = ai + bj + ck$ (3-D coordinate system), where a, b, and c are the scaling factors. Vectors can be represented usually in two forms: using components as in $r = ai + bj + ck$, or as coordinates in vector space as in $r = (a, b, c)$. In the coordinate notation, the unit vectors become $i = (1, 0, 0)$, $j = (0, 1, 0)$, $k = (0, 0, 1)$. The resultant of two vectors $p = ai + bj + ck$ and $q = di + ej + fk$ is equal to $r = (a + d)i + (b + e)j + (c + f)k$. Figure 6.3a shows that the vector $P = 3i + 4j$ can be split into two orthogonal components: $3i$ along the X-axis and $4j$ along the Y-axis. These components can in turn be represented as scaled unit vectors i.e. three times the i vector and four times the j vector. Figure 6.3b shows that the vector $P = 2i + 3j + 2k$ can be split into three orthogonal components: $2i$ along the X-axis, $3j$ along the Y-axis, and $2k$ along the Z-axis. These components can in turn be represented as scaled unit vectors i.e. two times the i vector, three times the j vector, and two times the k vector.

6.2 UNIT VECTOR

The magnitude of the vector $R = ai + bj + ck$ is given by:

$$|R| = \sqrt{\left(a^2 + b^2 + c^2\right)} \tag{6.1}$$

This is also known as the Euclidean length between the start and end points of the vector. The unit vector along the direction of R is given by:

$$r = \frac{R}{|R|} = \frac{(ai + bj + ck)}{\sqrt{\left(a^2 + b^2 + c^2\right)}} \tag{6.2}$$

Example 6.1

Find the magnitude and unit vector in the direction of a specified vector R = 2i – j + 2k.

From Equation (6.1): $|R| = \sqrt{\left(a^2 + b^2 + c^2\right)} = \sqrt{(4 + 1 + 4)} = 3$

From Equation (6.2): $r = (1/3) * (2i - j + 2k) = (2/3)i - (1/3)j + (2/3)k$ (Figure 6.4).

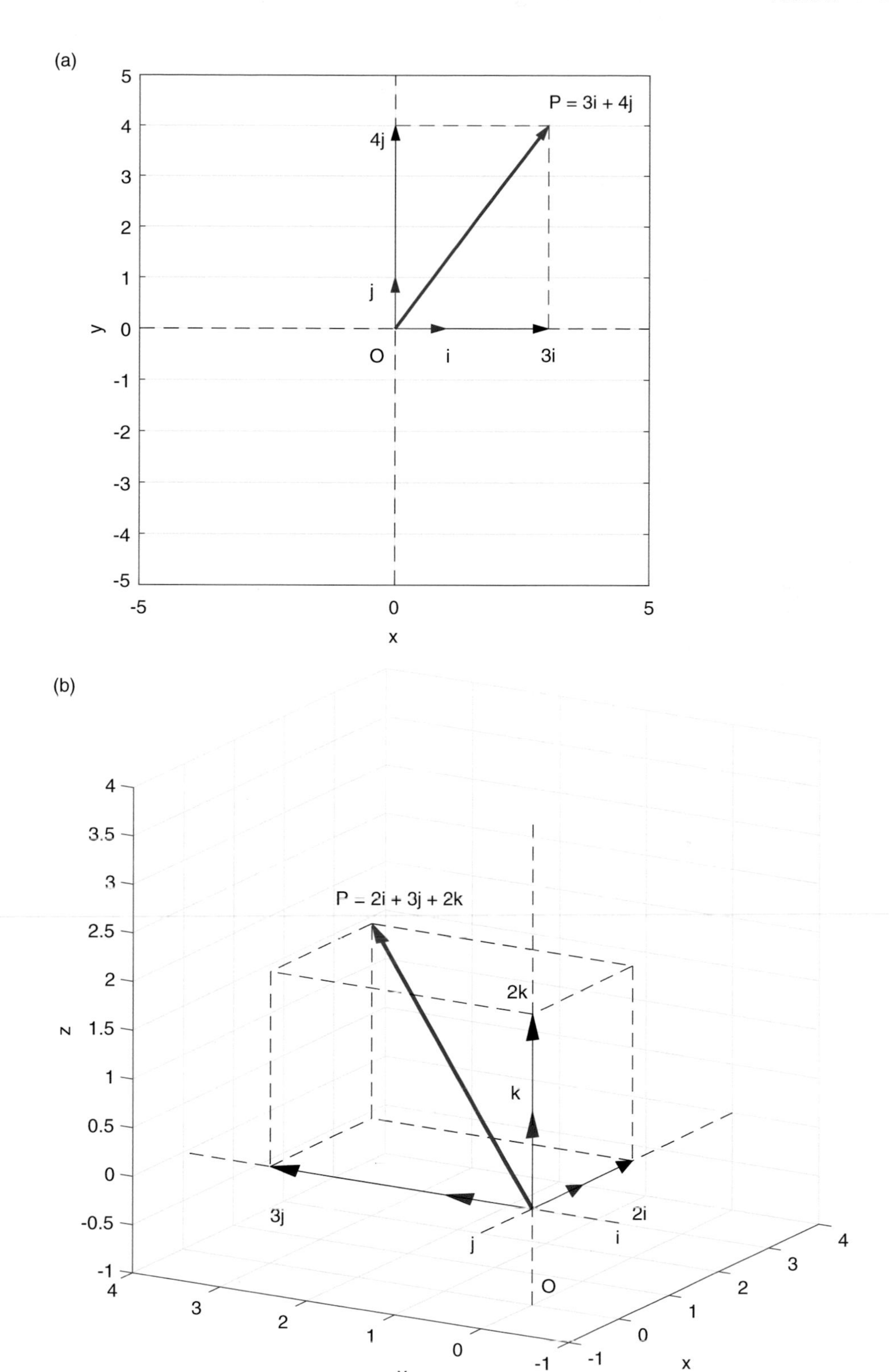

FIGURE 6.3 (a) and (b) Vector components (a) 2D plane (b) 3D space.

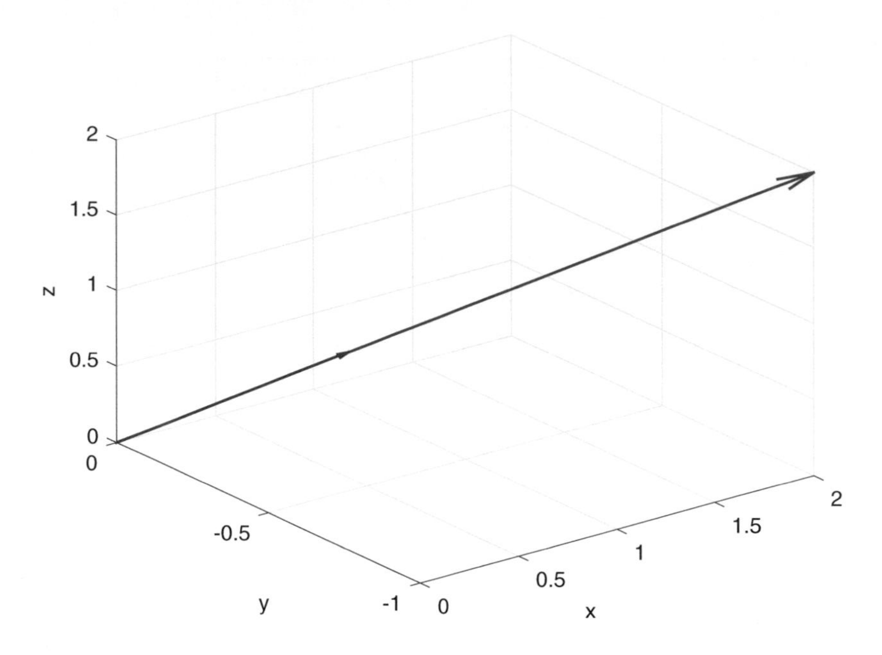

FIGURE 6.4 Plot for Example 6.1.

MATLAB® Code 6.1

```
clear all; clc;
R = [2, -1, 2];
nR = norm(R);
fprintf('Magnitude : %f \n', nR);
r = R/nR;
fprintf('Unit vector : (%f, %f, %f)\n', r(1), r(2), r(3));
quiver3(0, 0, 0, R(1), R(2), R(3), 1, 'b', 'LineWidth', 1.5);
hold on;
quiver3(0, 0, 0, r(1), r(2), r(3), 1, 'r', 'LineWidth', 1.5);
hold off;
xlabel('x'); ylabel('y'); zlabel('z');
```

NOTE

`norm`: calculates the magnitude or Euclidean length of a vector.
`quiver3`: depicts 3D vectors as arrows with direction and magnitude.

6.3 DIRECTION COSINES

Let A, B, and C be the angles made by the vector $R = ai + bj + ck$ with the three primary axes. The cosines of these three angles cos(A), cos(B), and cos(C) are known as direction cosines.

$$\cos A = \frac{a}{\sqrt{(a^2 + b^2 + c^2)}}$$

$$\cos B = \frac{b}{\sqrt{(a^2 + b^2 + c^2)}} \tag{6.3}$$

$$\cos C = \frac{c}{\sqrt{(a^2 + b^2 + c^2)}}$$

It follows from the above that: $\cos^2 A + \cos^2 B + \cos^2 C = 1$

Example 6.2

Find the direction cosines and angles, the vector $R = 3i + 5j - 2k$ makes with coordinate axes.

From Equation (6.3)

$$\cos A = \frac{a}{\sqrt{(a^2 + b^2 + c^2)}} = \frac{3}{\sqrt{38}} = 0.4867$$

$$\cos B = \frac{b}{\sqrt{(a^2 + b^2 + c^2)}} = \frac{5}{\sqrt{38}} = 0.8111$$

$$\cos C = \frac{c}{\sqrt{(a^2 + b^2 + c^2)}} = \frac{-2}{\sqrt{38}} = -0.3244$$

$A = \arccos(0.4867) = 60.87°$
$B = \arccos(0.8111) = 35.79°$
$C = \arccos(-0.3244) = 108.93°$

MATLAB Code 6.2

```
clear all;clc;
R = [3, 5, -2];
nR = norm(R);
cosA = R(1)/norm(R)
cosB = R(2)/norm(R)
cosC = R(3)/norm(R)
A = rad2deg(acos(cosA))
B = rad2deg(acos(cosB))
C = rad2deg(acos(cosC))
```

```
%alternatively
A = acosd(cosA)
B = acosd(cosB)
C = acosd(cosC)
```

NOTE

`rad2deg`: converts radian to degree values
`acos`: calculates inverse cosine in radians
`acosd`: calculates inverse cosine in degrees

6.4 DOT PRODUCT

Dot product of two vectors $A = ai + bj + ck$ and $P = pi + qj + rk$ with an angle θ between them is given by the following (Shirley, 2002):

$$A \bullet P = |A||P|\cos(\theta) \tag{6.4}$$

Also angle between the vectors:

$$\cos(\theta) = (A \bullet P)/(|A||P|) \tag{6.5}$$

Even though it is a product of two vector quantities, the product itself is a scalar number. To see how this is possible, we expand Equation (6.4)

$$A \bullet P = (ai + bj + ck)\cdot(pi + qj + rk) = ap + bq + cr \tag{6.6}$$

The above expression is true because $i \bullet i = j \bullet j = k \bullet k = 1$ as each has a magnitude of 1 and angle between them is 0. Also $i \bullet j = j \bullet k = k \bullet i = 0$ as angle between each pair is 90°.

Corollary 1: If vectors are parallel then $\theta = 0$ hence $A \bullet P = |A||P|$

Corollary 2: If vectors are perpendicular then $\theta = 90°$ hence $A \bullet P = 0$

Example 6.3

(a) Find if vectors $A = 3i + 5j - 2k$ and $B = 2i - 2j - 2k$ are perpendicular to each other.

(b) Find if vectors $A = 3i + 5j - 2k$ and $B = 0.5i + (5/6)j - 0.333k$ are parallel to each other.

(c) Find angle between vectors $A = 2i - 3j + k$ and $B = 4i + j - 3k$

(a)

From Equation (6.4)

$$A \bullet B = 3\times2 - 5\times2 + 2\times2 = 6 - 10 + 4 = 0$$

Thus, A and B are perpendicular.

(b)

From Equation (6.4)

$$A \bullet B = (3)(0.5) + (5)(5/6) + (-2)(-1/3) = 3/2 + 25/6 + 2/3 = (9 + 25 + 4)/6 = 38/6$$

From Equation (6.1)

$$|A| = \sqrt{}(9 + 25 + 4) = \sqrt{}38$$

$$|B| = \sqrt{}(1/4 + 25/36 + 1/9) = \sqrt{}(9 + 25 + 4)/\sqrt{}36 = \sqrt{}38/6$$

$$|A||B| = \sqrt{}38 \times \sqrt{}38/6 = 38/6 = A \bullet B$$

Thus, A and B are parallel.

(c)

From Equation (6.5)

$$\cos(\theta) = (A \bullet B)/(|A||B|)$$

$$A \bullet B = (2)(4) + (-3)(1) + (1)(-3) - 8 - 3 - 3 = 2$$

$$|A| = \sqrt{}(4 + 9 + 1) = \sqrt{}14$$

$$|B| = \sqrt{}(16 + 1 + 9) = \sqrt{}26$$

$$\cos(\theta) = (A \bullet B)/(|A||B|) = 2/(\sqrt{}14\sqrt{}26) = 2/19.0788 = 0.1048$$

$$\theta = \arccos(0.1048) = 83.98°$$

MATLAB Code 6.3

```
clear all; clc; format compact;

% (a)
A = [3, 5, -2];
B = [2, -2, -2];
C = dot(A, B);
if C == 0
fprintf('Perpendicular\n');
else
fprintf('Not perpendicular\n');
end
```

```
% (b)
clear all;
A = [3, 5, -2];
B = [0.5, 5/6, -0.333];
C = dot(A,B);
nA = norm(A);
nB = norm(B);
P = nA*nB;
if (C - P) < 0.001
fprintf('Parallel\n');
else
fprintf('Not parallel\n');
end

% (c)
clear all;
A = [2, -3, 1];
B = [4, 1, -3];
C = dot(A, B);
nA = norm(A);
nB = norm(B);
D = C/(nA*nB);
angle = acosd(D);
fprintf('Angle : %f deg\n', angle);
```

NOTE

dot: calculates dot product of vectors

6.5 CROSS PRODUCT

Cross (vector) product of two vectors $A = ai + bj + ck$ and $P = pi + qj + rk$ with an angle θ between them is given by the following (Shirley, 2002):

$$A \times P = |A||P|\sin(\theta) \cdot n \tag{6.7}$$

The resultant is a vector in the direction of n, which is a vector perpendicular to both A and P. The positive direction of n is governed by the right-handed corkscrew rule, which says that if a right-handed corkscrew is rotated from A to P then the positive direction of the n will be that along which the corkscrew advances.

Substituting the components in the above expression we get:

$$A \times P = (ai + bj + ck) \times (pi + qj + rk)$$

Remembering that $i \times i = j \times j = k \times k = 0$ and $i \times j = k, j \times k = i, k \times i = j$, it follows that $j \times i = -k, k \times j = -i, i \times k = -j$. The above expression can, therefore, be simplified as follows:

$$A \times P = (br - cq)i + (cp - ar)j + (aq - bp)k \tag{6.8}$$

Example 6.4

Find the vector perpendicular to both $A = 3i + 5j - 2k$ and $B = 2i - 2j - 2k$

From Equation (6.8) we get:

$$A \times B = (-10 - 4)i + (-4 + 6)j + (-6 - 10)k = -14i + 2j - 16k$$

MATLAB Code 6.4

```
clear all; clc;
A = [3, 5, -2];
B = [2, -2, -2];
C = cross(A, B)

%verification
dot(C,A)     % should be 0
dot(C,B)     % should be 0
```

> **NOTE**
>
> `cross`: calculates cross product of vectors

6.6 VECTOR EQUATION OF A LINE

Consider the dashed line is shown in Figure 6.5 whose vector equation needs to be computed. A vector equation of a line should be satisfied by any position vector on the line. Two parameters should be known regarding the line, first, a known point A on the line, and second, the direction of the line along a given vector b (Olive, 2003). Joining A with origin O, we obtain the position vector a of the point A. Let P be any other point on the line. Then r is the position vector of P. According to the triangle rule, $OA + AP = OP$ i.e. $a + AP = r$. Now AP being along the direction of b could be expressed as a scalar multiple of b. Let the scalar multiple be t so that $AP = t \cdot b$. Combining the above notations together vector equation of the line can be written as below:

$$r = a + t \cdot b \tag{6.9}$$

Any point on the line AP corresponds to some value of the scalar t and should satisfy the above equation (Shirley, 2002). Rewriting AP as $(r - a)$ and plugging this in Equation (6.9) we get: $r = a + t \cdot (r - a)$. Putting $t = 0$ gives us point A and putting $t = 1$ gives us point P. Putting $t = 0.5$ say, gives us the mid-point of line segment AP.

To convert the vector equation to Cartesian equation, the following steps are taken:

Expanding Equation (6.9) in terms of 3D components, $a = a_1 i + a_2 j + a_3 k$, $b = b_1 i + b_2 j + b_3 k$, and $r = xi + yj + zk$.

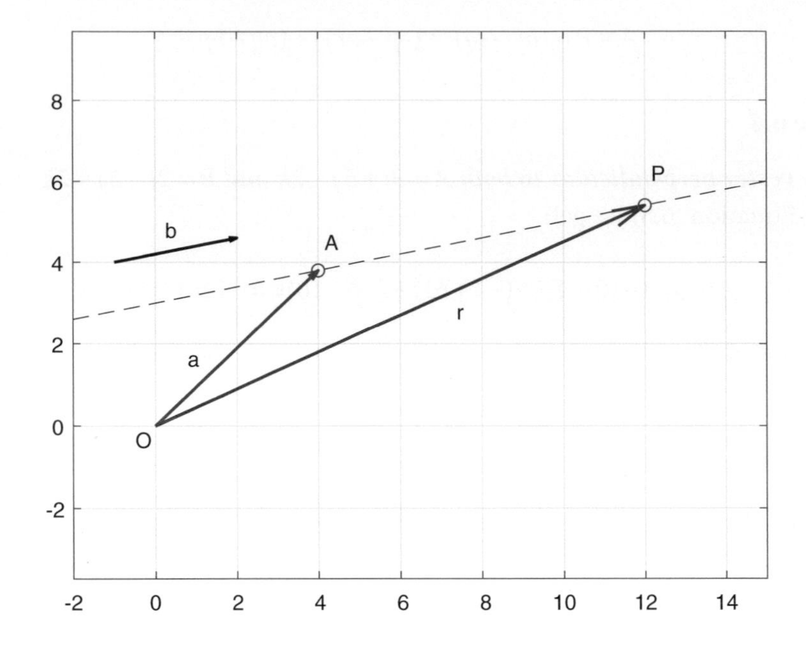

FIGURE 6.5 Deriving vector equation of a line.

$$xi + yj + zk = \left(a_1 i + a_2 j + a_3 k\right) + t \cdot \left(b_1 i + b_2 j + b_3 k\right)$$

For this to be true the $i, j,$ and k components should separately be equal. Thus

$$x = a_1 + t \cdot b_1$$

$$y = a_2 + t \cdot b_2$$

$$z = a_3 + t \cdot b_3$$

Rearranging terms

$$t = \frac{x - a_1}{b_1} = \frac{y - a_2}{b_2} = \frac{z - a_3}{b_3} \tag{6.10}$$

Equation (6.10) is the Cartesian representation of the line in 3D space as it is satisfied by any arbitrary point with coordinates (x, y, z) on the line.

Example 6.5

A straight line passes through points (6, 3, −5) and (2, 1, −4). Find its equation in vector form and Cartesian form

Position vectors $P = 6i + 3j - 5k, Q = 2i + j - 4k$

Direction vector along QP: $Q - P = (2i + j - 4k) - (6i + 3j - 5k) = -4i - 2j + k$

From Equation (6.9) vector equation:

$$r = P + t(Q - P) = (6i + 3j - 5k) + t(-4i - 2j + k) = (6 - 4t)i + (3 - 2t)j + (t - 5)k$$

From Equation (6.10)

$$t = \frac{x - 6}{2 - 6} = -x/4 + 3/2$$

$$t = \frac{y - 3}{1 - 3} = -y/2 + 3/2$$

$$t = \frac{z + 5}{-4 + 5} = z + 5$$

Cartesian equation: $\frac{-x + 6}{4} = \frac{-y + 3}{2} = \frac{z + 5}{1}$

Verification:

Substituting point (6, 3, –5) in the equation: $t = 0$ in all cases.
Substituting point (2, 1, –4) in the equation: $t = 1$ in all cases.

MATLAB Code 6.5

```
clear all; clc;

P = [6, 3, -5];
Q = [2, 1, -4];

syms t;
fprintf('Vector equation : \n');
r = P | t*(Q - P)

syms x, syms y, syms z;

x1 = P(1); x2 = Q(1);
y1 = P(2); y2 = Q(2);
z1 = P(3); z2 = Q(3);

dx = x2 - x1;
dy = y2 - y1;
dz = z2 - z1;

nx = (x - x1);
ny = (y - y1);
nz = (z - z1);

fprintf('\nCartesian equation : \n');
disp(nx/dx), disp('='), disp(ny/dy), disp('='), disp(nz/dz)
```

6.7 VECTOR EQUATION OF PLANE

To derive vector equation of a plane, two situations are to be considered: (a) the origin lies on the plane and (b) the origin lies outside the plane (Olive, 2003). For the first case, if the plane passes through the origin, let a and b be any two non-parallel vectors lying on the plane and let P be any arbitrary point on the plane (see Figure 6.6a). Then the position vector r of point P can be expressed as a combination of scaled versions of a and b

$$r = s \cdot a + t \cdot b \tag{6.11}$$

where s and t are the scaling factors for vectors a and b, respectively. This is the vector equation of the plane because any point P on the plane satisfies Equation (6.11) for different values of s and t.

For the second case, if the plane does not pass through the origin, then an additional vector c is required to define the position vector of point C, where the vectors a and b meet. Then the position vector r of the arbitrary point P on the plane can be expressed as a combination of scaled versions of a, b, and the vector c (see Figure 6.6b).

$$r = c + p = c + s \cdot a + t \cdot b \tag{6.12}$$

To derive the Cartesian equation of the plane consider Figure 6.7, where ON is the perpendicular from origin O onto the plane at N and let n be the unit vector along ON. Let length of ON be D. As before, let P be any arbitrary point on the plane and let r be its position vector. Also let the angle between ON and OP be θ.

Now from Equation (6.4)

$$r \bullet n = |r||n|\cos(\theta) = |r| \cdot 1 \cdot \cos\theta = D \tag{6.13}$$

In words, the above equation means that the dot product of the position vector of any point P on the plane and the unit normal vector equals the perpendicular distance D of the plane from the origin. Expanding into components if $n = n_1 i + n_2 j + n_3 k$ and $r = xi + yj + zk$ then

$$r \bullet n = x \cdot n_1 + y \cdot n_2 + z \cdot n_3 = D \tag{6.14}$$

The general form of the Cartesian equation of a plane is given by (Hearn and Baker, 1996):

$$Ax + By + Cz = D \tag{6.15}$$

Comparing Equations (6.14) and (6.15)

$$\begin{aligned} A &= n_1 \\ B &= n_2 \\ C &= n_3 \end{aligned} \tag{6.16}$$

Hence, the coefficients of the Cartesian equation of a plane are equal to the components of the unit normal vector from the origin onto the plane.

(a)

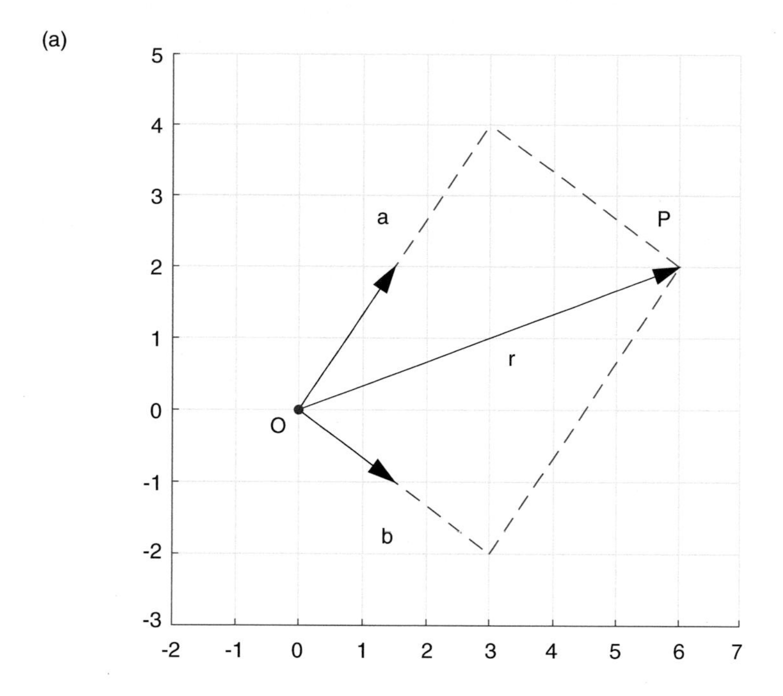

(b)

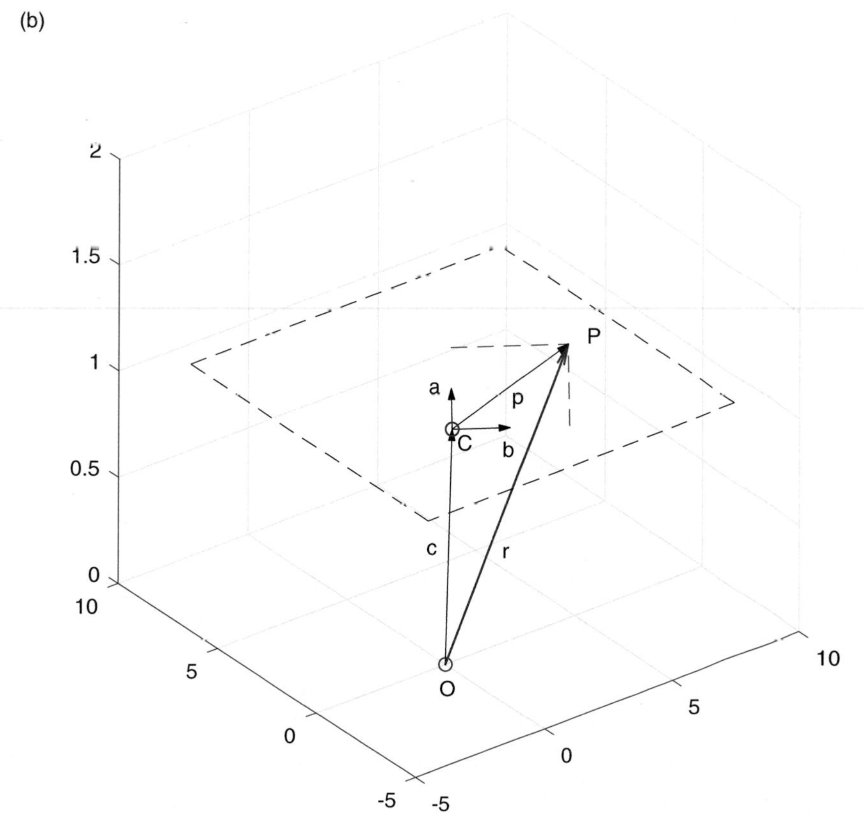

FIGURE 6.6 (a) and (b) Deriving vector equations of a plane (a) passing through origin (b) not passing through origin.

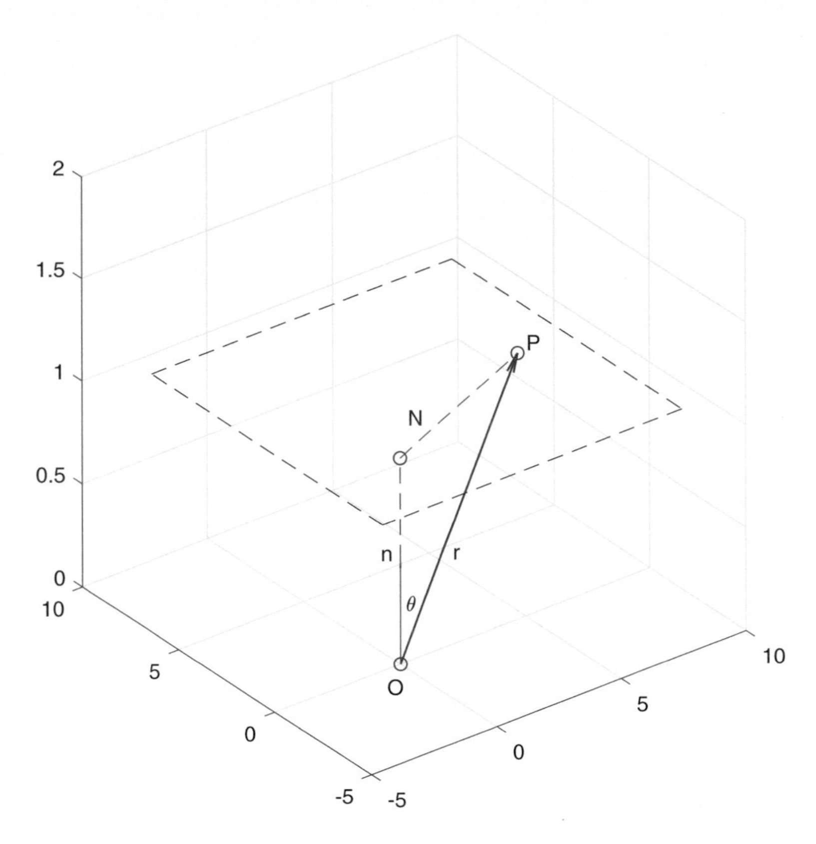

FIGURE 6.7 Deriving the Cartesian equation of a plane.

Example 6.6

Vector equation of a plane is $r = c + as + bt$, where $c = 2i + 4j - 3k$, $a = -3i + 3j - 3k$, $b = 4i + 2j + 3k$, and s, t are the scalars. Find the normal vector N and the normal distance D from the origin and Cartesian equation of the plane.

Let P be any point on the plane with $OP = r = xi + yj + zk$

Let N be normal vector of the plane

Then $N = a \times b = 15i - 3j - 18k$

Magnitude $|N| = \sqrt{\{225 + 9 + 324\}} = 23.622$

Unit vector $n = N/|N| = 0.6350i - 0.1270j - 0.7620k$

Hence, for the plane $Ax + By + Cz = D$, we have $A = 0.6350, B = -0.1270$, and $C = -0.7620$

Now for any point C on the plane, we must have $c \bullet n = D$ as per Equation (6.13)

Hence, $D = (2i + 4j - 3k) \bullet (0.6350i - 0.1270j - 0.7620k) = 3.048$

Thus, Cartesian equation of the plane is $0.6350x - 0.1270y - 0.7620z = 3.048$

Verification: since $C(2, 4, -3)$ is a point on the plane it should satisfy the plane equation i.e.

$$0.6350(2) - 0.1270(4) - 0.7620(-3) - 3.048 = 0$$

MATLAB Code 6.6

```
clear all; clc;
c = [2, 4, -3];
a = [-3, 3, -3];
b = [4, 2, 3];
N = cross(a, b);
fprintf('Normal vector : \n');
N
n = N/norm(N);
d = dot(c, n);
fprintf('Distance : \n');
d
n1 = n(1);
n2 = n(2);
n3 = n(3);
fprintf('Cartesian equation : \n (%.2f)x + (%.2f)y + (%.2f)z = %.2f\n', n1, n2, n3, d);

%verification
x = c(1); y = c(2); z = c(3);
f = n1*x + n2*y + n3*z - d;      % should be zero
```

6.8 VECTOR ALIGNMENT (2D)

A position vector $P = ai + bj$ can be aligned with a primary axis along positive or negative directions by rotating through the appropriate angle θ given by $\theta = \arctan(b/a)$. Depending on the orientation of the original vector and the primary axis along which it is to be aligned, the angle of rotation can be either positive or negative. If the angle of orientation with respect to +X axis, of vector $(ai + bj)$ be θ, then angle of orientation of vector $(-ai + bj)$ is $(180 - \theta)$, angle of orientation of vector $(-ai - bj)$ is $(180 + \theta)$, and angle of orientation of vector $(ai - bj)$ is $-\theta$. The rotation angles for aligning these vectors with the +X axis will be just the negative of these angles of orientation i.e. $-\theta, -(180 - \theta), -(180 + \theta), +\theta$, respectively. Angles are positive measured along counter-clockwise (CCW) direction and negative along the clockwise (CW) direction. The rotation matrix is calculated based on the latter set of values i.e. the rotation angles (see Figure 6.8).

Example 6.7

For each of the following vectors, find the angle it makes with the positive X-axis and the rotation matrix to align it with the positive X-axis: (a) $3i + 4j$, (b) $-3i + 4j$, (c) $-3i - 4j$, and (d) $3i - 4j$. Also for each case verify, the alignment by multiplying the vector with the rotation matrix.

(a)

$P = 3i + 4j = [3, 4]$
$\theta = \arctan(4/3) = 53.13°$
$R_1 = R(-53.13°)$
Verification: $Q = R_1 {*} P = [5, 0, 1]^T$

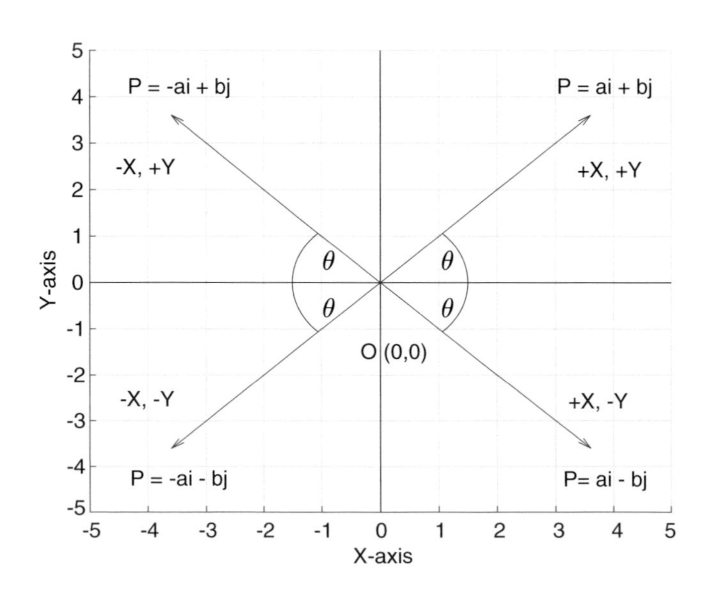

FIGURE 6.8 Orientation angles of vectors.

(b)

$P = -3i + 4j = [-3, 4]$

$\theta = 180° - \arctan(4/3) = 126.87°$

$R_1 = R(-126.87°)$

Verification: $Q = R_1 * P = [5, 0, 1]^T$

(c)

$P = -3i - 4j = [-3, -4]$

$\theta = 180° + \arctan(4/3) = 233.13°$

$R_1 = R(-233.13°)$

Verification: $Q = R_1 * P = [5, 0, 1]^T$

(d)

$P = 3i - 4j = [3, -4]$

$\theta = -\arctan(4/3) = -53.13°$

$R_1 = R(53.13°)$

Verification: $Q = R_1 * P = [5, 0, 1]^T$

MATLAB Code 6.7

```
clear; clc;
P = [3, 4]
B = atan(abs(P(2))/abs(P(1)));
ang = rad2deg(B)
A = -B;
R = [cos(A) -sin(A) 0 ; sin(A) cos(A) 0 ; 0 0 1];
Q = R * [P, 1]';
P = [-3, 4]
B = pi - atan(abs(P(2))/abs(P(1)));
ang = rad2deg(B)
A = -B;
R = [cos(A) -sin(A) 0 ; sin(A) cos(A) 0 ; 0 0 1];
Q = R * [P, 1]';

P = [-3, -4]
B = pi + atan(P(2)/P(1));
ang = rad2deg(B)
A = -B;
R = [cos(A) -sin(A) 0 ; sin(A) cos(A) 0 ; 0 0 1];
Q = R * [P, 1]';

P = [3, -4]
B = -atan(abs(P(2))/abs(P(1)));
ang = rad2deg(B)
A = -B;
R = [cos(A) -sin(A) 0 ; sin(A) cos(A) 0 ; 0 0 1];
Q = R * [P, 1]';
```

TABLE 6.1 Orientation Angles of Vectors

Vector	Quadrant	+X-axis	−X-axis	+Y-axis	−Y-axis
$P=ai+bj$	Q1	$+\theta$	$-(180-\theta)$	$-(90-\theta)$	$+(90+\theta)$
$P=-ai+bj$	Q2	$+(180-\theta)$	$-\theta$	$+(90-\theta)$	$-(90+\theta)$
$P=-ai-bj$	Q3	$+(180+\theta)$	$+\theta$	$+(90+\theta)$	$-(90-\theta)$
$P=ai-bj$	Q4	$-\theta$	$+(180-\theta)$	$-(90+\theta)$	$+(90-\theta)$

Apart from the $+X$ axis angles of orientation can also be calculated with respect to the other primary axes. If θ be the angle with the $+X$ axis, then angle with $+Y$ axis will be $(90 - \theta)$, angle with the $-X$ axis will be $(180 - \theta)$, and angle with the $-Y$ axis will be $(90 + \theta)$. The signs are calculated by observing whether we are moving in the CW or CCW direction. For example, for vector $(ai + bj)$ which is in Q1 quadrant, when angle is measured from $+X$ axis we are moving in CCW direction hence $+\theta$, when measuring from $-X$ axis we are moving in CW direction hence $-(180 - \theta)$, when measuring from $+Y$ axis we are moving in CW direction hence $-(90 - \theta)$, when measuring from $-Y$ axis we are moving in CCW direction hence $+(90 + \theta)$. Given in Table 6.1 are the possible cases for a and b positives. Column 1 shows the vectors, column 2 shows the quadrant in which the vector lies, and columns 3 to 6 indicate the angle and direction values (CCW +ve and CW −ve). It is to be noted that for actually aligning the vector to the corresponding axis, the sign of the angle will be just the negative to that given in the table. For example, for $(ai + bj)$ the angle from $-X$-axis is $-(180 - \theta)$ as it is measured in CW direction but to align the vector with the $-X$-axis, the vector needs to be rotated in the CCW direction, so angle for alignment is $+(180 - \theta)$. Similar for $-ai - bj$ which lies in Q3 the angle for aligning with the $+Y$ axis is $-(90 + \theta)$ since the vector would need to be rotated in CW direction.

Example 6.8

For each of the following vectors, find the rotation matrix to align it with the primary axis specified (a) $3i - 4j$ with positive Y-axis and (b) $-3i - 4j$ with negative X-axis. Also for each case, verify the alignment by multiplying the vector with the rotation matrix.

(a)

$P = 3i - 4j = [3, -4]$

$\theta = \arctan(4/3) = 53.13°$

To align with $+Y$ axis, the vector needs to be rotated by $(90 + 53.13)° = 143.13°$ in CCW direction.

$R_1 = R(143.13°)$

Verification: $Q = R_1 * P = [0, 5, 1]^T$

(b)

$P = -3i - 4j = [-3, -4]$

$\theta = \arctan(4/3) = 53.13°$

$\theta = 180° + \arctan(4/3) = 233.13°$
To align with $-X$ axis, the vector needs to be rotated by 53.13° CW direction.
$R_1 = R(-53.13°)$
Verification: $Q = R1*P = [-5, 0, 1]^T$

MATLAB Code 6.8

```
clear; clc;
P = [3, -4]
B = atan(abs(P(2))/abs(P(1)));
A = B + pi/2;
R = [cos(A) -sin(A) 0 ; sin(A) cos(A) 0 ; 0 0 1];
fprintf('verification \n');
Q = R * [P, 1]'

P = [-3, -4]
B = atan(abs(P(2))/abs(P(1)));
A = -B;
R = [cos(A) -sin(A) 0 ; sin(A) cos(A) 0 ; 0 0 1];
fprintf('verification \n');
Q = R * [P, 1]'
```

NOTE

Vector alignment in 3D has been dealt with in the next chapter.

6.9 VECTOR EQUATIONS IN HOMOGENEOUS COORDINATES (2D)

We know that Cartesian equation of a line is $ax + by + c = 0$. Let (X, Y, W) are homogeneous coordinates of point (x, y) i.e. $x = X/W,\ y = Y/W$. Substituting in line equation:

$$aX + bY + cW = 0 \tag{6.17}$$

This is called homogeneous line equation (Marsh, 2005) and the vector $\ell = (a, b, c)$ is denoted as line vector. If $P(X, Y, W)$ be a point on the line, we must have:

$$\ell \bullet P = (a, b, c) \bullet (X, Y, W) = aX + bY + cW = 0 \tag{6.18}$$

Thus for line $\ell = (a, b, c)$ passing through $P(X, Y, W)$ we must have $\ell \bullet P = 0$

Let the line ℓ through two given points $P_1(X_1, Y_1, W_1)$ and $P_2(X_2, Y_2, W_2)$. From the above equation, $\ell \bullet P_1 = 0$ and $\ell \bullet P_2 = 0$ which implies that vector ℓ is perpendicular to both P_1 and P_2. This can be represented as $\ell = P_1 \times P_2$.

Thus, for line $\ell = (a, b, c)$ passing through two points $P_1(X_1, Y_1, Z_1)$ and $P_2(X_2, Y_2, Z_2)$, we must have:

$$\ell = P_1 \times P_2 \tag{6.19}$$

Let two non-parallel lines ℓ_1 and ℓ_2 have an intersection point P.

Since P lies on both lines, we must have $\ell_1 \bullet P = 0$ and $\ell_2 \bullet P = 0$, which implies that vector P is perpendicular to both lines. This can be represented as $P = \ell_1 \times \ell_2$. Thus, intersection point P of two lines ℓ_1 and ℓ_2 is given by:

$$P = \ell_1 \times \ell_2 \tag{6.20}$$

Example 6.9

Solve the following using vector-based methods: (a) Find which of the points (0, 5), (2, 1), and (10/3, 0) lies on the line $3x + 4y - 10 = 0$. (b) Find the equation of the line passing through the points (1, 8) and (6, –7). (c) Find the point where the lines $x - 2y + 3 = 0$ and $4x - 5y + 6 = 0$ intersects.

(a)

Here, $L = [3, 4, -10]$, $P_1 = [0, 5, 1]$, $P_2 = [2, 1, 1]$, and $P_3 = [10/3, 0, 1]$

From Equation (6.18),

$L \bullet P_1 = (3)(0) + (4)(5) + (-10)(1) = 0 + 20 - 10 \neq 0$

Hence, point (0, 5) does not lie on line $3x + 64 - 10 = 0$

$L \bullet P_2 = (3)(2) + (4)(1) + (-10)(1) = 6 + 4 - 10 = 0$

Hence, point (2, 1) lies on line $3x + 64 - 10 = 0$

$L \bullet P_3 = (3)(10/3) + (4)(0) + (-10)(1) = 10 + 0 - 10 = 0$

Hence, point (10/3, 0) lies on line $3x + 64 - 10 = 0$

(b)

Here, $P_1 = [1, 8, 1]$ and $P_2 = [6, -7, 1]$

From Equation (6.19),

$L = P_1 \times P_2 = \{(8)(1) - (1)(7)\}i + \{(1)(6) - (1)(1)\}j + \{(1)(-7) - (8)(6)\}k = 15i + 5j - 55k$

Required equation of line: $15x + 5y - 55 = 0$

Verification: P_1 and P_2 are points on this line: $15(1) + 5(8) - 55 = 0$, $15(6) + 5(-7) - 55 = 0$

(c)

Here, $L_1 = [1, -2, 3]$ and $L_2 = [4, -5, 6]$

From Equation (6.20), $P = L_1 \times L_2 = [3, 6, 3]$ (in homogeneous coordinates)

Converting to Cartesian coordinates, intersection point $P = (3/3, 6/3) = (1, 2)$

Verification: P satisfies both lines: $1(1) + (-2)(2) + 3 = 0$, $4(1) + (-5)(2) + 6 = 0$

MATLAB Code 6.9

```
%(a)

clear all; clc;

L = [3, 4, -10];
P1 = [0, 5, 1];
P2 = [2, 1, 1];
P3 = [10/3, 0, 1];

D1 = dot(L, P1);
if D1 == 0 fprintf('Point P1 lies on line L\n'),
else fprintf('Point P1 does not lie on line L\n'), end;

D2 = dot(L, P2);
if D2 == 0 fprintf('Point P2 lies on line L\n'),
else fprintf('Point P2 does not lie on line L\n'), end;

D3 = dot(L, P3);
if D3 == 0 fprintf('Point P3 lies on line L\n'),
else fprintf('Point P3 does not lie on line L\n'), end;

%(b)

clear all;

P1 = [1, 8, 1];
P2 = [6, -7, 1];
```

```
L = cross(P1, P2);
fprintf('Equation of required line : (%.2f)x + (%.2f)y + (%.2f) = 0\n', L(1), L(2), L(3));

%(c)

clear all;

L1 = [1, -2, 3];
L2 = [4, -5, 6];
P = cross(L1, L2);
fprintf('Point in Cartesian coordinates : {(%.2f), (%.2f)}\n', P(1)/P(3), P(2)/P(3));
```

6.10 VECTOR EQUATIONS IN HOMOGENEOUS COORDINATES (3D)

We know that Cartesian equation of a plane in 3D space is $ax+by+cz+d=0$. Let (X, Y, Z, W) be the homogeneous coordinates of point (x, y, z) i.e. $x = X/W$, $y=Y/W$, $z=Z/W$

Substituting in plane equation:

$$aX+bY+cZ+dW=0 \tag{6.21}$$

This is called homogeneous plane equation (Marsh, 2005) and the vector $N=(a, b, c, d)$ is denoted as plane vector.

Let $P(X, Y, Z, W)$ be a point on the plane, hence we must have:

$$N \bullet P=(a, b, c, d) \bullet (X, Y, Z ,W)=aX+bY+cZ+dW=0 \tag{6.22}$$

Thus for plane $N=(a, b, c, d)$ passing through point $P(X, Y, Z, W)$ we must have $N \bullet P=0$.

Let the plane N pass through three given points $P_1(X_1, Y_1, Z_1, W_1)$, $P_2(X_2, Y_2, Z_2, W_2)$, and $P_3(X_3, Y_3, Z_3, W_3)$. From the above equation, $N \bullet P_1=0$, $N \bullet P_2=0$, $N \bullet P_3=0$, which implies that vector N is perpendicular to P_1, P_2, and P_3.

The condition for this to occur is given by the vector determinant:

$$N=\begin{bmatrix} e_1 & e_2 & e_3 & e_4 \\ X_1 & Y_1 & Z_1 & W_1 \\ X_2 & Y_2 & Z_2 & W_2 \\ X_3 & Y_3 & Z_3 & W_3 \end{bmatrix} \tag{6.23}$$

where $e_1 = (1, 0, 0, 0)$, $e_2 = (0, 1, 0, 0)$, $e_3 = (0, 0, 1, 0)$, and $e_4 = (0, 0, 0, 1)$ are the unit vectors along the four orthogonal directions.

Let three non-parallel planes $N_1(a_1, b_1, c_1, d_1)$, $N_2(a_2, b_2, c_2, d_2)$, and $N_3(a_3, b_3, c_3, d_3)$ have an intersection point P. Since P lies on all planes we must have $N_1 \bullet P = 0$, $N_2 \bullet P=0$, and $N_3 \bullet P=0$, which implies that vector P is perpendicular to all. The condition for this to occur is given by the following vector determinant, where e_1, e_2, e_3, and e_4 are the unit vectors along the four orthogonal directions:

$$P=\begin{bmatrix} e_1 & e_2 & e_3 & e_4 \\ a_1 & b_1 & c_1 & d_1 \\ a_2 & b_2 & c_2 & d_2 \\ a_3 & b_3 & c_3 & d_3 \end{bmatrix} \tag{6.24}$$

Example 6.10

Solve the following using vector-based methods: (a) Find the plane connecting the three points (1, 2, 3), (–4, –5, –6), and (7, 8, 9) (b) Find the point of intersection of the three planes $x+2y-3z+4=0, 3x+4y-2z+1=0,$ and $5x+6y-4z+3=0$.

(a)

Here, $P_1 = [1, 2, 3]$, $P_2 = [-4, -5, -6]$, and $P_3 = [7, 8, 9]$;
Let e_1, e_2, e_3, and e_4 be the unit vectors along the four orthogonal directions.
From Equation (6.23),

$$N = \begin{bmatrix} e_1 & e_2 & e_3 & e_4 \\ X_1 & Y_1 & Z_1 & W_1 \\ X_2 & Y_2 & Z_2 & W_2 \\ X_3 & Y_3 & Z_3 & W_3 \end{bmatrix} = \begin{bmatrix} e_1 & e_2 & e_3 & e_4 \\ 1 & 2 & 3 & 1 \\ -4 & -5 & -6 & 1 \\ 7 & 8 & 9 & 1 \end{bmatrix}$$

$$d1 = \det\left(\begin{bmatrix} 2 & 3 & 1 \\ -5 & -6 & 1 \\ 8 & 9 & 1 \end{bmatrix}\right) = 12$$

$$d2 = -\det\left(\begin{bmatrix} 1 & 3 & 1 \\ -4 & -6 & 1 \\ 7 & 9 & 1 \end{bmatrix}\right) = -24$$

$$d3 = \det\left(\begin{bmatrix} 1 & 2 & 1 \\ -4 & -5 & 1 \\ 7 & 8 & 1 \end{bmatrix}\right) = 12$$

$$d4 = -\det\left(\begin{bmatrix} 1 & 2 & 3 \\ -4 & -5 & -6 \\ 7 & 8 & 9 \end{bmatrix}\right) = 0$$

Equation of plane: $12x - 24y + 12z = 0$
Verification: P_1, P_2, and P_3 all satisfy this equation

(b)

Here, $N_1 = [1, 2, -3, 4]$, $N_2 = [3, 4, -2, 1]$, and $N_3 = [5, 6, -4, 3]$;
Let e_1, e_2, e_3, and e_4 be the unit vectors along the four orthogonal directions.
From Equation (6.24),

$$P = \begin{bmatrix} e_1 & e_2 & e_3 & e_4 \\ a_1 & b_1 & c_1 & d_1 \\ a_2 & b_2 & c_2 & d_2 \\ a_3 & b_3 & c_3 & d_3 \end{bmatrix} = \begin{bmatrix} e_1 & e_2 & e_3 & e_4 \\ 1 & 2 & -3 & 4 \\ 3 & 4 & -2 & 1 \\ 5 & 6 & -4 & 3 \end{bmatrix}$$

$$d1 = \det\left(\begin{bmatrix} 2 & -3 & 4 \\ 4 & -2 & 1 \\ 6 & -4 & 3 \end{bmatrix}\right) = -2$$

$$d2 = -\det\left(\begin{bmatrix} 1 & -3 & 4 \\ 3 & -2 & 1 \\ 5 & -4 & 3 \end{bmatrix}\right) = -2$$

$$d3 = \det\left(\begin{bmatrix} 1 & 2 & 4 \\ 3 & 4 & 1 \\ 5 & 6 & 3 \end{bmatrix}\right) = -10$$

$$d4 = -\det\left(\begin{bmatrix} 1 & 2 & -3 \\ 3 & 4 & -2 \\ 5 & 6 & -4 \end{bmatrix}\right) = -6$$

Point in homogeneous coordinates: (–2,–2,–10,–6)
Point in Cartesian coordinates: (1/3, 1/3, 5/3)
Verification: point P satisfy all three plane equations

MATLAB Code 6.10

```
%(a)

clear all; clc;

syms e1 e2 e3 e4;

P1 = [1, 2, 3];
P2 = [-4, -5, -6];
P3 = [7, 8, 9];

N = [e1, e2, e3, e4 ; P1(1), P1(2), P1(3), 1 ; P2(1), P2(2), P2(3), 1 ; P3(1), P3(2), P3(3), 1];

d1 = det([N(2,2), N(2,3), N(2,4) ; N(3,2), N(3,3), N(3,4) ; N(4,2), N(4,3), N(4,4)]);
d2 = -det([N(2,1), N(2,3), N(2,4) ; N(3,1), N(3,3), N(3,4) ; N(4,1), N(4,3), N(4,4)]);
d3 = det([N(2,1), N(2,2), N(2,4) ; N(3,1), N(3,2), N(3,4) ; N(4,1), N(4,2), N(4,4)]);
d4 = -det([N(2,1), N(2,2), N(2,3) ; N(3,1), N(3,2), N(3,3) ; N(4,1), N(4,2), N(4,3)]);

d1 = double(d1); d2 = double(d2); d3 = double(d3); d4 = double(d4);
fprintf('Equation of plane : (%.2f)x + (%.2f)y + (%.2f)z + (%0.2f) = 0\n', d1, d2, d3, d4)

%(b)

clear all;

syms e1 e2 e3 e4;

N1 = [1, 2, -3, 4];
N2 = [3, 4, -2, 1];
N3 = [5, 6, -4, 3];
```

```
N = [e1, e2, e3, e4 ; N1(1), N1(2), N1(3), N1(4) ; ...
    N2(1), N2(2), N2(3), N2(4) ; N3(1), N3(2), N3(3), N3(4)];

d1 = det([N(2,2), N(2,3), N(2,4) ; N(3,2), N(3,3), N(3,4) ;
... N(4,2), N(4,3), N(4,4)]);
d2 = -det([N(2,1), N(2,3), N(2,4) ; N(3,1), N(3,3), N(3,4) ;
... N(4,1), N(4,3), N(4,4)]);
d3 = det([N(2,1), N(2,2), N(2,4) ; N(3,1), N(3,2), N(3,4) ;
... N(4,1), N(4,2), N(4,4)]);
d4 = -det([N(2,1), N(2,2), N(2,3) ; N(3,1), N(3,2), N(3,3) ;
... N(4,1), N(4,2), N(4,3)]);

d1 = double(d1); d2 = double(d2); d3 = double(d3); d4 =
     double(d4);
fprintf('Point in Cartesian coordinates : {(%.2f), (%.2f), (%.2f)}\n', d1/d4, d2/d4, d3/d4);
```

NOTE

det: calculates determinant of a matrix

6.11 NORMAL VECTOR AND TANGENT VECTOR

Given a function in implicit form $w = f(x,y)$, the gradient ∇f at a point of the curve $f(x,y)=k$ is perpendicular to the curve at that point and gives the normal vector (Shirley, 2002). Let $P(x_0, y_0)$ be a point on the curve so that $(x_0, y_0)=k$. Let the parametric representation of the curve be $g(t)=f(x(t),y(t))=k$. Also let at P, $t=t_0$

Differentiating with respect to t at P we get:

$$\frac{dg}{dt}=\left.\frac{\partial f}{\partial x}\right|_P \cdot \left.\frac{dx}{dt}\right|_{t_0}+\left.\frac{\partial f}{\partial y}\right|_P \cdot \left.\frac{dy}{dt}\right|_{t_0}=0 \tag{6.25}$$

Rewriting in vector form:

$$\left(\left.\frac{\partial f}{\partial x}\right|_P, \left.\frac{\partial f}{\partial y}\right|_P\right) \bullet \left(\left.\frac{dx}{dt}\right|_{t_0}, \left.\frac{dy}{dt}\right|_{t_0}\right)=0 \tag{6.26}$$

The second term in the above equation gives the tangent vector and since the dot product is zero the gradient vector is the normal vector perpendicular to the curve. Note that the tangent vector can also be obtained by rotating the normal vector by 90°.

Example 6.11

For the curve $x^2 + y^2 = 4$, find the normal vector and tangent vector at a point on the curve. Also derive the equation of the tangent line through that point.

Let $f(x,y)=x^2+y^2-4$ and let $P(1, \sqrt{3})$ be a point on the curve.

Differentiating, $\nabla f = \left(\frac{\partial f}{\partial x}, \frac{\partial f}{\partial y}\right)=(2x,2y)$

Gradient at P: $\nabla f|_P = (2, 2\sqrt{3}) = N_P(n_1, n_2)$ which is the normal vector at P i.e. $N_P = 2i + 3.46j$

The tangent vector is obtained by rotating the normal by 90° i.e.

$$T_P = R(90) * N = \begin{bmatrix} 0 & -1 & 0 \\ 1 & 0 & 0 \\ 0 & 0 & 1 \end{bmatrix}\begin{bmatrix} n_1 \\ n_2 \\ 1 \end{bmatrix} = \begin{bmatrix} -n_2 \\ n_1 \\ 1 \end{bmatrix} = \begin{bmatrix} -2\sqrt{3} \\ 2 \\ 1 \end{bmatrix} \text{i.e. } T_P = -3.46i + 2j$$

The tangent line equation can be obtained as the line through P with slope same as T_P

$$\frac{(y-y_0)}{(x-x_0)}=\frac{n_1}{-n_2}$$

Equation of the tangent line at P is given by T_{LP}: $2x + 2\sqrt{3}y - 8 = 0$.

Verification: Point P satisfies the tangent line equation (Figure 6.9)

MATLAB Code 6.11

```
clear all; clc; format compact;
syms x y;
f = x^2 + y^2 - 4;
df = [diff(f, x), diff(f, y)];
p = [1, sqrt(3)];
nv = subs(df, [x, y], [p(1), p(2)]);     % normal vector
r90 = [cosd(90), -sind(90), 0 ; sind(90), cosd(90), 0 ; 0 0 1];
tv = r90*[nv 1]';    % tangent vector
tl = nv(1)*(x - p(1)) + nv(2)*(y - p(2));    % tangent line

fprintf('Normal vector : (%.2f)i + (%.2f)j \n', eval(nv(1)), eval(nv(2)));
fprintf('Tangent vector : (%.2f)i + (%.2f)j \n', eval(tv(1)), eval(tv(2)));
fprintf('Tangent line : ');
disp(simplify(tl))

%plotting
ezplot(f); hold on; grid;
plot(p(1), p(2), 'ro');
quiver(p(1), p(2), nv(1), nv(2));
ezplot(tl);
quiver(p(1), p(2), tv(1), tv(2));
axis equal; hold off;
```

NOTE

sind: calculates sine of an angle in degrees
cosd: calculates cosine of an angle in degrees

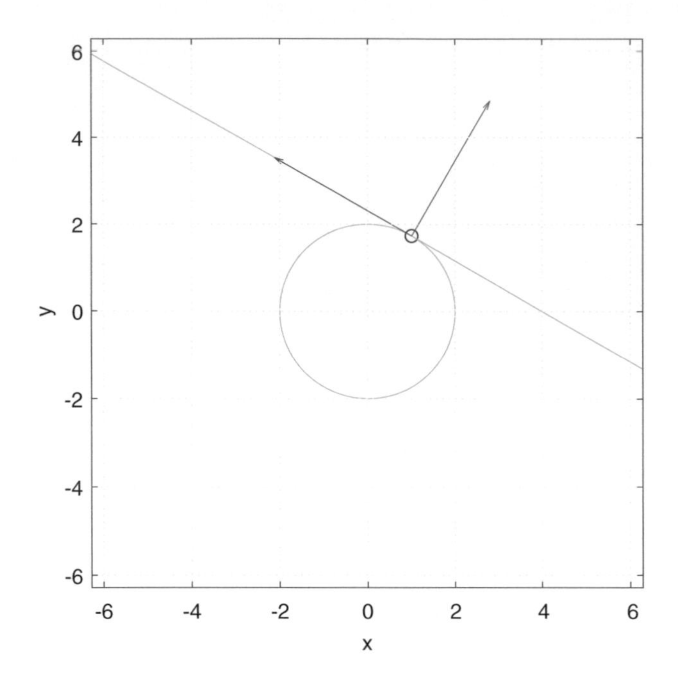

FIGURE 6.9 Plot for Example 6.11.

Similar to the case of a 2D curve, the gradient ∇f at a point of the surface $f(x, y, z) = k$ is perpendicular to the surface at that point and gives the normal vector (Shirley, 2002). In vector form

$$\left(\left.\frac{\partial f}{\partial x}\right|_P, \left.\frac{\partial f}{\partial y}\right|_P, \left.\frac{\partial f}{\partial z}\right|_P \right) \bullet \left(\left.\frac{dx}{dt}\right|_{t_0}, \left.\frac{dy}{dt}\right|_{t_0}, \left.\frac{dz}{dt}\right|_{t_0} \right) = 0 \tag{6.27}$$

The first set of terms in the above equation gives the normal vector $N(n_1, n_2, n_3)$ and the second set of terms gives the tangent vector $T(t_1, t_2, t_3)$ at point (x_0, y_0, z_0), whose dot product is zero. The equation of the tangent plane through $P(x_0, y_0, z_0)$ is given by:

$$n_1(x - x_0) + n_2(y - y_0) + n_3(z - z_0) = 0 \tag{6.28}$$

Example 6.12

For the surface $x^2 + y^2 + z^2 = 12$, find the normal vector and tangent plane at a point on the surface.

Let $f(x, y) = x^2 + y^2 + z^2 - 12$ and $P(2, 2, 2)$ be a point on the surface.

Differentiating, $\nabla f = \left(\frac{\partial f}{\partial x}, \frac{\partial f}{\partial y}, \frac{\partial f}{\partial z} \right) = (2x, 2y, 2z)$

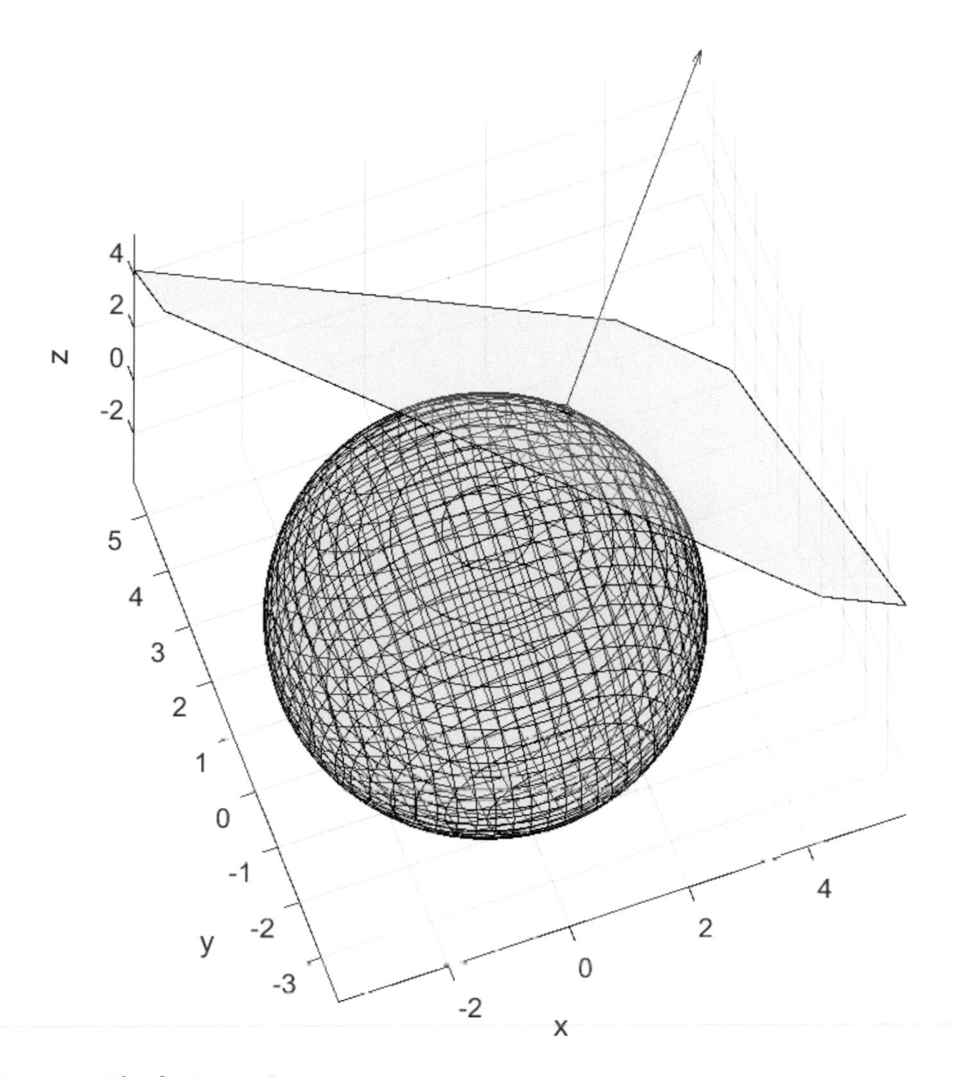

FIGURE 6.10 Plot for Example 6.12.

Gradient at P: $\nabla f\big|_P = (4, 4, 4) = N_P\left(n_1, n_2, n_3\right)$ which is the normal vector at P
Equation of the tangent plane T_{PP} through P is given by (Figure 6.10):

$$4(x-2)+4(y-2)+4(z-2)=0$$

Simplifying we get T_{PP}: $4x+4y+4z-24=0$

MATLAB Code 6.12

```
clear all; clc;
syms x y z;
f = x^2 + y^2 + z^2 - 12;
df = [diff(f, x), diff(f, y), diff(f, z)];
p = [2, 2, 2];
n = subs(df, [x, y, z], [p(1), p(2) p(3)]);
t = n(1)*(x - p(1)) + n(2)*(y - p(2)) + n(3)*(z - p(3));
fprintf('Normal vector : (%.2f)i + (%.2f)j + (%.2f)k \n', eval(n(1)), eval(n(2)), eval(n(3)));
fprintf('Tangent plane : ');
disp(t)

%plotting
fimplicit3(f, 'FaceColor', 'y', 'FaceAlpha',0.3);
axis square; hold on;
plot3(p(1), p(2), p(3), 'ro')
quiver3(p(1), p(2), p(3), n(1), n(2), n(3));
fimplicit3(t, 'MeshDensity', 2, 'FaceColor', 'y', 'FaceAlpha',0.3);
view(-20, 66);
xlabel('x');ylabel('y');zlabel('z');
hold off;
```

NOTE

`fimplicit3`: generates a 3D plot of an implicit function
`plot3`: creates 3D graphical plots from a set of values
`view`: specifies the horizontal and vertical angles for viewing a 3D scene

6.12 CHAPTER SUMMARY

The following points summarize the topics discussed in this chapter:

- Vectors have both magnitude and direction.
- Multiplying a vector by a scalar only changes its magnitude but keeps its direction same.
- Triangle law, parallelogram law, and polygon law define how multiple vectors can be combined.
- A unit vector is derived by dividing a vector by its own magnitude.
- Direction cosines are cosines of the angles made by a vector with the primary axes.
- Dot product of two vectors is the product of their magnitudes and cosine of the angle between them.
- Cross product of two vectors is a vector perpendicular to both of them.
- Vector equation of a line can be derived from a point on the line and a direction vector.
- Vector equation of a plane can be derived if two non-parallel vectors on the plane are known.
- Expressed in homogeneous coordinates, for line L passing through point P, $L \bullet P = 0$.
- Expressed in homogeneous coordinates, for line L passing through points P_1 and P_2, $L = P_1 \times P_2$.
- Expressed in homogeneous coordinates, for two lines L_1 and L_2 intersecting at point P, $P = L_1 \times L_2$.
- In homogeneous coordinates, for normal N of a plane and point P on the plane, $N \bullet P = 0$.
- For a plane with normal N passing through points P_1, P_2, and P_3; $N \bullet P_1 = 0$, $N \bullet P_2 = 0$, and $N \bullet P_3 = 0$.
- For three non-parallel planes N_1, N_2, and N_3 intersecting at P, $N_1 \bullet P = 0$, $N_2 \bullet P = 0$, and $N_3 \bullet P = 0$.

6.13 REVIEW QUESTIONS

1. For two vectors to be equal what conditions need to be fulfilled?
2. How can multiple vectors be combined to produce a resultant?
3. How can the magnitude and direction of a vector be computed?
4. How can the dot product of two vectors be used to determine whether they are orthogonal?

5. If the cross product of two vectors is zero what does it indicate?

6. For three vectors A, B, and C specify the result: (a) $A \bullet B \times C$ and (b) $A \times B \bullet C$.

7. Why is the vector equation of a line or a plane not unique?

8. If N is the normal to a plane and r any point on the plane, what is the value of $N \bullet r$?

9. If θ be the angle of orientation of a vector with $+X$ axis what is its angle with $-X$, $+Y$, and $-Y$ axes?

10. If $ai + bj$ is oriented at θ with $+X$ axis, what are the angles of $ai - bj$, $-ai + bj$, and $-ai - bj$?

6.14 PRACTICE PROBLEMS

1. Find relation between the vector $p = ai + bj + ck$ and $= (\cos A)i + (\cos B)j + (\cos C)k$, where A, B, and C are the angles made by the vector p with the three primary axes.

2. Convert the Cartesian equation of the line $3x + 4y = 12$ to a vector equation.

3. Consider two lines $A = i - j + 4k + s(i - j + k)$ and $B = 2i + 4j + 7k + t(2i + j + 3k)$, where s and t are the scaling factors. For what values of s and t do the lines intersect?

4. Find Cartesian equation of a line with vector equation $r = (2i + 3j - 4k) + t(3i - j + 2k)$.

5. Consider the plane $3x + 4y + 5z = 12$. Find the equation of the line going through the point of intersection of the plane with the X- and Y-axes.

6. Find out if the line $L: r = (3, 2, 5) + a \cdot (5, -5, 1)$ is perpendicular to the plane $P: r = (3, -2, 5) + b \cdot (3, 2, -5) + c \cdot (2, 3, 5)$, where a, b, and c are the scalars.

7. Where does the line $r = (1, 3, 5) + t(2, -4, 6)$ meet the plane $x - 2y + 3z = -4$?

8. Find out if the line $L: r = (1, 3, 8) + u(-2, 5, 7)$ is parallel to the plane $P: r = (0.3, 0.25, -0.5) + s(4, -1, 2) + t(6, -15, -21)$.

9. Find vector equation of line along which two planes $P_1: r = (2, 0, 0) + s_1(2, -3, 0) + t_1(2, 0, -4)$ and $P_2: r = (5, 0, 0) + s_2(5, 1, 0) + t_2(5, 0, 4)$ meet.

10. Find the transformation that aligns the position vector (a) $-4i + 5j$ with positive X-axis and (b) $4i + 5j$ with negative Y-axis.

CHAPTER 7

3D Transformations

7.1 INTRODUCTION

Three dimensional transformations enable us to change the location, orientation, and shapes of splines in 3D space. These transformations are translation, rotation, scaling, reflection, and shear applied individually or in combination of two or more. Given known coordinates of a point, each of these transformations is represented by a matrix which when multiplied to the original coordinates give us a new set of coordinates. Similar to the case of 2D transformations, we use homogeneous coordinates to derive transformation matrices. Coordinates of points are measured using a right-handed coordinate system. Here, the location of each point is measured by three numbers representing coordinates along an *X*-, *Y*-, and *Z* axes mutually at right angles or 90°. The positive directions of the axes are defined using the right-handed rule, which states that if the thumb, the fore-finger, and the middle-finger of the right-hand are stretched so that they are mutually at right angles to each other, then the thumb denotes the positive direction of the *X*-axis, the fore-finger denotes the positive direction of the *Y*-axis, and the middle-finger denotes the positive direction of the *Z*-axis (O'Rourke, 2003). Angles are considered positive when measured in the counter-clockwise (CCW) direction observed from the tip of a primary axis and negative in the clockwise (CW) direction. Along with the three primary axes, there are three primary planes which together divide the coordinate space into eight octants (see Figure 7.1). The *X–Y* plane (shown in green) is located at $Z = 0$ and divides the space into top and bottom segments, the *Y–Z* plane (shown in red) is located at $X = 0$ and divides the space into left and right segments, the *X–Z* plane (shown in yellow) is located at $Y = 0$ and divides the space into front and rear segments. The three primary axes and the three primary planes meet at the origin.

7.2 TRANSLATION

A translation operation changes the location of points and graphic objects by adding translation factors (t_x, t_y, t_z) to the *X*-, *Y*-, *Z*-coordinates of each point of the object (Hearn and Baker, 1996), (Shirley, 2002). If the factors are positive, the object moves along the positive direction of coordinate axes, if they are negative, the object moves along the negative direction.

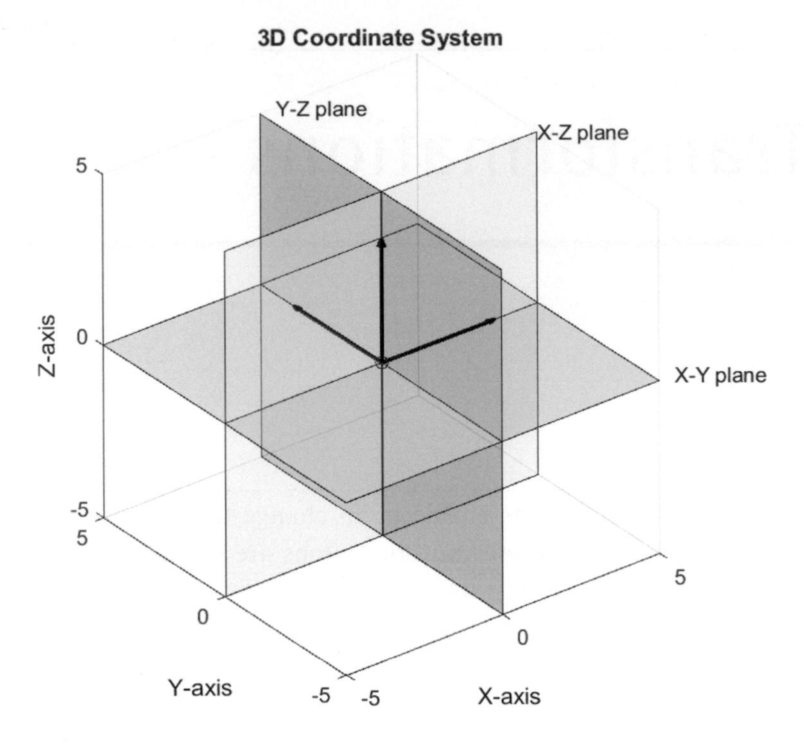

FIGURE 7.1 3D coordinate system.

A point $P(x_1, y_1, z_1)$ when translated by amounts (t_x, t_y, t_z) has new coordinates $Q(x_2, y_2, z_2)$ given by:

$$\begin{bmatrix} x_2 \\ y_2 \\ z_2 \\ 1 \end{bmatrix} = \begin{bmatrix} 1 & 0 & 0 & t_x \\ 0 & 1 & 0 & t_y \\ 0 & 0 & 1 & t_z \\ 0 & 0 & 0 & 1 \end{bmatrix} \begin{bmatrix} x_1 \\ y_1 \\ z_1 \\ 1 \end{bmatrix} \tag{7.1}$$

The inverse transformation is computed by taking the inverse of the matrix as below:

$$\begin{bmatrix} x_1 \\ y_1 \\ z_1 \\ 1 \end{bmatrix} = \begin{bmatrix} 1 & 0 & 0 & t_x \\ 0 & 1 & 0 & t_y \\ 0 & 0 & 1 & t_z \\ 0 & 0 & 0 & 1 \end{bmatrix}^{-1} \begin{bmatrix} x_2 \\ y_2 \\ z_2 \\ 1 \end{bmatrix} \tag{7.2}$$

It can be verified that the inverse of the matrix is equal to the negative of the arguments.

$$\begin{bmatrix} 1 & 0 & 0 & t_x \\ 0 & 1 & 0 & l_y \\ 0 & 0 & 1 & t_z \\ 0 & 0 & 0 & 1 \end{bmatrix}^{-1} = \begin{bmatrix} 1 & 0 & 0 & -t_x \\ 0 & 1 & 0 & -t_y \\ 0 & 0 & 1 & -t_z \\ 0 & 0 & 0 & 1 \end{bmatrix} \tag{7.3}$$

Symbolically, if T denotes the forward translation operation with arguments (t_x, t_y) and T' denotes the reverse translation then the above can be written as:

$$T'(t_x, t_y, t_z) = T(-t_x, -t_y, -t_z)$$

As before, this is the convention followed throughout this book i.e. the operations themselves would be denoted by single letters such as *T, S, R*, and so on for translation, scaling, and rotation while a specific matrix would be denoted with a letter with a subscript e.g. T_1. For example:

$$T_1 = T(3, -4, 5) = \begin{bmatrix} 1 & 0 & 0 & 3 \\ 0 & 1 & 0 & -4 \\ 0 & 0 & 1 & 5 \\ 0 & 0 & 0 & 1 \end{bmatrix}$$

Example 7.1

A cube with center at origin and vertices at (–1, 1, 1), (1, 1, 1), (1, –1, 1), (–1, –1, 1), (–1, 1, –1), (1, 1, –1), (1, –1, –1), and (–1, –1, –1) is translated by amounts (–2, –1, 3). Find its new vertices.

$$\text{Original coordinate matrix: } C = \begin{bmatrix} -1 & 1 & 1 & -1 & -1 & 1 & 1 & -1 \\ 1 & 1 & -1 & -1 & 1 & 1 & -1 & -1 \\ 1 & 1 & 1 & 1 & -1 & -1 & -1 & -1 \\ 1 & 1 & 1 & 1 & 1 & 1 & 1 & 1 \end{bmatrix}$$

$$\text{Translation matrix: } T_1 = T(-2, -1, 3) = \begin{bmatrix} 1 & 0 & 0 & -2 \\ 0 & 1 & 0 & -1 \\ 0 & 0 & 1 & 3 \\ 0 & 0 & 0 & 1 \end{bmatrix}$$

From Equation (7.1), new coordinate matrix:

$$D = T_1 * C = \begin{bmatrix} -3 & -1 & -1 & -3 & -3 & -1 & -1 & -3 \\ 0 & 0 & -2 & -2 & 0 & 0 & -2 & -2 \\ 4 & 4 & 4 & 4 & 2 & 2 & 2 & 2 \\ 1 & 1 & 1 & 1 & 1 & 1 & 1 & 1 \end{bmatrix}$$

New coordinates are (–3, 0, 4), (–1, 0, 4), (–1, –2, 4), (–3, –2, 4), (–3, 0, 2), (–1, 0, 2), (–1, –2, 2), and (–3, –2, 2) (Figure 7.2).

MATLAB® Code 7.1

```
clear all; clc;

p1 = [-1,1,1];
p2 = [1,1,1];
p3 = [1,-1,1];
p4 = [-1,-1,1];
p5 = [-1,1,-1];
p6 = [1,1,-1];
p7 = [1,-1,-1];
p8 = [-1,-1,-1];
C = [p1' p2' p3' p4' p5' p6' p7' p8' ;
    1 1 1 1 1 1 1 1 ]
tx = -2; ty = -1; tz = 3;
T1 = [1 0 0 tx ; 0 1 0 ty ; 0 0 1 tz ; 0 0 0 1];
D  = T1*C

fprintf('New vertices : \n');
for i=1:8
    fprintf('(%.2f, %.2f, %.2f) \n', D(1,i), D(2,i), D(3,i));
end;

%plotting
C = [p1' p2' p3' p4' p1' p5' p6' p7' p8' p5' p8' p4' p3' p7' p6' p2' ;
    1 1 1 1 1 1 1 1 1 1 1 1 1 1 1 1];
D  = T1*C;

plot3(C(1,:), C(2,:), C(3,:), 'b'); hold on;
plot3(D(1,:), D(2,:), D(3,:), 'r');
xlabel('x'); ylabel('y'); zlabel('z');
legend('original', 'new'); axis equal; grid; hold off;
```

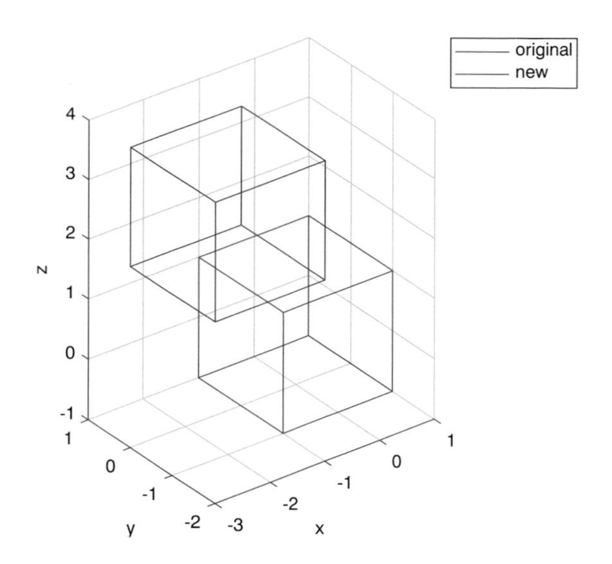

FIGURE 7.2 Plot for Example 7.1.

7.3 SCALING

A scaling operation alters the size of graphic objects by multiplying the X-, Y-, Z-coordinates of each point of the object by scaling factors s_x, s_y, s_z. If scaling factors are less than 1, they reduce the size of the object, if they are more than 1, they increase the size and if they are equal to 1 they keep the size unaltered (Hearn and Baker, 1996), (Shirley, 2002). If the factors are positive, the size increases along the positive direction of coordinate axes, if they are negative, the size increases along the negative direction. If all factors are equal then scaling is uniform otherwise non-uniform.

A point $P(x_1, y_1, z_1)$ when scaled by amounts (s_x, s_y, s_z) has new coordinates $Q(x_2, y_2, z_2)$ given by:

$$\begin{bmatrix} x_2 \\ y_2 \\ z_2 \\ 1 \end{bmatrix} = \begin{bmatrix} s_x & 0 & 0 & 0 \\ 0 & s_y & 0 & 0 \\ 0 & 0 & s_z & 0 \\ 0 & 0 & 0 & 1 \end{bmatrix} \begin{bmatrix} x_1 \\ y_1 \\ z_1 \\ 1 \end{bmatrix} \tag{7.4}$$

It can be verified that the inverse of the matrix is equal to the reciprocal of the arguments.

$$\begin{bmatrix} s_x & 0 & 0 & 0 \\ 0 & s_y & 0 & 0 \\ 0 & 0 & s_z & 0 \\ 0 & 0 & 0 & 1 \end{bmatrix}^{-1} = \begin{bmatrix} \frac{1}{s_x} & 0 & 0 & 0 \\ 0 & \frac{1}{s_y} & 0 & 0 \\ 0 & 0 & \frac{1}{s_z} & 0 \\ 0 & 0 & 0 & 1 \end{bmatrix} \tag{7.5}$$

Symbolically: $S'(s_x, s_y, s_z) = S\left(\frac{1}{s_x}, \frac{1}{s_y}, \frac{1}{s_z}\right)$

The scaling operation pertaining to the above matrix is always with respect to the origin.

Example 7.2

A cube with center at origin and vertices at (–1, 1, 1), (1, 1, 1), (1, –1, 1), (–1, –1, 1), (–1, 1, –1), (1, 1, –1), (1, –1, –1), and (–1, –1, –1) is scaled by amounts (2, 1, 3). Find its new vertices.

$$\text{Original coordinate matrix: } C = \begin{bmatrix} -1 & 1 & 1 & -1 & -1 & 1 & 1 & -1 \\ 1 & 1 & -1 & -1 & 1 & 1 & -1 & -1 \\ 1 & 1 & 1 & 1 & -1 & -1 & -1 & -1 \\ 1 & 1 & 1 & 1 & 1 & 1 & 1 & 1 \end{bmatrix}$$

$$\text{Scaling matrix: } S_1 = S(2, 1, 3) = \begin{bmatrix} 2 & 0 & 0 & 0 \\ 0 & 1 & 0 & 0 \\ 0 & 0 & 3 & 0 \\ 0 & 0 & 0 & 1 \end{bmatrix}$$

From Equation (7.4), new coordinate matrix:

$$D = S_1 * C = \begin{bmatrix} -2 & 2 & 2 & -2 & -2 & 2 & 2 & -2 \\ 1 & 1 & -1 & -1 & 1 & 1 & -1 & -1 \\ 3 & 3 & 3 & 3 & -3 & -3 & -3 & -3 \\ 1 & 1 & 1 & 1 & 1 & 1 & 1 & 1 \end{bmatrix}$$

New coordinates are (–2, 1, 3), (2, 1, 3), (2, –1, 3), (–2, –1, 3), (–2, 1, –3), (2, 1, –3), (2, –1, –3), and (–2, –1, –3) (Figure 7.3).

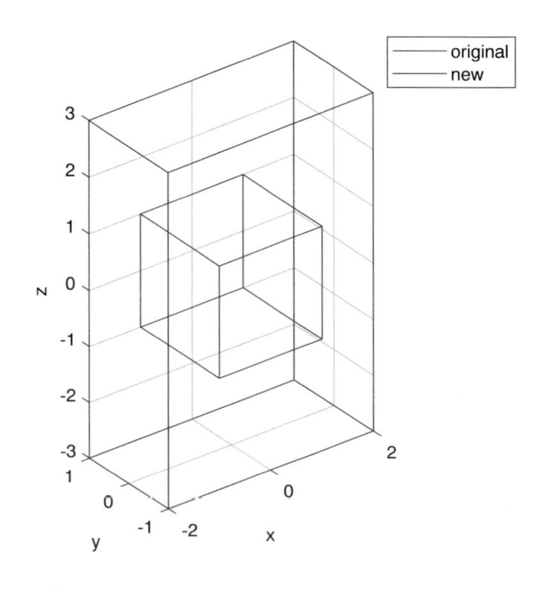

FIGURE 7.3 Plot for Example 7.2.

MATLAB Code 7.2

```
clear all; clc;

p1 = [-1,1,1];
p2 = [1,1,1];
p3 = [1,-1,1];
p4 = [-1,-1,1];
p5 = [-1,1,-1];
p6 = [1,1,-1];
p7 = [1,-1,-1];
p8 = [-1,-1,-1];
C = [p1' p2' p3' p4' p5' p6' p7' p8' ;
    1 1 1 1 1 1 1 1 ]

sx = 2; sy = 1; sz = 3;
S1 =[sx 0 0 0 ; 0 sy 0 0 ; 0 0 sz 0 ; 0 0 0 1];
D = S1*C

fprintf('New vertices : \n');
for i=1:8
    fprintf('(%.2f, %.2f, %.2f) \n', D(1,i), D(2,i), D(3,i));
end;

%plotting
C = [p1' p2' p3' p4' p1' p5' p6' p7' p8' p5' p8' p4' p3' p7' p6' p2' ;
    1 1 1 1 1 1 1 1 1 1 1 1 1 1 1 1];
D  = S1*C;
plot3(C(1,:), C(2,:), C(3,:), 'b'); hold on;
plot3(D(1,:), D(2,:), D(3,:), 'r');
xlabel('x'); ylabel('y'); zlabel('z');
legend('original', 'new'); axis equal; grid; hold off;
```

7.4 ROTATION

A rotation operation moves a point along the circumference of a circle centered at the origin and radius equal to the distance of the point from the origin. Rotation is considered positive when it is in the CCW direction and negative along the CW direction. Unlike the 2D case where there is a single rotation matrix, for 3D there are three different rotation matrices depending on which of the three primary axes is the axis of rotation (Hearn and Baker, 1996), (Shirley, 2002).

Rotation about X-axis:

$$R_x(\theta)=\begin{bmatrix} 1 & 0 & 0 & 0 \\ 0 & \cos\theta & -\sin\theta & 0 \\ 0 & \sin\theta & \cos\theta & 0 \\ 0 & 0 & 0 & 1 \end{bmatrix} \tag{7.6}$$

Rotation about Y-axis:

$$R_y(\theta)=\begin{bmatrix} \cos\theta & 0 & \sin\theta & 0 \\ 0 & 1 & 0 & 0 \\ -\sin\theta & 0 & \cos\theta & 0 \\ 0 & 0 & 0 & 1 \end{bmatrix} \tag{7.7}$$

Rotation about Z-axis:

$$R_z(\theta)=\begin{bmatrix} \cos\theta & -\sin\theta & 0 & 0 \\ \sin\theta & \cos\theta & 0 & 0 \\ 0 & 0 & 1 & 0 \\ 0 & 0 & 0 & 1 \end{bmatrix} \tag{7.8}$$

Rotation by default is always with respect to the origin around any of the three primary axes.

Example 7.3

A cube with center at origin and vertices at (–1, 1, 1), (1, 1, 1), (1, –1, 1), (–1, –1, 1), (–1, 1, –1), (1, 1, –1), (1, –1, –1), and (–1, –1, –1) is rotated by 45° about the X-axis. Find its new vertices.

$$\text{Original coordinate matrix: } C=\begin{bmatrix} -1 & 1 & 1 & -1 & -1 & 1 & 1 & -1 \\ 1 & 1 & -1 & -1 & 1 & 1 & -1 & -1 \\ 1 & 1 & 1 & 1 & -1 & -1 & -1 & -1 \\ 1 & 1 & 1 & 1 & 1 & 1 & 1 & 1 \end{bmatrix}$$

Rotation matrix:

$$R_x(45)=\begin{bmatrix} 1 & 0 & 0 & 0 \\ 0 & \cos 45 & -\sin 45 & 0 \\ 0 & \sin 45 & \cos 45 & 0 \\ 0 & 0 & 0 & 1 \end{bmatrix}=\begin{bmatrix} 1 & 0 & 0 & 0 \\ 0 & 0.71 & -0.71 & 0 \\ 0 & 0.71 & 0.71 & 0 \\ 0 & 0 & 0 & 1 \end{bmatrix}$$

New coordinate matrix:

$$D=R_x(45)^{*}C=\begin{bmatrix} -1 & 1 & 1 & -1 & -1 & 1 & 1 & -1 \\ 0 & 0 & -1.41 & -1.41 & 1.41 & 1.41 & 0 & 0 \\ 1.41 & 1.41 & 0 & 0 & 0 & 0 & -1.41 & -1.41 \\ 1 & 1 & 1 & 1 & 1 & 1 & 1 & 1 \end{bmatrix}$$

New vertices: (–1.00, 0, 1.41), (1.00, 0, 1.41), (1.00, –1.41, 0), (–1.00, –1.41, 0), (–1.00, 1.41, 0), (1.00, 1.41, 0), (1.00, 0, –1.41), and (–1.00, 0, –1.41) (Figure 7.4)

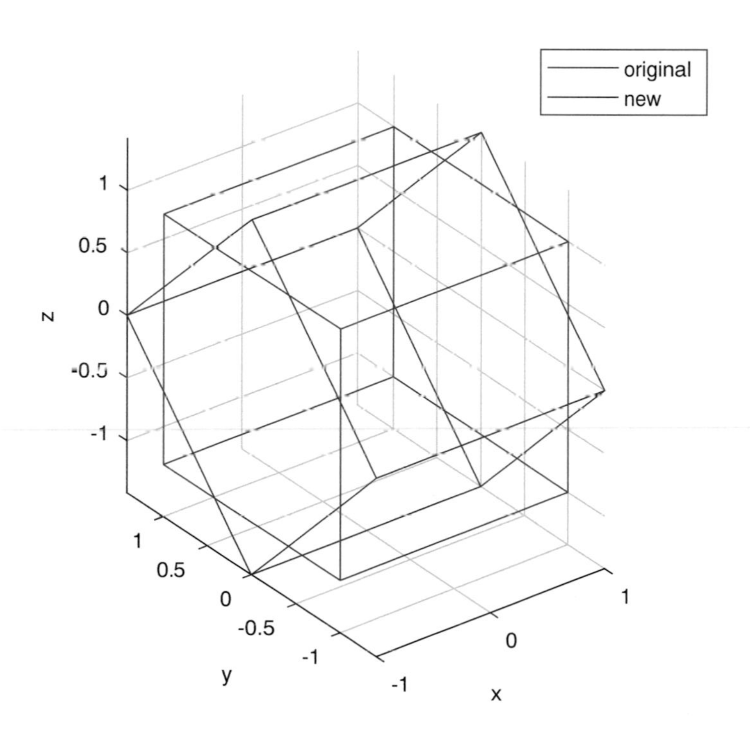

FIGURE 7.4 Plot for Example 7.3.

MATLAB Code 7.3

```
clear all; clc;

p1 = [-1,1,1];
p2 = [1,1,1];
p3 = [1,-1,1];
p4 = [-1,-1,1];
p5 = [-1,1,-1];
p6 = [1,1,-1];
p7 = [1,-1,-1];
p8 = [-1,-1,-1];
C = [p1' p2' p3' p4' p5' p6' p7' p8' ;
    1 1 1 1 1 1 1 1 ];

A = deg2rad(45);
R1 = [1 0 0 0 ; 0 cos(A) -sin(A) 0 ; 0 sin(A) cos(A) 0 ; 0 0 0 1];
D = R1*C;

fprintf('New vertices : \n');
for i=1:8
    fprintf('(%.2f, %.2f, %.2f) \n', D(1,i), D(2,i), D(3,i));
end;

%plotting

C = [p1' p2' p3' p4' p1' p5' p6' p7' p8' p5' p8' p4' p3' p7' p6' p2' ;
    1 1 1 1 1 1 1 1 1 1 1 1 1 1 1 1];
D = R1*C;

plot3(C(1,:), C(2,:), C(3,:), 'b'); hold on;
plot3(D(1,:), D(2,:), D(3,:), 'r');
xlabel('x'); ylabel('y'); zlabel('z');
legend('original', 'new'); axis equal; grid; hold off;
```

7.5 FIXED-POINT SCALING

As mentioned previously, a scaling operation is by default about the origin. For a general scaling operation with respect to a fixed point (x_f, y_f, z_f), the following steps are taken:

- Translate object so that fixed point moves to origin: $T_1 = T(-x_f, -y_f, -z_f)$
- Scale object about origin: $S_1 = S(s_x, s_y, s_z)$
- Reverse translate the object to original location: $T_2 = T(x_f, y_f, z_f)$
- Compute composite transformation matrix: $M = T_2 * S_1 * T_1$

Example 7.4

A cube with center at origin and vertices at (–1, 1, 1), (1, 1, 1), (1, –1, 1), (–1, –1, 1), (–1, 1, –1), (1, 1, –1), (1, –1, –1), and (–1, –1, –1) is scaled by amounts (2, 1, 3) with respect to its vertex (–1, –1, –1). Find its new vertices.

$$\text{Original coordinate matrix: } C - \begin{bmatrix} -1 & 1 & 1 & -1 & -1 & 1 & 1 & -1 \\ 1 & 1 & -1 & -1 & 1 & 1 & -1 & -1 \\ 1 & 1 & 1 & 1 & -1 & -1 & -1 & -1 \\ 1 & 1 & 1 & 1 & 1 & 1 & 1 & 1 \end{bmatrix}$$

$$\text{Forward translation: } T_1 = T(1, 1, 1) = \begin{bmatrix} 1 & 0 & 0 & 1 \\ 0 & 1 & 0 & 1 \\ 0 & 0 & 1 & 1 \\ 0 & 0 & 0 & 1 \end{bmatrix}$$

$$\text{Scaling: } S_1 = S(2, 1, 3) = \begin{bmatrix} 2 & 0 & 0 & 0 \\ 0 & 1 & 0 & 0 \\ 0 & 0 & 3 & 0 \\ 0 & 0 & 0 & 1 \end{bmatrix}$$

$$\text{Reverse translation: } T_2 = T(-1, -1, -1) = \begin{bmatrix} 1 & 0 & 0 & -1 \\ 0 & 1 & 0 & -1 \\ 0 & 0 & 1 & -1 \\ 0 & 0 & 0 & 1 \end{bmatrix}$$

$$\text{Composite transformation: } M = T_2 \cdot S_1 \cdot T_1 = \begin{bmatrix} 2 & 0 & 0 & 1 \\ 0 & 1 & 0 & 0 \\ 0 & 0 & 3 & 2 \\ 0 & 0 & 0 & 1 \end{bmatrix}$$

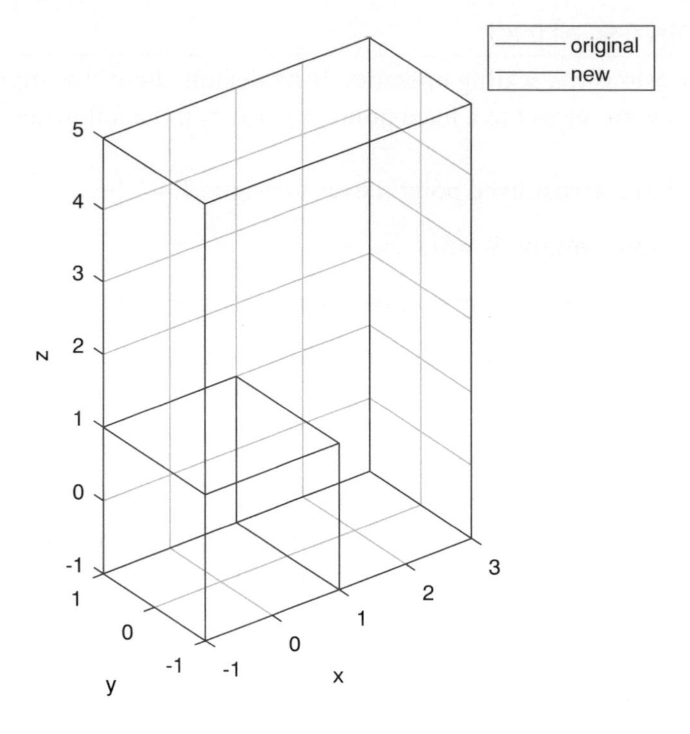

FIGURE 7.5 Plot for Example 7.4.

New coordinate matrix:

$$D = M \cdot C = \begin{bmatrix} -1 & 3 & 3 & -1 & -1 & 3 & 3 & -1 \\ 1 & 1 & -1 & -1 & 1 & 1 & -1 & -1 \\ 5 & 5 & 5 & 5 & -1 & -1 & -1 & -1 \\ 1 & 1 & 1 & 1 & 1 & 1 & 1 & 1 \end{bmatrix}$$

New vertices: (–1.00, 1.00, 5.00), (3.00, 1.00, 5.00), (3.00, –1.00, 5.00), (–1.00, –1.00, 5.00), (–1.00, 1.00, –1.00), (3.00, 1.00, –1.00), (3.00, –1.00, –1.00), and (–1.00, –1.00, –1.00) (Figure 7.5).

MATLAB Code 7.4

```
clear all; clc;

p1 = [-1,1,1];
p2 = [1,1,1];
p3 = [1,-1,1];
p4 = [-1,-1,1];
p5 = [-1,1,-1];
p6 = [1,1,-1];
p7 = [1,-1,-1];
p8 = [-1,-1,-1];
C = [p1' p2' p3' p4' p5' p6' p7' p8' ;
    1 1 1 1 1 1 1 1 ];

tx = 1; ty = 1; tz = 1;
T1 = [1 0 0 tx ; 0 1 0 ty ; 0 0 1 tz ; 0 0 0 1];
sx = 2; sy = 1; sz = 3;
S1 = [sx 0 0 0 ; 0 sy 0 0 ; 0 0 sz 0 ; 0 0 0 1];
T2 = inv(T1);
M = T2*S1*T1;
D  = M*C;

fprintf('New vertices : \n');
for i=1:8
    fprintf('(%.2f, %.2f, %.2f) \n', D(1,i), D(2,i), D(3,i));
end;

%plotting
C = [p1' p2' p3' p4' p1' p5' p6' p7' p8' p5' p8' p4' p3' p7' p6' p2' ;
    1 1 1 1 1 1 1 1 1 1 1 1 1 1 1 1];
D  = M*C;

plot3(C(1,:), C(2,:), C(3,:), 'b'); hold on;
plot3(D(1,:), D(2,:), D(3,:), 'r');
xlabel('x'); ylabel('y'); zlabel('z');
legend('original', 'new'); axis equal; grid; hold off;
```

7.6 FIXED-POINT ROTATION

As mentioned previously, a rotation operation is by default with respect to the origin. For a general rotation operation with respect to a fixed point (x_f, y_f, z_f), the following steps are taken:

- Translate object so that fixed point moves to origin: $T_1 = T(-x_f, -y_f, -z_f)$
- Rotate with respect to the origin about a primary axis: $R_1 = R_x(\theta)$ or $R_1 = R_y(\theta)$ or $R_1 = R_z(\theta)$
- Reverse translate the object to original location: $T_2 = T(x_f, y_f, z_f)$
- Compute composite transformation matrix: $M = T_2 \times R_1 \times T_1$

Example 7.5

A cube with center at origin and vertices at (–1, 1, 1), (1, 1, 1), (1, –1, 1), (–1, –1, 1), (–1, 1, –1), (1, 1, –1), (1, –1, –1), and (–1, –1, –1) is rotated by 45° with respect to its vertex (–1, –1, –1) around Z-axis. Find its new vertices.

$$\text{Original coordinate matrix: } C = \begin{bmatrix} -1 & 1 & 1 & -1 & -1 & 1 & 1 & -1 \\ 1 & 1 & -1 & -1 & 1 & 1 & -1 & -1 \\ 1 & 1 & 1 & 1 & -1 & -1 & -1 & -1 \\ 1 & 1 & 1 & 1 & 1 & 1 & 1 & 1 \end{bmatrix}$$

$$\text{Forward translation: } T_1 = T(1, 1, 1) = \begin{bmatrix} 1 & 0 & 0 & 1 \\ 0 & 1 & 0 & 1 \\ 0 & 0 & 1 & 1 \\ 0 & 0 & 0 & 1 \end{bmatrix}$$

$$\text{Rotation about } Z\text{-axis: } R_1 = R_z(\theta) = \begin{bmatrix} \cos 45 & -\sin 45 & 0 & 0 \\ \sin 45 & \cos 45 & 0 & 0 \\ 0 & 0 & 1 & 0 \\ 0 & 0 & 0 & 1 \end{bmatrix}$$

$$\text{Reverse translation: } T_2 = T(-1, -1, -1) = \begin{bmatrix} 1 & 0 & 0 & -1 \\ 0 & 1 & 0 & -1 \\ 0 & 0 & 1 & -1 \\ 0 & 0 & 0 & 1 \end{bmatrix}$$

Composite transformation: $M = T_2 \cdot R_1 \cdot T_1$
New coordinate matrix: $D = M \cdot C$

New vertices: (–2.41, 0.41, 1.00), (–1.00, 1.83, 1.00), (0.41, 0.41, 1.00), (–1.00, –1.00, 1.00), (–2.41, 0.41, –1.00), (–1.00, 1.83, –1.00), (0.41, 0.41, –1.00), and (–1.00, –1.00, –1.00) (Figure 7.6).

MATLAB Code 7.5

```
clear all; clc;
p1 = [-1,1,1];
p2 = [1,1,1];
p3 = [1,-1,1];
p4 = [-1,-1,1];
p5 = [-1,1,-1];
p6 = [1,1,-1];
p7 = [1,-1,-1];
p8 = [-1,-1,-1];
C = [p1' p2' p3' p4' p5' p6' p7' p8' ;
    1 1 1 1 1 1 1 1 ];
tx = 1; ty = 1; tz = 1;
T1 = [1 0 0 tx ; 0 1 0 ty ; 0 0 1 tz ; 0 0 0 1];
A = 45;
R1 = [cosd(A) -sind(A) 0 0 ; sind(A) cosd(A) 0 0 ; 0 0 1 0 ; 0 0 0 1];
T2 = inv(T1);
M = T2*R1*T1;
D  = M*C;
fprintf('New vertices : \n');
for i=1:8
    fprintf('(%.2f, %.2f, %.2f) \n', D(1,i), D(2,i), D(3,i));
end;
%plotting
C = [p1' p2' p3' p4' p1' p5' p6' p7' p8' p5' p8' p4' p3' p7' p6' p2' ;
    1 1 1 1 1 1 1 1 1 1 1 1 1 1 1 1];
D  = M*C;
plot3(C(1,:), C(2,:), C(3,:), 'b'); hold on;
plot3(D(1,:), D(2,:), D(3,:), 'r');
xlabel('x'); ylabel('y'); zlabel('z');
legend('original', 'new'); axis equal; grid; hold off;
```

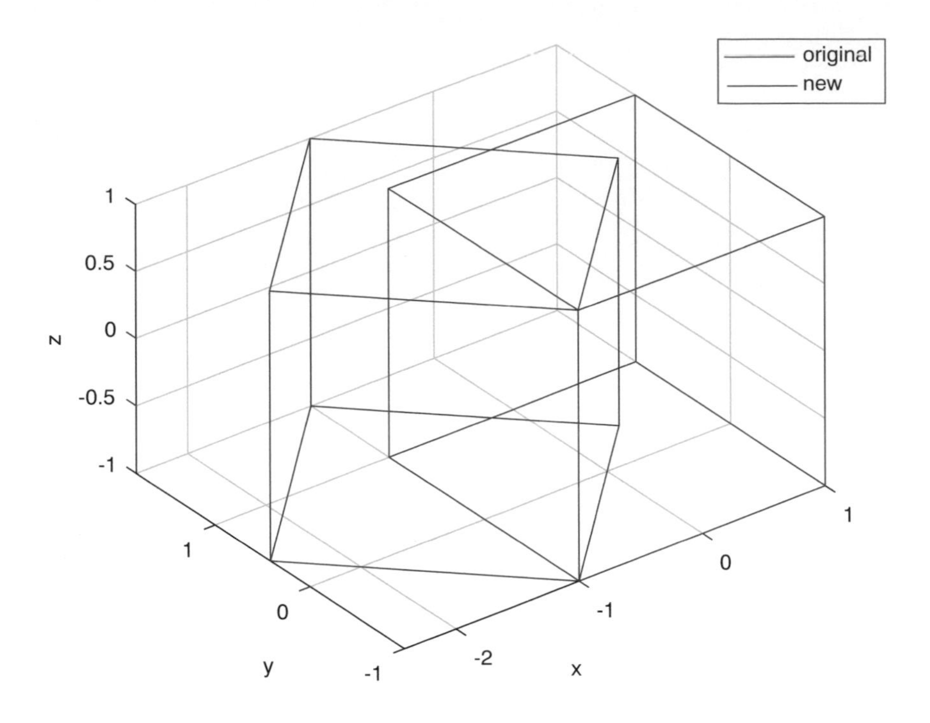

FIGURE 7.6 Plot for Example 7.5.

7.7 ROTATION PARALLEL TO PRIMARY AXES

Consider a straight line parallel to the Y-axis joining points $P(a, b, c)$ and $Q(a, 0, c)$ (see Figure 7.7). To derive the matrix for rotation by angle θ in the CCW direction about this line, the following steps are followed:

- Translate line such that Q coincides with origin: $T_1 = T(-a, 0, -c)$
- Rotate with respect to the origin around Y-axis by angle θ: $R_1 = R_y(\theta)$
- Reverse translate back to original location: $T_2 = T(a, 0, c)$
- Compute composite transformation: $M = T(a, 0, c) \cdot R_y(\theta) \cdot T(-a, 0, -c)$

Example 7.6

A point C(1, 1, 1) is to be rotated by 180° around a line parallel to the Y-axis joining points P(5, 2, 3) and Q(5, 0, 3). Find its new coordinates.

Original coordinate matrix: $C = [1, 1, 1, 1]^T$
Forward translation to coincide with Y-axis: $T_1 = T(-5, 0, -3)$
Rotation around Y-axis: $R_1 = R_y(180)$
Reverse translation to original location: $T_2 = T(5, 0, 3)$
Composite transformation: $M = T(5, 0, 3) \cdot R_y(180) \cdot T(-5, 0, -3)$
New coordinate matrix $D = \text{M} \cdot \text{C} = [9, 1, 5, 1]^T$
New coordinates: (9, 1, 5) (Figure 7.8)

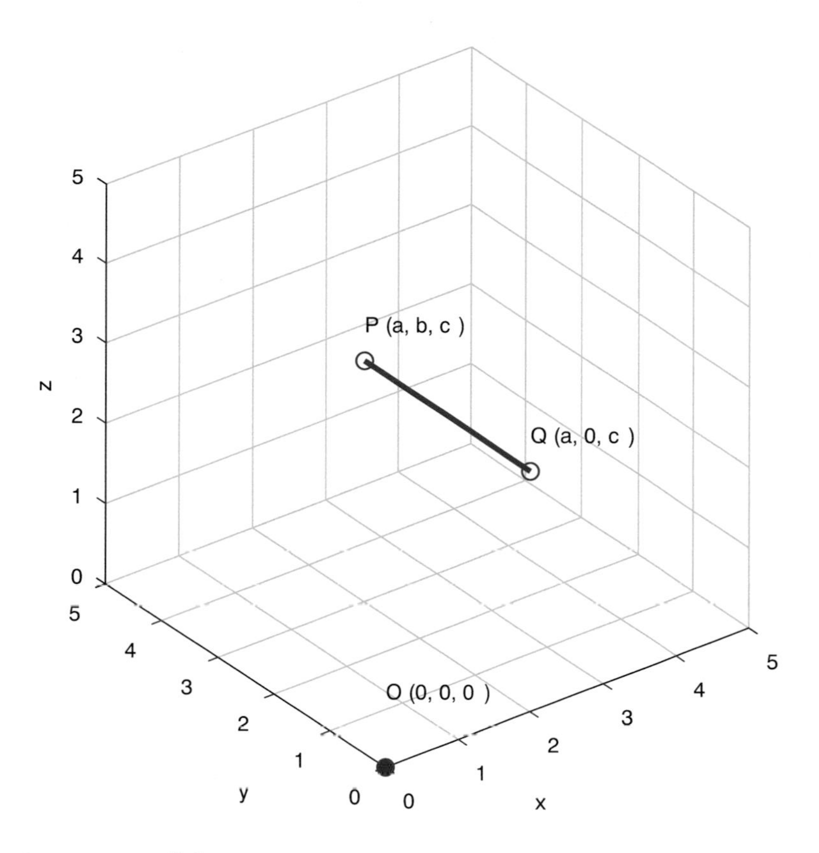

FIGURE 7.7 Rotation parallel to *Y*-axis.

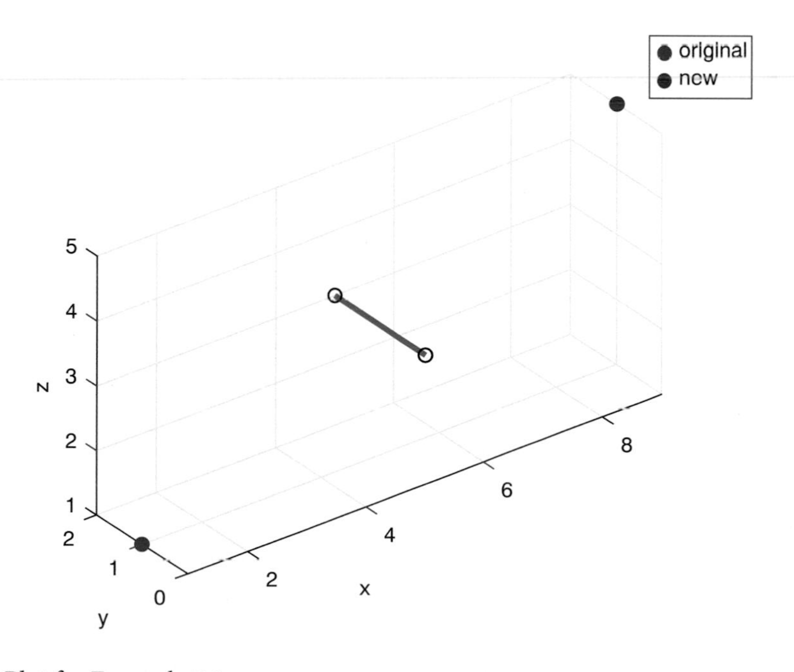

FIGURE 7.8 Plot for Example 7.6.

MATLAB Code 7.6

```
clear all; clc;
p = [1,1,1];
C = [p' ; 1];
P = [5, 2, 3]; Q = [5, 0, 3];
tx = -Q(1); ty = -Q(2); tz = -Q(3);
T1 = [1 0 0 tx ; 0 1 0 ty ; 0 0 1 tz ; 0 0 0 1];
A = deg2rad(180);
R1 = [cos(A), 0, sin(A), 0 ; 0, 1, 0, 0 ; -sin(A), 0, cos(A), 0 ; 0, 0, 0, 1];
T2 = inv(T1);
M = T2*R1*T1;
D  = M*C;
fprintf('New vertices : \n')
fprintf('(%.2f, %.2f, %.2f) \n', D(1,1), D(2,1), D(3,1));

%plotting
plot3(C(1,:), C(2,:), C(3,:), 'bo', 'MarkerFaceColor', 'b'); hold on;
plot3(D(1,:), D(2,:), D(3,:), 'ro', 'MarkerFaceColor', 'r'); grid;
line([5, 5], [2, 0], [3, 3], 'LineWidth', 2);
plot3(P(1), P(2), P(3), 'ko');
plot3(Q(1), Q(2), Q(3), 'ko');
xlabel('x'); ylabel('y'); zlabel('z');
legend('original', 'new'); axis equal; hold off;
```

NOTE

`line`: Draws a line from one point to another

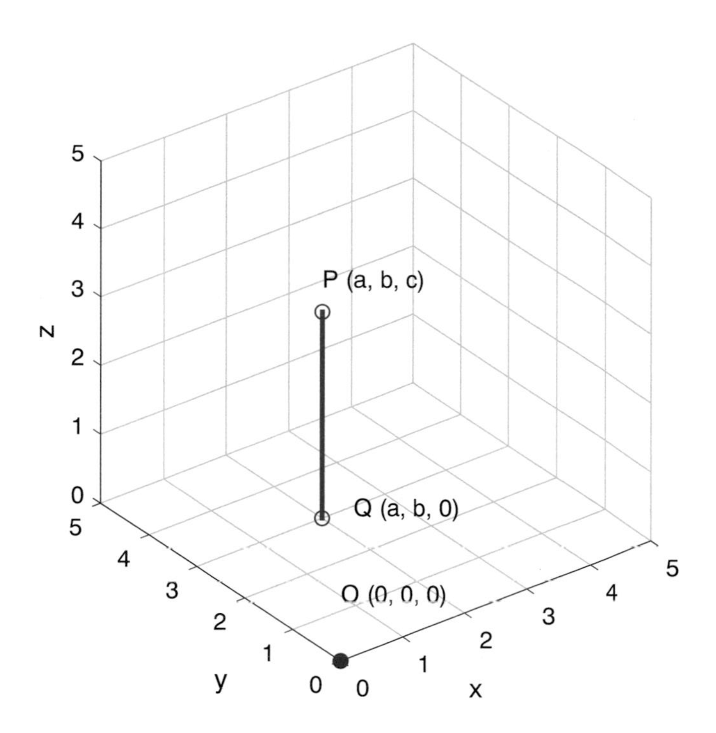

FIGURE 7.9 Rotation parallel to Z-axis.

Next consider a straight line parallel to the Z-axis joining points $P(a, b, c)$ and $Q(a, b, 0)$ (see Figure 7.9). To derive the matrix for rotation by angle θ in the CCW direction about this line, the following steps are followed:

- Translate line such that Q coincides with origin: $T_1 = T(-a, -b, 0)$
- Rotate with respect to the origin around Z-axis by angle θ: $R_1 = R_z(\theta)$
- Reverse translate back to original location: $T_2 = T(a, b, 0)$
- Compute composite transformation: $M = T(a, b, 0) \cdot R_z(\theta) \cdot T(-a, -b, 0)$

Example 7.7

A point C(1, 1, 1) is to be rotated by 180° around a line parallel to the Z-axis joining points P(5, 2, 3) and Q(5, 2, 0). Find its new coordinates.

Original coordinate matrix: $C = [1, 1, 1, 1]^T$
Forward translation to coincide with Y-axis: $T_1 = T(-5, -2, 0)$
Rotation around Y-axis: $R_1 = R_z(180)$
Reverse translation to original location: $T_2 = T(5, 2, 0)$
Composite transformation: $M = T(5, 2, 0) \cdot R_z(180) \cdot T(-5, -2, 0)$
New coordinate matrix $D = M \cdot C = [9, 3, 1, 1]^T$
New coordinates: (9, 3, 1) (Figure 7.10)

MATLAB Code 7.7

```
clear all; clc;
p = [1,1,1];
C = [p' ; 1];
P = [5, 2, 3]; Q = [5, 2, 0];
tx = -Q(1); ty = -Q(2); tz = -Q(3);
T1 = [1 0 0 tx ; 0 1 0 ty ; 0 0 1 tz ; 0 0 0 1];
A = 180;
R1 = [cosd(A) -sind(A) 0 0 ; sind(A) cosd(A) 0 0 ; 0 0 1 0 ; 0 0 0 1];
T2 = inv(T1);
M = T2*R1*T1;
D  = M*C;

fprintf('New vertices : \n')
fprintf('(%.2f, %.2f, %.2f) \n', D(1,1), D(2,1), D(3,1));

%plotting
plot3(C(1,:), C(2,:), C(3,:), 'bo', 'MarkerFaceColor', 'b'); hold on;
plot3(D(1,:), D(2,:), D(3,:), 'ro', 'MarkerFaceColor', 'r'); grid;
line([5, 5], [2, 2], [3, 0], 'LineWidth', 2);
plot3(P(1), P(2), P(3), 'ko');
plot3(Q(1), Q(2), Q(3), 'ko');
xlabel('x'); ylabel('y'); zlabel('z');
legend('original', 'new'); axis equal; hold off;
```

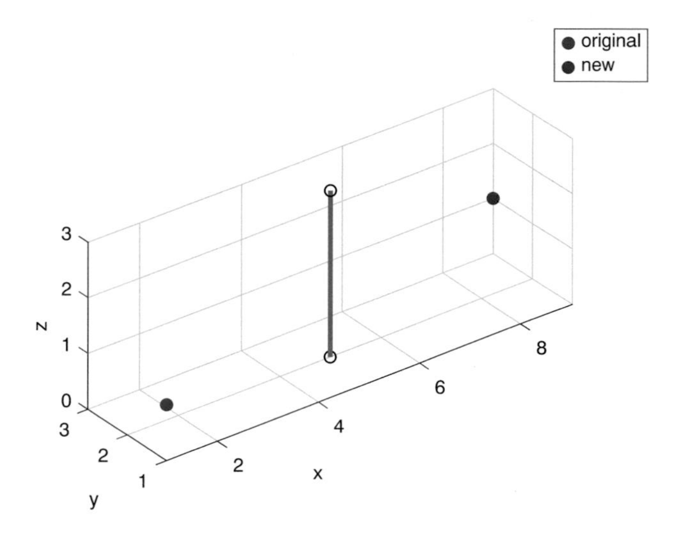

FIGURE 7.10 Plot for Example 7.7.

Finally, consider a straight line parallel to the X-axis joining points $P(a, b, c)$ and $Q(0, b, c)$ (see Figure 7.11). To derive the matrix for rotation by angle θ in the CCW direction about this line, the following steps are followed:

- Translate line such that Q coincides with origin: $T_1 = T(0, -b, -c)$
- Rotate with respect to the origin around X-axis by angle θ: $R_1 = R_x(\theta)$
- Reverse translate back to original location: $T_2 = T(0, b, c)$
- Compute composite transformation: $M = T(0, b, c) \cdot R_x(\theta) \cdot T(0, -b, -c)$

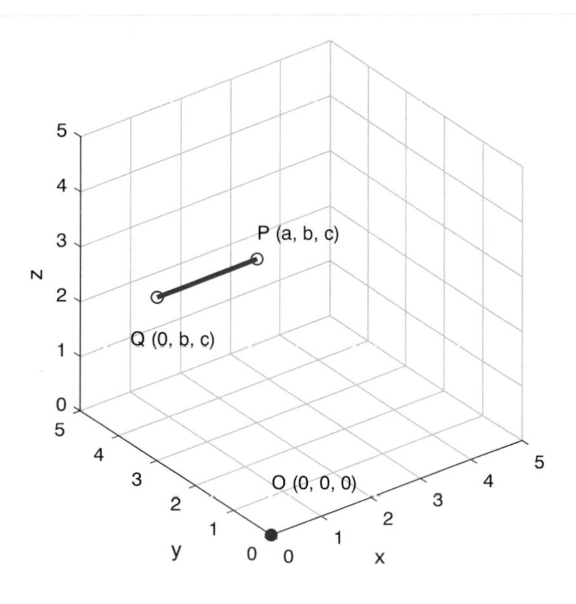

FIGURE 7.11 Rotation parallel to X-axis.

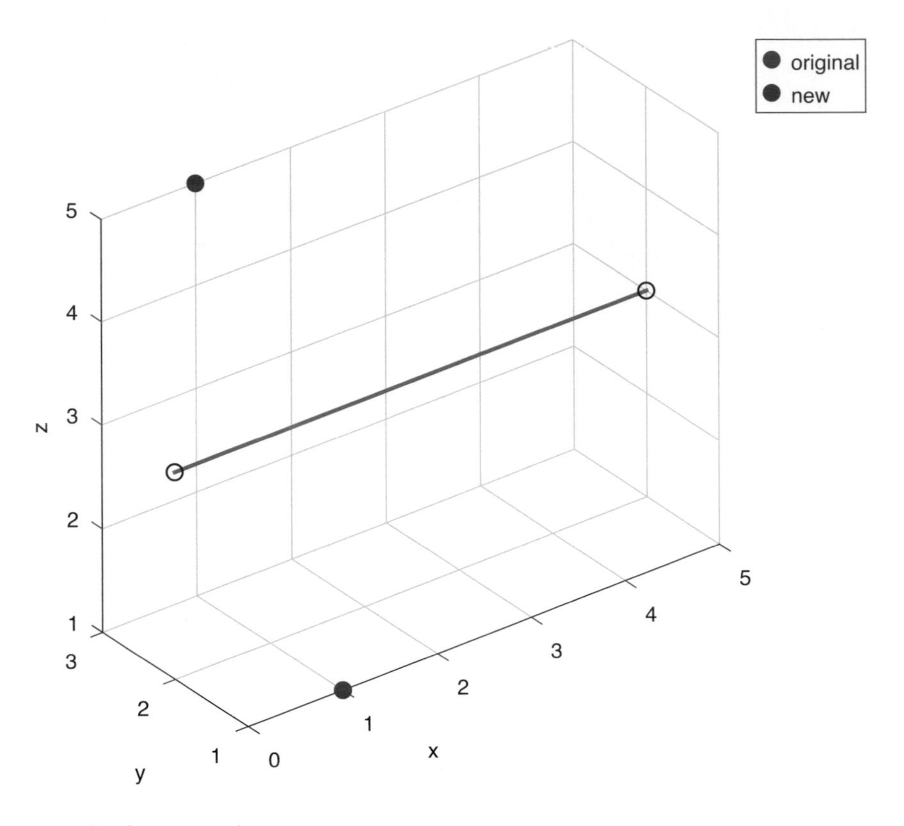

FIGURE 7.12 Plot for Example 7.8.

Example 7.8

A point C(1, 1, 1) is to be rotated by 180° around a line parallel to the X-axis joining points P(5, 2, 3) and Q(0, 2, 3). Find its new coordinates.

Original coordinate matrix: $C = [1, 1, 1, 1]^T$

Forward translation to coincide with *Y*-axis: $T_1 = T(0, -2, -3)$

Rotation around *Y*-axis: $R_1 = R_x(180)$

Reverse translation to original location: $T_2 = T(0, 2, 3)$

Composite transformation: $M = T(0, 2, 3) \cdot R_x(180) \cdot T(0, -2, -3)$

New coordinate matrix: $D = \text{M} \cdot \text{C} = [1, 3, 5, 1]^T$

New coordinates: (1, 3, 5) (Figure 7.12)

MATLAB Code 7.8

```
clear all; clc;
p = [1,1,1];
C = [p' ; 1];
P = [5, 2, 3]; Q = [0, 2, 3];
tx = -Q(1); ty = -Q(2); tz = -Q(3);
T1 = [1 0 0 tx ; 0 1 0 ty ; 0 0 1 tz ; 0 0 0 1];
A = deg2rad(180);
R1 = [1 0 0 0 ; 0 cos(A) -sin(A) 0 ; 0 sin(A) cos(A) 0 ; 0 0 0 1];
T2 = inv(T1);
M = T2*R1*T1;
D  = M*C;

fprintf('New vertices : \n');
fprintf('(%.2f, %.2f, %.2f) \n', D(1,1), D(2,1), D(3,1));

%plotting
plot3(C(1,:), C(2,:), C(3,:), 'bo', 'MarkerFaceColor', 'b'); hold on;
plot3(D(1,:), D(2,:), D(3,:), 'ro', 'MarkerFaceColor', 'r'); grid;
line([5, 0], [2, 2], [3, 3], 'LineWidth', 2);
plot3(P(1), P(2), P(3), 'ko');
plot3(Q(1), Q(2), Q(3), 'ko');
xlabel('x'); ylabel('y'); zlabel('z');
legend('original', 'new'); axis equal; hold off;
```

7.8 VECTOR ALIGNMENT (3D)

Consider a position vector $P = ai + bj + ck$ from origin O to point P, which is to be aligned along the positive Z-axis (see Figure 7.13). To derive the transformation matrix, the following steps are followed (Chakraborty, 2010):

Step 1:

Rotate vector OP by angle α CCW around X-axis to lie on X–Z plane at OQ: $R_x(\alpha)$

Construction: To find the value of α in terms of (a, b, c), the following set of constructions are done.

- Project $P(a, b, c)$ onto X–Z plane at $A(a, 0, c)$
- Project $A(a, 0, c)$ along X–Z plane onto Z-axis at $B(0, 0, c)$
- Project $P(a, b, c)$ onto Y–Z plane at $C(0, b, c)$
- Join BC and OC

We now observe the following:

Angle POQ between OP and OQ is α.

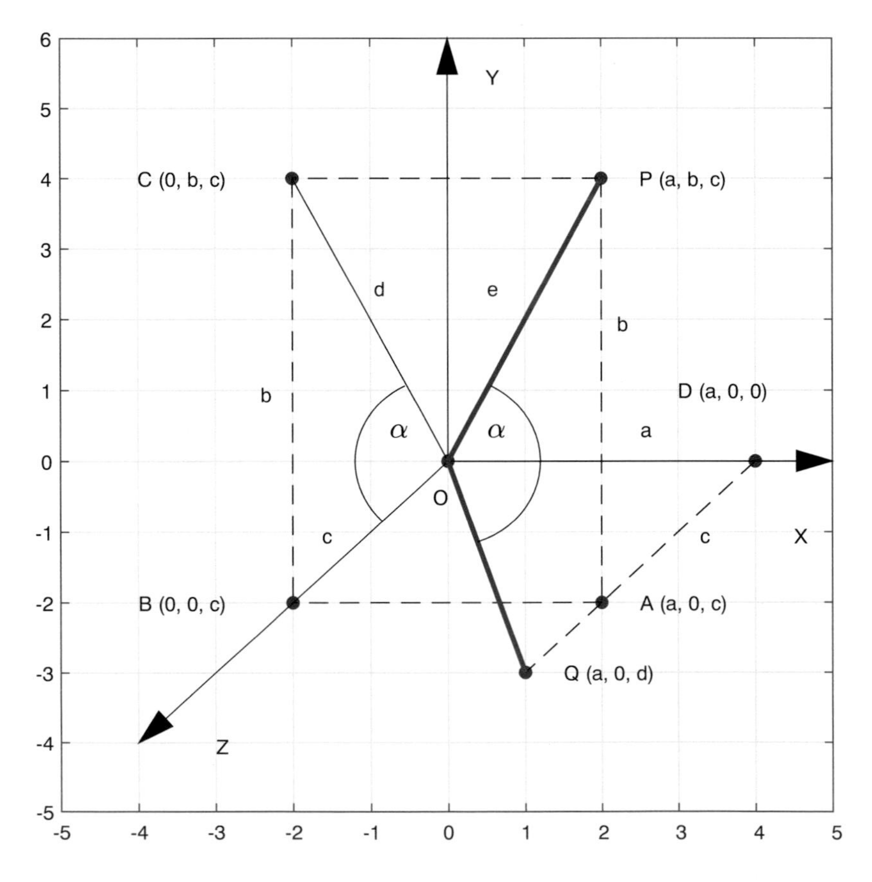

FIGURE 7.13 Vector alignment: Step 1.

Since, *OP* and *OQ* are projected parallel to *X–Z* plane onto the *Y–Z* plane at *OC* and *OB*, angle *BOC* is also α.

Since, *C* has coordinates (0, *b*, *c*) and *B* has coordinates (0, 0, *c*), length *BC* equals *b*.

Also since, *B* has coordinates (0, 0, *c*) length *OB* equals *c*.

Let length *OC* be $d=\sqrt{b^2+c^2}$

Thus in triangle *OBC*, $\cos(\alpha)=OB/OC=c/d$ and $\sin(\alpha)=BC/OC=b/d$

$$\text{Hence } R_x(\alpha)=\begin{bmatrix} 1 & 0 & 0 & 0 \\ 0 & \cos\alpha & -\sin\alpha & 0 \\ 0 & \sin\alpha & \cos\alpha & 0 \\ 0 & 0 & 0 & 1 \end{bmatrix}=\begin{bmatrix} 1 & 0 & 0 & 0 \\ 0 & c/d & -b/d & 0 \\ 0 & b/d & c/d & 0 \\ 0 & 0 & 0 & 1 \end{bmatrix}$$

$$\text{Coordinates of } Q=R_x(\alpha)\cdot P=\begin{bmatrix} 1 & 0 & 0 & 0 \\ 0 & c/d & -b/d & 0 \\ 0 & b/d & c/d & 0 \\ 0 & 0 & 0 & 1 \end{bmatrix}\begin{bmatrix} a \\ b \\ c \\ 1 \end{bmatrix}=\begin{bmatrix} a \\ 0 \\ d \\ 1 \end{bmatrix} \text{ i.e. } (a, 0, d)$$

Step 2:
Rotate vector *OQ* by angle φ CW around *Y*-axis to coincide with *Z*-axis at *R*: $R_y(-\varphi)$.
Note: CW rotation is considered negative (see Figure 7.14).
Construction: To find the value of φ in terms of (*a*, *b*, *c*), the following set of constructions are done.

- Project *Q*(*a*, 0, *d*) along *X–Z* plane onto *Z*-axis at *S*(0, 0, *d*)

We now observe the following:
Let length of *OP* = length of *OQ* = length of *OR* = $e=\sqrt{a^2+b^2+c^2}$.
Since coordinates of *Q* are (*a*, 0, *d*), length *QS* equals *a*.
Since coordinates of *S* are (0, 0, *d*), length *OS* equals *d*.
In triangle *OQS*, angle *QOS* equals. Thus $\sin(\varphi)=a/e$, $\cos(\varphi)=d/e$

$$\text{Hence } R_y(-\varphi)=\begin{bmatrix} \cos(-\varphi) & 0 & \sin(-\varphi) & 0 \\ 0 & 1 & 0 & 0 \\ -\sin(-\varphi) & 0 & \cos(-\varphi) & 0 \\ 0 & 0 & 0 & 1 \end{bmatrix}=\begin{bmatrix} d/e & 0 & -a/e & 0 \\ 0 & 1 & 0 & 0 \\ a/e & 0 & d/e & 0 \\ 0 & 0 & 0 & 1 \end{bmatrix}$$

$$\text{Coordinates of } R=R_y(-\varphi)\cdot Q=\begin{bmatrix} d/e & 0 & -a/e & 0 \\ 0 & 1 & 0 & 0 \\ a/e & 0 & d/e & 0 \\ 0 & 0 & 0 & 1 \end{bmatrix}\begin{bmatrix} a \\ 0 \\ d \\ 1 \end{bmatrix}=\begin{bmatrix} 0 \\ 0 \\ e \\ 1 \end{bmatrix} \text{ i.e. } (0, 0, e)$$

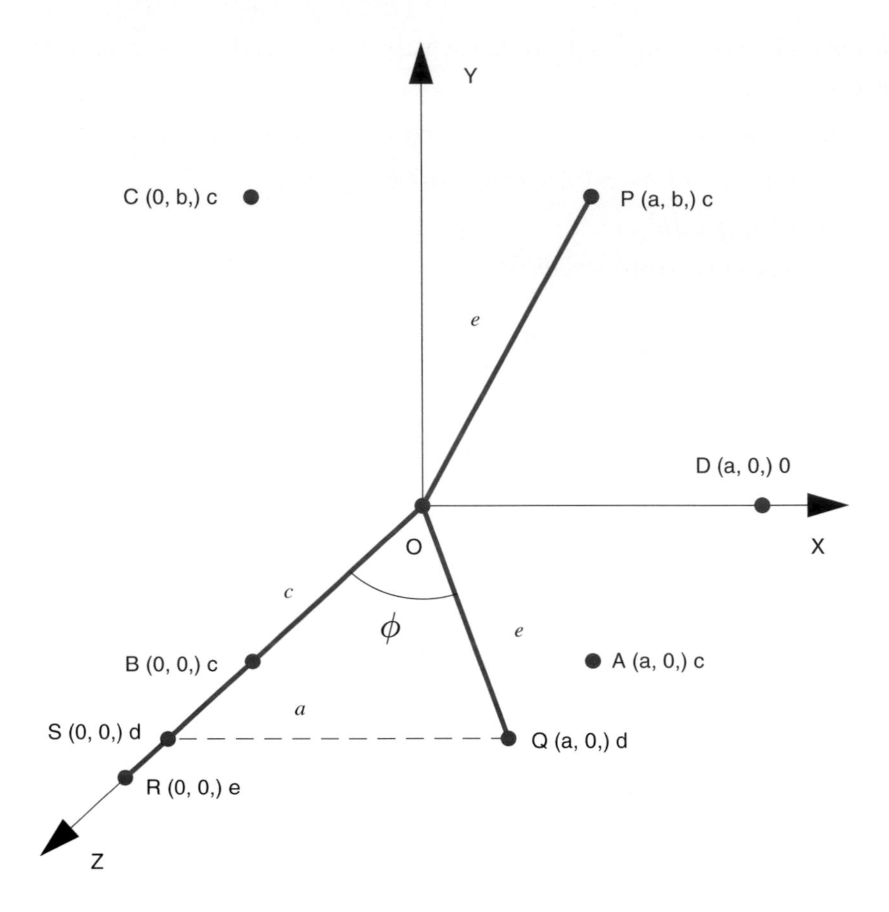

FIGURE 7.14 Vector alignment: Step 2.

This matches our expectation since the length of vector OP is $\sqrt{a^2+b^2+c^2}=e$ hence when it is aligned along the Z-axis the tip of the vector should have coordinates (0, 0, e).

Composite transformation: $M = R_y(-\varphi)\cdot R_x(\alpha)$

It can be verified that coordinates of the final vector OR can also be obtained by multiplying the original vector OP with the composite transformation matrix M i.e. $R = M \cdot P$. This is left as an exercise for the readers.

Example 7.9

Find the transformation for aligning the vector $P = 2i + j + 2k$ with the positive Z-axis. Also find the new vector after alignment.

For the given problem,

$$a=2,\ b=1,\ c=2,\ d=\sqrt{b^2+c^2}=2.2361,\ e=\sqrt{a^2+b^2+c^2}=3$$

$$\sin(\alpha)=\frac{b}{d}=0.4472,\quad \cos(\alpha)=\frac{c}{d}=0.8944$$

$$\sin(\varphi)=\frac{a}{e}=0.6667,\quad \cos(\varphi)=\frac{d}{e}=0.7454$$

$$R_x(\alpha)=\begin{bmatrix}1 & 0 & 0 & 0\\ 0 & \cos\alpha & -\sin\alpha & 0\\ 0 & \sin\alpha & \cos\alpha & 0\\ 0 & 0 & 0 & 1\end{bmatrix}=\begin{bmatrix}1 & 0 & 0 & 0\\ 0 & 0.8944 & -0.4472 & 0\\ 0 & 0.4472 & 0.8944 & 0\\ 0 & 0 & 0 & 1\end{bmatrix}$$

$$R_y(-\varphi)=\begin{bmatrix}\cos(-\varphi) & 0 & \sin(-\varphi) & 0\\ 0 & 1 & 0 & 0\\ -\sin(-\varphi) & 0 & \cos(-\varphi) & 0\\ 0 & 0 & 0 & 1\end{bmatrix}=\begin{bmatrix}0.7454 & 0 & -0.6667 & 0\\ 0 & 1 & 0 & 0\\ 0.6667 & 0 & 0.7454 & 0\\ 0 & 0 & 0 & 1\end{bmatrix}$$

$$\text{Composite transformation } M=R_y(-\varphi)\cdot R_x(\alpha)=\begin{bmatrix}0.7454 & -0.2981 & -0.5963 & 0\\ 0 & 0.8944 & -0.4472 & 0\\ 0.6667 & 0.3333 & 0.6667 & 0\\ 0 & 0 & 0 & 1\end{bmatrix}$$

Original coordinate matrix: $P=[2, 1, 2, 1]^T$
New coordinate matrix $Q=M\cdot C=[0, 0, 3, 1]^T$
New coordinates of Q: (0, 0, 3)
New vector is $Q=3k$

MATLAB Code 7.9

```
clear all; clc;

P = [2; 1; 2; 1];

a = P(1);
b = P(2);
c = P(3);

d = sqrt(b^2 + c^2);
A = asin(b/d);
A = acos(c/d);
R1 = [1 0 0 0; 0 cos(A) -sin(A) 0; 0 sin(A) cos(A) 0; 0 0 0 1];

e = sqrt(a^2 + d^2);
B = asin(a/e);
R2 = [cos(B) 0 -sin(B) 0; 0 1 0 0; sin(B) 0 cos(B) 0; 0 0 0 1];
fprintf('Transformation matrix : \n');
M = R2*R1
fprintf('New vector : \n');
Q = M*P
```

TABLE 7.1 Alignment of Vector $P = ai + bj + ck$ with Primary Axes (Here, $e = \sqrt{a^2 + b^2 + c^2}$)

Primary Axis	d	$\cos\alpha$	$\sin\alpha$	$\cos\varphi$	$\sin\varphi$	M
Z-axis	$d = \sqrt{b^2 + c^2}$	c/d	b/d	d/e	a/e	$R_y(-\varphi) \cdot R_x(\alpha)$
X-axis	$d = \sqrt{b^2 + c^2}$	c/d	b/d	a/e	d/e	$R_y(\varphi) \cdot R_x(\alpha)$
Y-axis	$d = \sqrt{a^2 + c^2}$	a/d	c/d	b/e	d/e	$R_z(\varphi) \cdot R_y(\alpha)$

In a similar fashion alignment with the X- and Y-axes can be likewise analyzed. These are left for the reader as exercises. The final results are summarized below in Table 7.1 for convenience.

NOTE

`asin`: calculates inverse sine in radians.

To account for negative rotations in CW direction do one of the following:
Either (a) put a negative sign in the angle argument for the rotation matrix e.g. $R_y(-\varphi)$,
Or (b) put a negative sign before the sine component e.g. $\sin\varphi = -a/e$,
but **not** both. Cosine components are not affected by the negative sign i.e. $\cos\varphi = \cos(-\varphi)$

Example 7.10

Find the transformation for aligning the following vectors with the positive X-axis. Also find the new vector after alignment. (a) $2i + j + 2k$ and (b) $2i - j - 2k$.

(a)

$$a = 2, b = 1, c = 2$$

$$d = \sqrt{b^2 + c^2} = 2.2361, \; e = \sqrt{a^2 + b^2 + c^2} = 3$$

$$\sin(\alpha) = b/d = 0.4472, \; \cos(\alpha) = c/d = 0.8944$$

$$\sin(\varphi) = d/e = 0.7454, \; \cos(\varphi) = a/e = 0.6667$$

$$M = R_y(\varphi) \cdot R_x(\alpha) = \begin{bmatrix} 0.6667 & 0.3333 & 0.6667 & 0 \\ 0 & 0.8944 & -0.4472 & 0 \\ -0.7454 & 0.2981 & 0.5963 & 0 \\ 0 & 0 & 0 & 1 \end{bmatrix}$$

Original coordinate matrix: $C = [2, 1, 2, 1]^T$
New coordinate matrix $D = M \cdot C = [3, 0, 0, 1]^T$
New vector is $Q = 3i$

(b)

$$a = 2, b = -1, c = -2$$

$$d = \sqrt{b^2 + c^2} = 2.2361; e = \sqrt{a^2 + b^2 + c^2} = 3$$

$$\sin(\alpha) = b/d = -0.4472, \cos(\alpha) = c/d = -0.8944$$

$$\sin(\varphi) = d/e = 0.7454, \cos(\varphi) = a/e = 0.6667$$

$$M = R_y(\varphi) \cdot R_x(\alpha) = \begin{bmatrix} 0.6667 & -0.3333 & -0.6667 & 0 \\ 0 & -0.8944 & 0.4472 & 0 \\ -0.7454 & -0.2981 & -0.5963 & 0 \\ 0 & 0 & 0 & 1 \end{bmatrix}$$

Original coordinate matrix: $C = [2, 1, 2, 1]^T$
New coordinate matrix $D = M \cdot C = [3, 0, 0, 1]^T$
New vector is $Q = 3i$

MATLAB Code 7.10

```
%(a)

clear all; clc;
P = [2 ; 1 ; 2 ; 1];
a = 2; b = 1; c = 2;
d = sqrt(b^2 + c^2); e = sqrt(a^2 + b^2 + c^2);
sinA = b/d; cosA = c/d;
sinB = d/e; cosB = a/e;
Rx = [1 0 0 0 ; 0 cosA -sinA 0 ; 0 sinA cosA 0 ; 0 0 0 1];
Ry = [cosB 0 sinB 0 ; 0 1 0 0 ; -sinB 0 cosB 0 ; 0 0 0 1];
fprintf('Transformation matrix : \n');
M = Ry*Rx
fprintf('New vector : \n');
Q = M*P

%(b)

clear all;
P = [2 ; -1 ; -2 ; 1];
a = 2; b = -1; c = -2;
d = sqrt(b^2 + c^2); e = sqrt(a^2 + b^2 + c^2);
sinA = b/d; cosA = c/d;
sinB = d/e; cosB = a/e;
Rx = [1 0 0 0 ; 0 cosA -sinA 0 ; 0 sinA cosA 0 ; 0 0 0 1];
Ry = [cosB 0 sinB 0 ; 0 1 0 0 ; -sinB 0 cosB 0 ; 0 0 0 1];
fprintf('Transformation matrix : \n');
M = Ry*Rx
fprintf('New vector : \n');
Q = M*P
```

7.9 ROTATION AROUND A VECTOR

The transformation matrix for rotation around a vector $P = ai + bj + ck$ by a specified angle, is derived by the following steps:

Step 1: Align the vector along a primary axis (see Section 7.8)

Step 2: Rotate around that primary axis by the specified angle (see Section 7.4)

Step 3: Reverse align vector to its original location

Example 7.11

A point P(1, 2, 3) is to be rotated around vector V = 12i + 3j + 4k by 90° in CCW direction. Find its new coordinates. Verify the result by aligning the vector with each of the three primary axes.

$$V = [12, 3, 4],\ P = [1, 2, 3],\ \theta = 90^\circ$$

Here, $a = 12$, $b = 3$, $c = 4$, $e = \sqrt{a^2 + b^2 + c^2} = 13$

Aligning vector V along X-axis:

$$d = \sqrt{b^2 + c^2} = 5$$

$$\sin\alpha = \frac{b}{d} = \frac{3}{5}, \cos\alpha = \frac{c}{d} = \frac{4}{5}, \sin\varphi = \frac{d}{e} = \frac{5}{13}, \cos\varphi = \frac{a}{e} = \frac{12}{13}$$

$$R_x(\alpha) = \begin{bmatrix} 1 & 0 & 0 & 0 \\ 0 & \cos\alpha & -\sin\alpha & 0 \\ 0 & \sin\alpha & \cos\alpha & 0 \\ 0 & 0 & 0 & 1 \end{bmatrix}, R_y(\varphi) = \begin{bmatrix} \cos\varphi & 0 & \sin\varphi & 0 \\ 0 & 1 & 0 & 0 \\ -\sin\varphi & 0 & \cos\varphi & 0 \\ 0 & 0 & 0 & 1 \end{bmatrix},$$

$$R_x(\theta) = \begin{bmatrix} 1 & 0 & 0 & 0 \\ 0 & \cos\theta & -\sin\theta & 0 \\ 0 & \sin\theta & \cos\theta & 0 \\ 0 & 0 & 0 & 1 \end{bmatrix}$$

$$M = R_x(-\alpha) \cdot R_y(-\varphi) \cdot R_x(\theta) \cdot R_y(\varphi) \cdot R_x(\alpha)$$

$$Q = M \cdot P \rightarrow (2.2071, -1.9290, 2.3254)$$

Aligning vector V along Y-axis:

$$d = \sqrt{a^2 + c^2} = 4\sqrt{10}$$

$$\sin\alpha = \frac{c}{d} = \frac{1}{\sqrt{10}}, \cos\alpha = \frac{a}{d} = \frac{3}{\sqrt{10}}, \sin\varphi = \frac{d}{e} = \frac{4\sqrt{10}}{13}, \cos\varphi = \frac{b}{e} = \frac{3}{13}$$

$$R_y(\alpha) = \begin{bmatrix} \cos\alpha & 0 & \sin\alpha & 0 \\ 0 & 1 & 0 & 0 \\ -\sin\alpha & 0 & \cos\alpha & 0 \\ 0 & 0 & 0 & 1 \end{bmatrix}, R_z(\varphi) = \begin{bmatrix} \cos\varphi & -\sin\varphi & 0 & 0 \\ \sin\varphi & \cos\varphi & 0 & 0 \\ 0 & 0 & 1 & 0 \\ 0 & 0 & 0 & 1 \end{bmatrix},$$

$$R_y(\theta) = \begin{bmatrix} \cos\theta & 0 & \sin\theta & 0 \\ 0 & 1 & 0 & 0 \\ -\sin\theta & 0 & \cos\theta & 0 \\ 0 & 0 & 0 & 1 \end{bmatrix}$$

$$M = R_y(-\alpha)\cdot R_z(-\varphi)\cdot R_y(\theta)\cdot R_z(\varphi)\cdot R_y(\alpha)$$

$$Q = M\cdot P \rightarrow (2.2071, -1.9290, 2.3254)$$

Aligning vector V along Z-axis:

$$\sqrt{b^2 + c^2} = 5$$

$$\sin\alpha = \frac{b}{d} = \frac{3}{5}, \cos\alpha = \frac{c}{d} = \frac{4}{5}, \sin\varphi = -\frac{a}{e} = -\frac{12}{13}, \cos\varphi = \frac{d}{e} = \frac{5}{13}$$

$$R_x(\alpha) = \begin{bmatrix} 1 & 0 & 0 & 0 \\ 0 & \cos\alpha & -\sin\alpha & 0 \\ 0 & \sin\alpha & \cos\alpha & 0 \\ 0 & 0 & 0 & 1 \end{bmatrix}, R_y(-\varphi) = \begin{bmatrix} \cos\varphi & 0 & \sin\varphi & 0 \\ 0 & 1 & 0 & 0 \\ -\sin\varphi & 0 & \cos\varphi & 0 \\ 0 & 0 & 0 & 1 \end{bmatrix},$$

$$R_z(\theta) = \begin{bmatrix} \cos\theta & -\sin\theta & 0 & 0 \\ \sin\theta & \cos\theta & 0 & 0 \\ 0 & 0 & 1 & 0 \\ 0 & 0 & 0 & 1 \end{bmatrix}$$

$$M = R_x(-\alpha)\cdot R_y(-\varphi)\cdot R_z(\theta)\cdot R_y(\varphi)\cdot R_x(\alpha)$$

$$Q = M\cdot P \rightarrow (2.2071, -1.9290, 2.3254)$$

MATLAB Code 7.11

```
clear all; clc; format compact;
V = [12 ; 3 ; 4 ; 1];
P = [1 ; 2 ; 3 ; 1];
a = V(1); b = V(2); c = V(3);
e = sqrt(a^2 + b^2 + c^2);
C = pi/2;

fprintf('Aligning vector V along X-axis :\n');

d = sqrt(b^2 + c^2);
sinA = b/d; cosA = c/d;
sinB = d/e; cosB = a/e;
R1 = [1 0 0 0 ; 0 cosA -sinA 0 ; 0 sinA cosA 0 ; 0 0 0 1];
R2 = [cosB 0 sinB 0 ; 0 1 0 0 ; -sinB 0 cosB 0 ; 0 0 0 1];
Rx = [1 0 0 0 ; 0 cos(C) -sin(C) 0 ; 0 sin(C) cos(C) 0 ; 0 0 0 1];
R4 = inv(R2);
R5 = inv(R1);
Mx = R5*R4*Rx*R2*R1;
fprintf('New coordinates : \n');
Qx = Mx*P

fprintf('Aligning vector V along Z-axis :\n');

d = sqrt(b^2 + c^2);
sinA = b/d; cosA = c/d;
sinB = -a/e; cosB = d/e;
R1 = [1 0 0 0 ; 0 cosA -sinA 0 ; 0 sinA cosA 0 ; 0 0 0 1];
R2 = [cosB 0 sinB 0 ; 0 1 0 0 ; -sinB 0 cosB 0 ; 0 0 0 1];
Rz = [cos(C) -sin(C) 0 0 ; sin(C) cos(C) 0 0 ; 0 0 1 0 ; 0 0 0 1];
R4 = inv(R2);
R5 = inv(R1);
Mz = R5*R4*Rz*R2*R1;
fprintf('New coordinates : \n');
Qz = Mz*P

fprintf('Aligning vector V along Y-axis :\n');

d = sqrt(a^2 + c^2);
sinA = c/d; cosA = a/d;
sinB = d/e; cosB = b/e;
R1 = [cosA 0 sinA 0 ; 0 1 0 0 ; -sinA 0 cosA 0 ; 0 0 0 1];
R2 = [cosB -sinB 0 0 ; sinB cosB 0 0 ; 0 0 1 0 ; 0 0 0 1];
Ry = [cos(C) 0 sin(C) 0 ; 0 1 0 0 ; -sin(C) 0 cos(C) 0 ; 0 0 0 1];
R4 = inv(R2);
R5 = inv(R1);
```

```
My = R5*R4*Ry*R2*R1;
fprintf('New coordinates : \n');
Qy = My*P
```

7.10 ROTATION AROUND AN ARBITRARY LINE

The rotation matrix about an arbitrary line joining points $P(x_1, y_1, z_1)$ and $Q(x_2, y_2, z_2)$, in the CCW direction by an angle θ along a plane perpendicular to the line, is derived by the following steps:

Step 1: Translate line so that one end coincides with origin (see Section 7.2)

Step 2: Align the resulting vector along a primary axis (see Section 7.8)

Step 3: Rotate around that primary axis by the given amount (see Section 7.4)

Step 4: Reverse align vector to its original location

Step 5: Reverse translate line to original location

Example 7.12

A cube with center at origin and vertices at (−1, 1, 1), (1, 1, 1), (1, −1, 1), (−1, −1, 1), (−1, 1, −1), (1, 1, −1), (1, −1, −1), (−1, −1, −1) is rotated by 45° CCW around an arbitrary line joining points P(2, 1, −2) and Q(3, 3, 2) along a plane perpendicular to the line. Find its new vertices.

Here, $x_1 = 2,\ y_1 = 1,\ z_1 = -2,\ x_2 = 3,\ y_2 = 3,\ z_2 = 2$, angle of rotation $\theta = 45°$

Translate axis of rotation such that P coincides with the origin:

Translation matrix $T_1 = T(-x_1, -y_1, -z_1)$

Align the resulting vector along X-axis:

The coordinates of the vector tip: $a = x_2 - x_1 = 1,\ b = y_2 - y_1 = 2,\ c = z_2 - z_1 = 4$

Then $d = \sqrt{b^2 + c^2} = 4.4721, e = \sqrt{a^2 + b^2 + c^2} = 4.5826$

$$\sin\alpha = \frac{b}{d} = 0.4472, \cos\alpha = \frac{c}{d} = 0.8944, \sin\varphi = \frac{d}{e} = 0.9759, \cos\varphi = \frac{a}{e} = 0.2182$$

$$R_1 = R_x(\alpha) = \begin{bmatrix} 1 & 0 & 0 & 0 \\ 0 & \cos\alpha & -\sin\alpha & 0 \\ 0 & \sin\alpha & \cos\alpha & 0 \\ 0 & 0 & 0 & 1 \end{bmatrix}, R_2 = R_y(\varphi) = \begin{bmatrix} \cos\varphi & 0 & \sin\varphi & 0 \\ 0 & 1 & 0 & 0 \\ -\sin\varphi & 0 & \cos\varphi & 0 \\ 0 & 0 & 0 & 1 \end{bmatrix}$$

$$\text{Rotate about the origin around } X\text{-axis by angle } \theta: R_x(\theta) = \begin{bmatrix} 1 & 0 & 0 & 0 \\ 0 & \cos\theta & -\sin\theta & 0 \\ 0 & \sin\theta & \cos\theta & 0 \\ 0 & 0 & 0 & 1 \end{bmatrix}$$

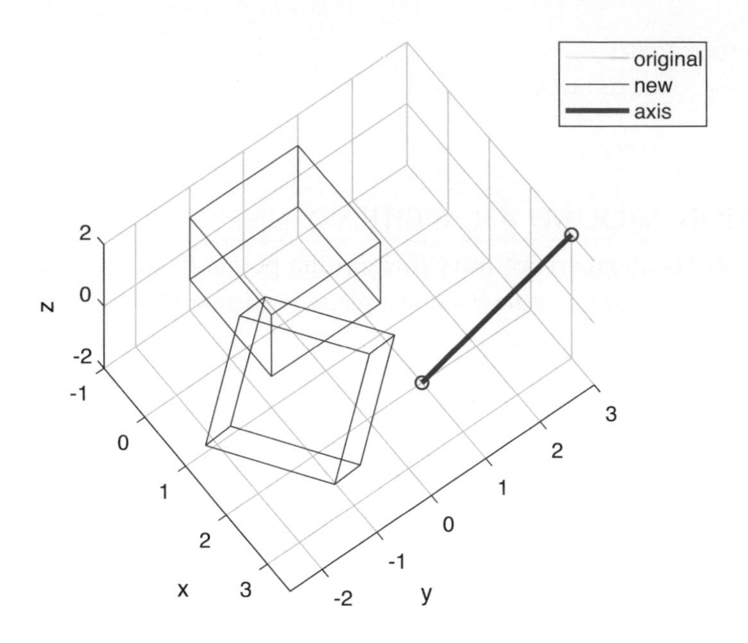

FIGURE 7.15 Plot for Example 7.12.

Reverse align vector to its original location: $R_4 = R_y(-\varphi), R_5 = R_x(-\alpha)$

Reverse translate line to original location: $T_2 = T\left(x_1, y_1, z_1\right)$

Composite transformation matrix: $M = T_2 \cdot R_5 \cdot R_4 \cdot R_x(\theta) \cdot R_2 \cdot R_1 \cdot T_1$

Original coordinate matrix: $C = \begin{bmatrix} -1 & 1 & 1 & -1 & -1 & 1 & 1 & -1 \\ 1 & 1 & -1 & -1 & 1 & 1 & -1 & -1 \\ 1 & 1 & 1 & 1 & -1 & -1 & -1 & -1 \\ 1 & 1 & 1 & 1 & 1 & 1 & 1 & 1 \end{bmatrix}$

New coordinate matrix: $D = M \cdot C$

New vertices: (0.93, −1.06, 1.55), (2.37, 0.23, 1.04), (3.55, −1.30, 0.51), (2.11, −2.59, 1.02), (0.20, −0.98, −0.31), (1.64, 0.31, −0.82), (2.82, −1.21, −1.35), and (1.38, −2.50, −0.84) (Figure 7.15)

MATLAB Code 7.12

```
clear all; clc;
p1 = [-1,1,1];
p2 = [1,1,1];
p3 = [1,-1,1];
p4 = [-1,-1,1];
p5 = [-1,1,-1];
p6 = [1,1,-1];
p7 = [1,-1,-1];
p8 = [-1,-1,-1];

x1 = 2; y1 = 1; z1 = -2;
x2 = 3; y2 = 3; z2 = 2;
a = x2 - x1 ; b = y2 - y1 ; c = z2 - z1;
d = sqrt(b^2 + c^2);
e = sqrt(a^2 + b^2 + c^2);
N = pi/4;
tx = -x1; ty = -y1 ; tz = -z1;
T1 = [1, 0, 0, tx ; 0, 1, 0, ty ; 0, 0, 1, tz ; 0, 0, 0, 1];

sinA = b/d; cosA = c/d;
sinB = d/e; cosB = a/e;
R1 = [1 0 0 0 ; 0 cosA -sinA 0 ; 0 sinA cosA 0 ; 0 0 0 1];
R2 = [cosB 0 sinB 0 ; 0 1 0 0 ; -sinB 0 cosB 0 ; 0 0 0 1];
Rx = [1 0 0 0 ; 0 cos(N) -sin(N) 0 ; 0 sin(N) cos(N) 0 ; 0 0 0 1];
R4 = inv(R2);
R5 = inv(R1);
T2 = inv(T1);
M = T2*R5*R4*Rx*R2*R1*T1;
C = [p1' p2' p3' p4' p5' p6' p7' p8' ; 1 1 1 1 1 1 1 1 ];
D = M*C;
```

```
fprintf('New vertices : \n')
for i=1:8
    fprintf('(%.2f, %.2f, %.2f) \n',D(1,i), D(2,i), D(3,i));
end

%plotting

C = [p1' p2' p3' p4' p1' p5' p6' p7' p8' p5' p8' p4' p3' p7' p6' p2' ;
    1 1 1 1 1 1 1 1 1 1 1 1 1 1 1 1];
D = M*C;
plot3(C(1,:), C(2,:), C(3,:), 'b'); hold on;
plot3(D(1,:), D(2,:), D(3,:), 'r'); grid;
xlabel('x'); ylabel('y'); zlabel('z');
plot3([x1, x2],[y1, y2],[z1, z2], 'b', 'LineWidth', 2);
plot3(x1, y1, z1, 'bo', x2, y2, z2, 'bo');
legend('original', 'new', 'axis'); axis equal;
view(53, 63); hold off;
```

7.11 REFLECTION

Reflection about a primary plane reverses the coordinate value along an axis perpendicular to the plane (Hearn and Baker, 1996). Reflection along X-axis is equivalent to reflection about YZ-plane and reverses the x-coordinate of a point. The corresponding matrix is given by:

$$F_x = \begin{bmatrix} -1 & 0 & 0 & 0 \\ 0 & 1 & 0 & 0 \\ 0 & 0 & 1 & 0 \\ 0 & 0 & 0 & 1 \end{bmatrix} \tag{7.9}$$

Reflection along Y-axis is equivalent to reflection about XZ-plane and reverse the y-coordinate of a point. The corresponding matrix is given by:

$$F_y = \begin{bmatrix} 1 & 0 & 0 & 0 \\ 0 & -1 & 0 & 0 \\ 0 & 0 & 1 & 0 \\ 0 & 0 & 0 & 1 \end{bmatrix} \tag{7.10}$$

Reflection along Z-axis is equivalent to reflection about XY-plane and reverse the z-coordinate of a point. The corresponding matrix is given by:

$$F_z = \begin{bmatrix} 1 & 0 & 0 & 0 \\ 0 & 1 & 0 & 0 \\ 0 & 0 & -1 & 0 \\ 0 & 0 & 0 & 1 \end{bmatrix} \tag{7.11}$$

If the plane of reflection is not a primary plane then it is first made to coincide with one of the primary planes and then the above formulas are applied.

Example 7.13

A cube with center at origin and vertices at (–1, 1, 1), (1, 1, 1), (1, –1, 1), (–1, –1, 1), (–1, 1, –1), (1, 1, –1), (1, –1, –1), (–1, –1, –1) is to be reflected about a plane z = 3 parallel to the X–Y plane along the Z-axis. Find its new vertices.

Translate plane so as to coincide with XY-plane: $T_1 = T(0, 0, -3)$
Reflect about XY-plane i.e. along Z-axis: F_z
Reverse translate plane to original location: $T_2 = T(0, 0, 3)$
Composite transformation: $M = T_2 \cdot F_z \cdot T_1$

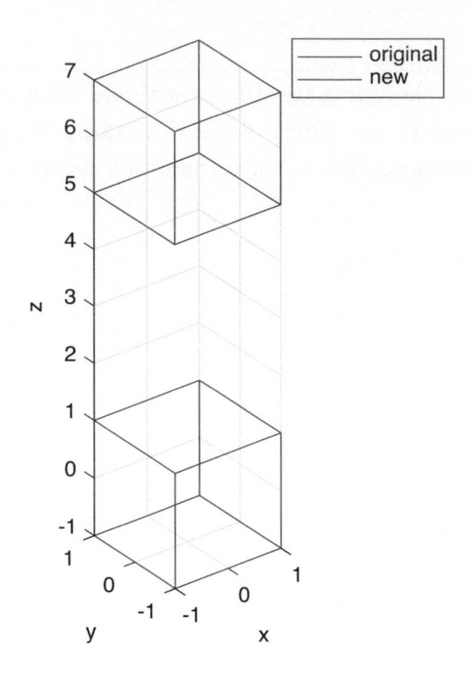

FIGURE 7.16 Plot for Example 7.13.

$$\text{Original coordinate matrix: } C = \begin{bmatrix} -1 & 1 & 1 & -1 & -1 & 1 & 1 & -1 \\ 1 & 1 & -1 & -1 & 1 & 1 & -1 & -1 \\ 1 & 1 & 1 & 1 & -1 & -1 & -1 & -1 \\ 1 & 1 & 1 & 1 & 1 & 1 & 1 & 1 \end{bmatrix}$$

New coordinate matrix: $D = M \cdot C$

New coordinates: (–1, 1, 5), (1, 1, 5), (1, –1, 5), (–1, –1, 5), (–1, 1, 7), (1, 1, 7), (1, –1, 7), and (–1, –1, 7) (Figure 7.16).

MATLAB Code 7.13

```
clear all; clc;
p1 = [-1,1,1];
p2 = [1,1,1];
p3 = [1,-1,1];
p4 = [-1,-1,1];
p5 = [-1,1,-1];
p6 = [1,1,-1];
p7 = [1,-1,-1];
p8 = [-1,-1,-1];
C = [p1' p2' p3' p4' p5' p6' p7' p8' ; 1 1 1 1 1 1 1 1 ];

tx = 0; ty = 0; tz = -3;
T1 = [1, 0, 0, tx ; 0, 1, 0, ty ; 0, 0, 1, tz ; 0, 0, 0, 1];
Fz = [1, 0, 0, 0 ; 0, 1, 0, 0 ; 0, 0, -1, 0 ; 0, 0, 0, 1];
T2 = inv(T1);
M = T2 * Fz * T1;
D = M*C;

fprintf('New vertices : \n');
for i=1:8
    fprintf('(%.2f, %.2f, %.2f) \n',D(1,i), D(2,i), D(3,i));
end

%plotting

C = [p1' p2' p3' p4' p1' p5' p6' p7' p8' p5' p8' p4' p3' p7' p6' p2' ;
    1 1 1 1 1 1 1 1 1 1 1 1 1 1 1 1];
D = M*C;
plot3(C(1,:), C(2,:), C(3,:), 'b'); hold on;
plot3(D(1,:), D(2,:), D(3,:), 'r'); grid;
xlabel('x'); ylabel('y'); zlabel('z')
legend('original', 'new'); axis equal; hold off;
```

7.12 SHEAR

Involves changing coordinate values along one axis by adding an amount proportional to coordinate values along another axis (Hearn and Baker, 1996). Shear along each axis can be of two types:

Shear along X-axis can be parallel to $Y = 0$ plane (front and back faces) governed by the relation $x_2 = x_1 + h \cdot y$, where h is the constant of proportionality. The corresponding transformation matrix is:

$$H_{xy} = \begin{bmatrix} 1 & h & 0 & 0 \\ 0 & 1 & 0 & 0 \\ 0 & 0 & 1 & 0 \\ 0 & 0 & 0 & 1 \end{bmatrix} \tag{7.12}$$

Shear along X-axis can also be parallel to $Z = 0$ plane (top and bottom faces) governed by the relation $x_2 = x_1 + h \cdot z$. The corresponding transformation matrix is:

$$H_{xz} = \begin{bmatrix} 1 & 0 & h & 0 \\ 0 & 1 & 0 & 0 \\ 0 & 0 & 1 & 0 \\ 0 & 0 & 0 & 1 \end{bmatrix} \tag{7.13}$$

Figure 7.17 illustrates the two types of shear along X-axis. These types are, however, not mutually exclusive, both can occur simultaneously. The transformation matrix in that case becomes:

$$H_{x} = \begin{bmatrix} 1 & h_1 & h_2 & 0 \\ 0 & 1 & 0 & 0 \\ 0 & 0 & 1 & 0 \\ 0 & 0 & 0 & 1 \end{bmatrix} \tag{7.14}$$

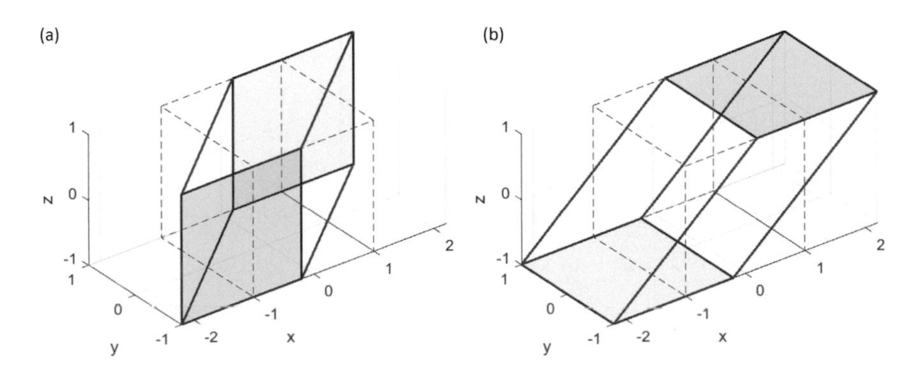

FIGURE 7.17 Shear along X-axis parallel to (a) front and back faces and (b) top and bottom faces.

In a likewise manner, shear along Y-axis can be along $X = 0$ plane and $Z = 0$ plane. The general transformation matrix is of the form:

$$H_y = \begin{bmatrix} 1 & 0 & 0 & 0 \\ h_1 & 1 & h_2 & 0 \\ 0 & 0 & 1 & 0 \\ 0 & 0 & 0 & 1 \end{bmatrix} \tag{7.15}$$

Shear along Z-axis can be along $X = 0$ plane and $Y = 0$ plane. The general transformation matrix is of the form:

$$H_z = \begin{bmatrix} 1 & 0 & 0 & 0 \\ 0 & 1 & 0 & 0 \\ h_1 & h_2 & 1 & 0 \\ 0 & 0 & 0 & 1 \end{bmatrix} \tag{7.16}$$

Example 7.14

A cube with center at origin and vertices at (–1, 1, 1), (1, 1, 1), (1, –1, 1), (–1, –1, 1), (–1, 1, –1), (1, 1, –1), (1, –1, –1), and (–1, –1, –1) is subjected to a shear along the Z-axis with parameters (1.2, 2.3). Find its new vertices.

From Equation (7.16), $H_z = \begin{bmatrix} 1 & 0 & 0 & 0 \\ 0 & 1 & 0 & 0 \\ 1.2 & 2.3 & 1 & 0 \\ 0 & 0 & 0 & 1 \end{bmatrix}$

Original coordinate matrix: $C = \begin{bmatrix} -1 & 1 & 1 & -1 & -1 & 1 & 1 & -1 \\ 1 & 1 & -1 & -1 & 1 & 1 & -1 & -1 \\ 1 & 1 & 1 & 1 & -1 & -1 & -1 & -1 \\ 1 & 1 & 1 & 1 & 1 & 1 & 1 & 1 \end{bmatrix}$

New coordinate matrix: $D = H_z \cdot C$

New vertices: (–1.00, 1.00, 2.10), (1.00, 1.00, 4.50), (1.00, –1.00, –0.10), (–1.00, –1.00, –2.50), (–1.00, 1.00, 0.10), (1.00, 1.00, 2.50), (1.00, –1.00, –2.10), and (–1.00, –1.00, –4.50) (Figure 7.18)

MATLAB Code 7.14

```
clear all; clc;
h1 = 1.2; h2 = 2.3;
H = [1, 0, 0, 0 ; 0, 1, 0, 0 ; h1, h2, 1, 0 ; 0, 0, 0, 1];
p1 = [-1,1,1];
p2 = [1,1,1];
p3 = [1,-1,1];
p4 = [-1,-1,1];
p5 = [-1,1,-1];
p6 = [1,1,-1];
p7 = [1,-1,-1];
p8 = [-1,-1,-1];
C = [p1' p2' p3' p4' p5' p6' p7' p8' ; 1 1 1 1 1 1 1 1 ];
D = H*C;

fprintf('New vertices : \n');
for i=1:8
    fprintf('(%.2f, %.2f, %.2f) \n',D(1,i), D(2,i), D(3,i));
end

%plotting

C = [p1' p2' p3' p4' p1' p5' p6' p7' p8' p5' p8' p4' p3' p7' p6' p2' ;
    1 1 1 1 1 1 1 1 1 1 1 1 1 1 1 1];
D = H*C;
plot3(C(1,:), C(2,:), C(3,:), 'b--'); hold on;
plot3(D(1,:), D(2,:), D(3,:), 'b-', 'LineWidth', 1.5); grid;
xlabel('x'); ylabel('y'); zlabel('z');
legend('original', 'new');
axis([-5 5 -5 5 -5 5]); hold off;
```

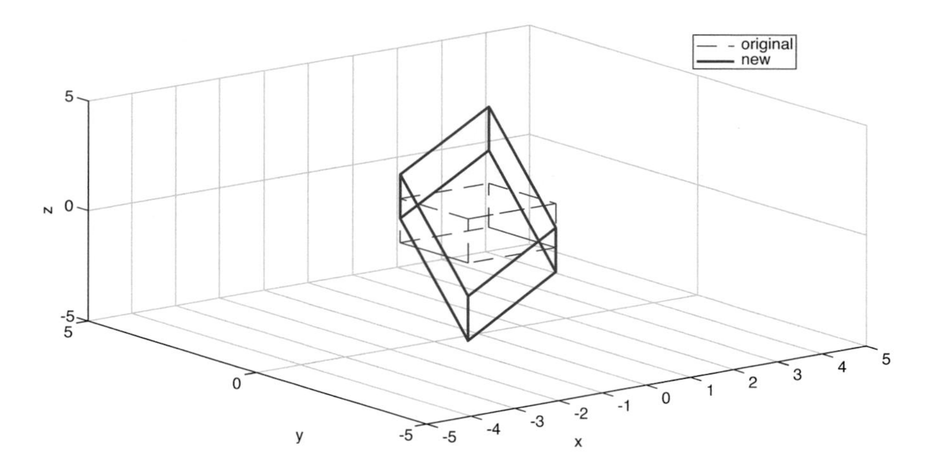

FIGURE 7.18 Plot for Example 7.14.

7.13 CHAPTER SUMMARY

The following points summarize the topics discussed in this chapter:

- Coordinates in 3D space are measured using the right-handed coordinate system.
- Angles are considered positive along the CCW direction and negative along the CW direction.
- The *X*-, *Y*-, and *Z*-axes are the primary axes mutually at right-angles meeting at the origin.
- The *X*–*Y*, *Y*–*Z*, and *Z*–*X* planes divide the 3D space into eight octants.
- Translation operation adds positive or negative increments to the coordinates of a point.
- Scaling operation multiples coordinates of a point by positive or negative scaling factors.
- Rotation operation moves a point along an arc about any of the three primary axes.
- Scaling and rotation by default are calculated with respect to the origin.
- Scaling and rotation operations with respect to an arbitrary point involves translating the point to the origin, performing the specified operation, and reverse translation to its original location.
- Rotations parallel to a primary axis are associated with a forward and reverse translation to and from a primary axis.
- To align a vector to a primary axis, requires in general two rotation operations.

- Rotation around a vector involves aligning the vector to a primary axis, rotating about that axis and reverse aligning the vector to its original location.
- Rotation around an arbitrary line involve translating it to the origin, aligning it to a primary axis, rotating about that axis, reverse aligning, and reverse translating to its original location.
- Reflection about a primary plane reverses the coordinate value along an axis perpendicular to the plane.
- Shear involves changing coordinate values along one axis by adding an amount proportional to coordinate values along other axes.

7.14 REVIEW QUESTIONS

1. What is meant by the right-handed coordinate system for 3D space?
2. Why are there three different rotation matrices for 3D space?
3. What are meant by positive and negative translation factors?
4. What are meant by positive and negative scaling factors?
5. What is considered as a positive direction of rotation around any of the primary axes?
6. How are scaling and rotations with respect to an arbitrary point calculated?
7. How is the rotation matrix parallel to a primary axis computed?
8. How is a vector aligned to a primary axis?
9. How is reflection parallel to a primary plane calculated?
10. Why does shear operation along a primary axis involve two different options?

7.15 PRACTICE PROBLEMS

1. A point with coordinates (2, 2, 2) is to be rotated by 45° about the Y-axis and then by 60° about the Z-axis. Find its new coordinates. Check if the final position of the point is same or different if it is first rotated by 60° about the Z-axis and then by 45° about the Y-axis.
2. Find the new coordinates of a point $P(k, -k, k)$ when rotated clockwise about the origin on the X–Z plane by 30°, where k is a constant.
3. A line is parallel to the Y-axis and joins point $P(k, -k, k)$ and $Q(k, 0, k)$, where k is a constant. Derive the transformation matrix for rotation by angle θ about the line. Assume $\sin(\theta) = 0.76$ and $\cos(\theta) = 0.65$.
4. A point $P(1, 2, 3)$ is to be rotated by 77° CW about a line joining points M (2, 1, 0) and N (3, 3, 1). Calculate the new coordinates of the point.

5. Find transformation matrix of reflection with respect to the plane passing through the origin and having normal vector $N = i + j + k$.

6. Find transformation matrix of reflection with respect to the plane: $2x - y + 2z - 2 = 0$.

7. Consider a cube with a vertex at the origin and length of side 1. It is first subjected to a shear $x_2 = x_1 + ay_1$ and then to another shear $y_2 = y_1 + bx_1$, where $a = 2$ and $b = 3$. Find the final coordinates of the vertices of the cube.

8. Find the transformations that align the following vectors with the positive Z-axis: (a) $2i + j + 2k$, (b) $2i - j + 2k$, and (c) $2i - j - 2k$.

9. Obtain a transformation that aligns the vector k with the vector $i + j + k$.

10. Show that new coordinates of a point $P(k, -2k, \sqrt{3}k)$ when rotated about Y-axis by $30°$ and then about Z-axis by $-30°$ is given by the following, where k is a constant: $Q(k/2, -3\sqrt{3}k/2, k)$.

CHAPTER 8

Surfaces

8.1 INTRODUCTION

Surfaces define the shape and contour of the 3D graphical objects and models. The most basic type of surface is a flat surfaces commonly represented by a plane. Planes have been discussed in Chapter 6. In this chapter, we will focus on curved surfaces. Curved surfaces are usually modeled using two splines u and v in orthogonal directions. Depending on the types of splines used surfaces may be named accordingly e.g. Bezier surfaces or B-spline surfaces. Spline-based surfaces often have a grid of control points associated with them using which the shape of the surface may be modified (Foley et al., 1995). Figure 8.1 shows a surface composed of a quadratic curve along u and a cubic curve along v. Apart from the surface structure, this chapter also takes a look at surface appearances. Surface appearances are determined by texture and lighting. Texture mapping provides a realistic look

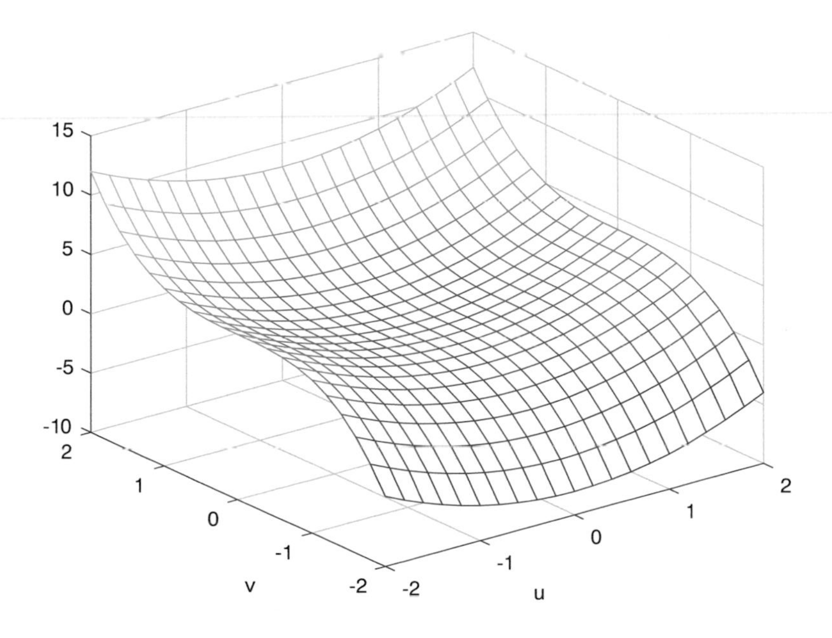

FIGURE 8.1 Spline-based surface.

to a surface and is extensively used in 3D graphical models e.g. wooden table or metal plate. Lighting determine the brightness and shading aspects of a surface. Surfaces can be categorized into two broad types based on their mathematical representation: implicit and parametric. Implicit surfaces have equations of the form $f(x, y, z)=0$ while parametric surfaces are represented using parametric variables viz. $\{x(u, v), y(u, v), z(u, v)\}$. Surfaces can also be categorized based on methods of generation e.g. extrusion and revolution.

8.2 PARAMETRIC SURFACES

Recall from Chapter 1 that a 2D curve in parametric form is represented as $C(t)=\{x(t), y(t)\}$, where t is the parametric variable defined over a specified interval. This essentially implies that as t takes on different values over this range, functions $x(t)$ and $y(t)$ generate values along the X- and Y-axes of a 2D graph, both of which together determine the locus of the curve $C(t)$. For example, $C(t)=\{r\cdot\cos t, r\cdot\sin t\}$ represents a circle of radius r on a 2D plane as t varies from 0 to 2π. A parametric surface is an extension of this concept to 3D. Instead of a single variable t now we have two parametric variables u, v and instead of two functions $x(t)$, $y(t)$, we have three functions $x(u, v)$, $y(u, v)$, and $z(u, v)$, which together generate a surface $S(u, v)$ in 3D space. In some books, the parametric variables are referred to as s and t instead of u and v.

$$S(u, v)=\{x(u, v), y(u, v), z(u, v)\} \tag{8.1}$$

Any point on the surface is therefore decided by three values along three orthogonal axes. If the functions are of degree 1 then the resulting surface is flat; otherwise, they are curves. To visualize a parametric surface consider the functions: $x=u+v, y=u-v, z=abs(u+v)$. A 2D plot of the X–Y plane is shown in Figure 8.2a for the range $-2\le u, v\le 2$, which depicts a plane surface in the shape of a quadrilateral. For each point on the plane surface, if the third value is plotted along the Z-axis then the resulting 3D plot is shown in Figure 8.2b. It depicts two planes intersecting on the X–Y plane.

If the functions are of degree more than 1 then the resulting surface will in general be curves. Figure 8.3 shows the surface generated from the parametric functions: $x=u^2, y=2uv, z=v^2$.

Example 8.1

A parametric surface $S=\left(u-v, u+v, u^2-u^2\right)$ is translated using $T(3, 5, 4)$ and then rotated about the Z-axis by 90°. Find the parametric representation of the resulting surface.

Translation: $T_1=T(3, 5, 4)$

Rotation: $R_1=R_z(90)$

Composite transformation: $M=R_1\cdot T_1=\begin{bmatrix} 0 & -1 & 0 & -5 \\ 1 & 0 & 0 & 3 \\ 0 & 0 & 1 & 4 \\ 0 & 0 & 0 & 1 \end{bmatrix}$

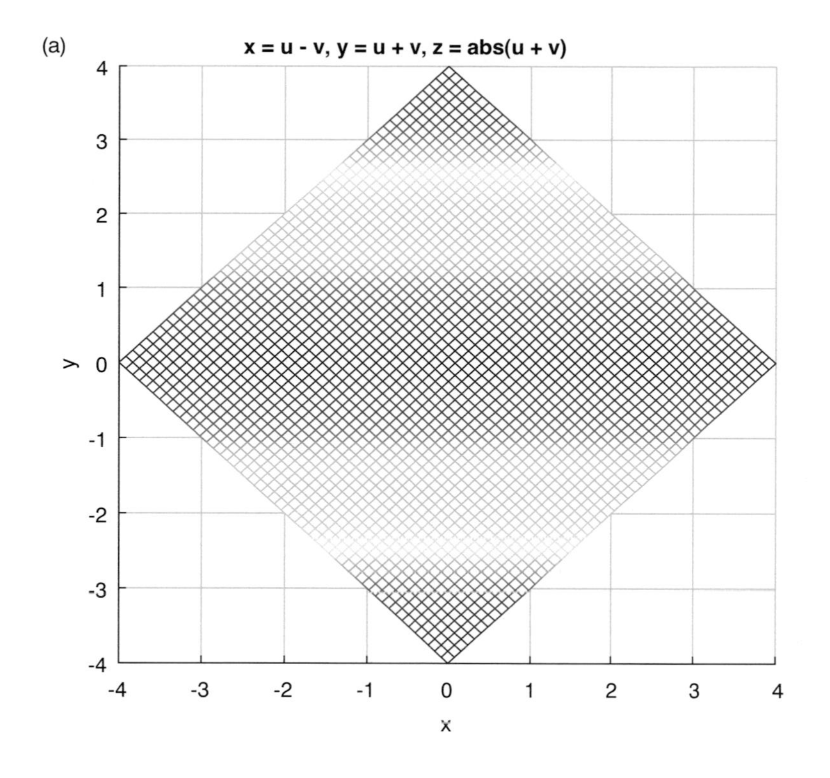

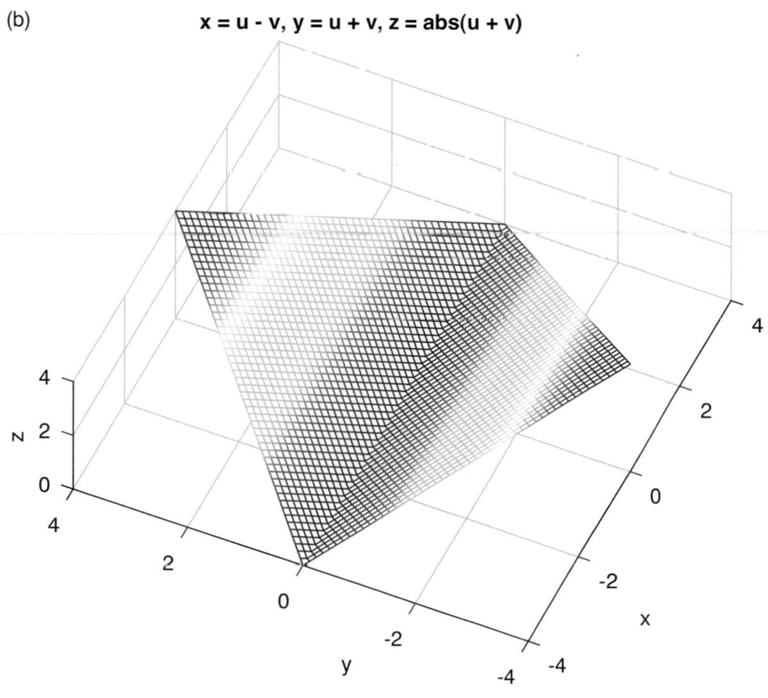

FIGURE 8.2 Linear parametric surface: (a) 2D view; (b) 3D view.

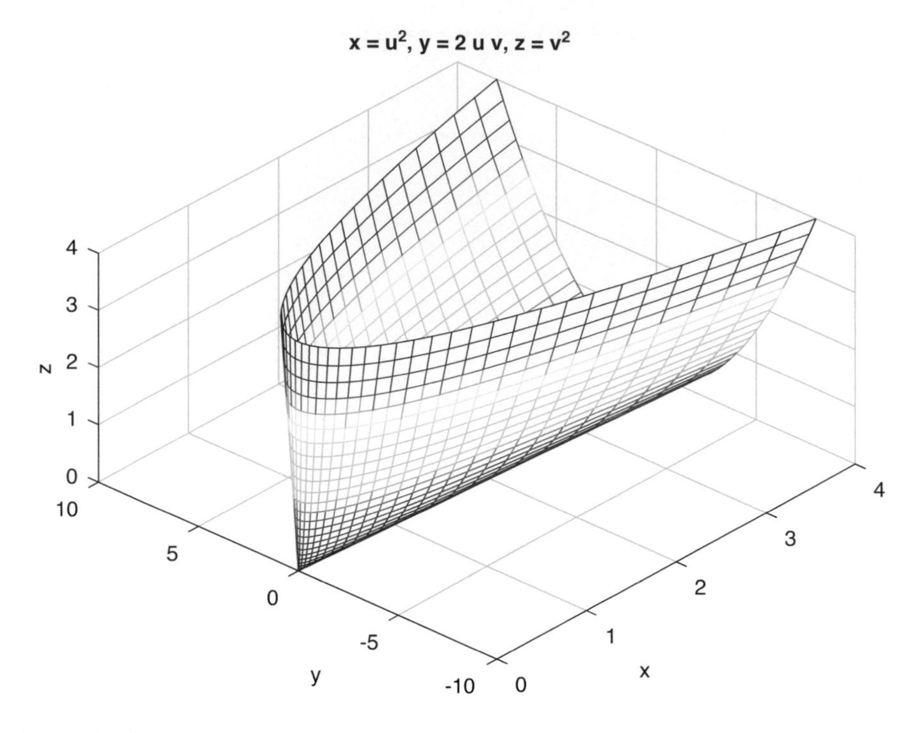

FIGURE 8.3 Non-linear parametric surface.

$$\text{Original curve: } P = \begin{bmatrix} u - v \\ u + v \\ u^2 - v^2 \\ 1 \end{bmatrix}$$

$$\text{Modified curve: } Q = M * P = \begin{bmatrix} u - v - 5 \\ u - v + 3 \\ u^2 - v^2 + 4 \\ 1 \end{bmatrix}$$

Required curve equation: $S(u, v) = (u - v - 5, u - v + 3, u^2 - v^2 + 4)$ (Figure 8.4)

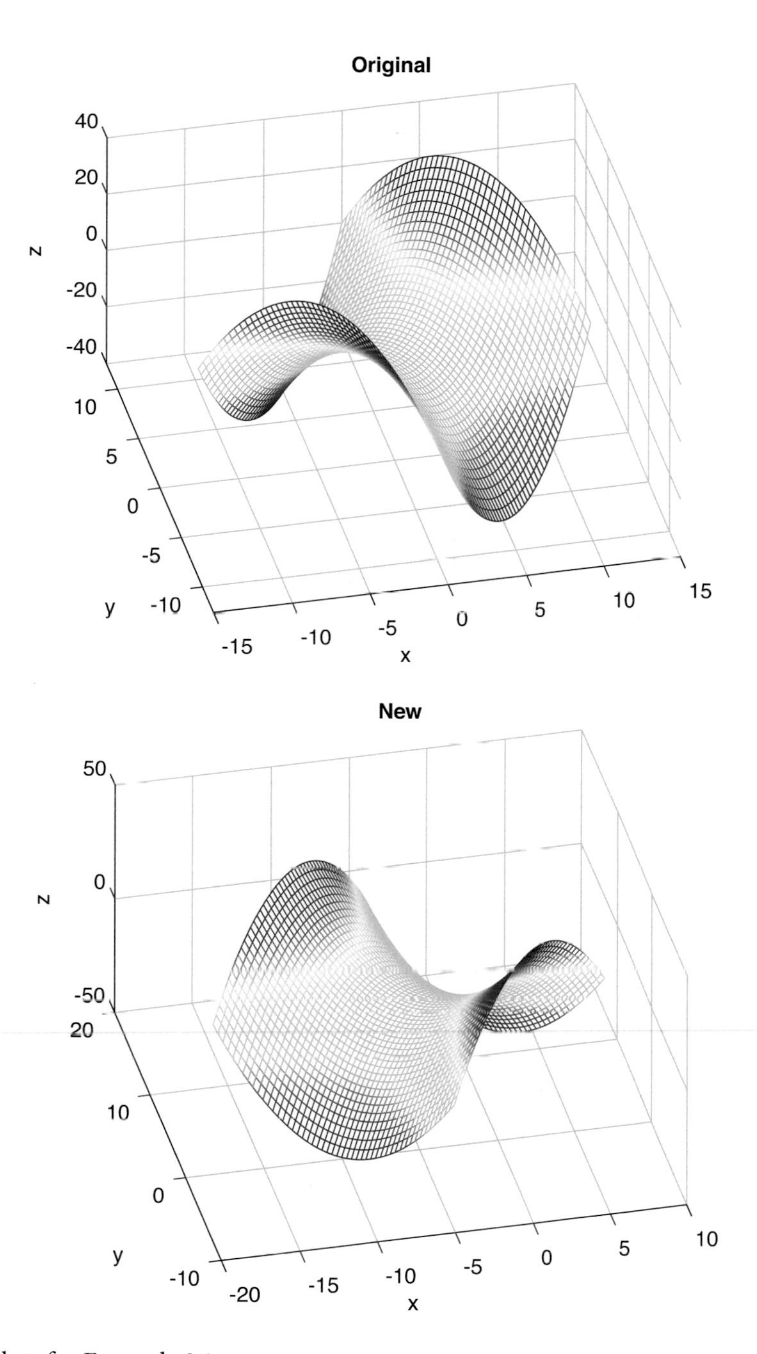

FIGURE 8.4 Plots for Example 8.1.

MATLAB® Code 8.1

```
clear all; clc;
syms u v;
tx = 3; ty = 5; tz = 4;
A = deg2rad(90);
P = [u - v ; u + v ; u^2 - v^2 ; 1]
T1 = [1, 0, 0, tx ; 0, 1, 0, ty ; 0, 0, 1, tz ; 0, 0, 0, 1];
R1 = [cos(A), -sin(A), 0, 0 ; sin(A), cos(A), 0, 0 ; 0, 0, 1, 0 ; 0, 0, 0, 1];
M = R1*T1;
Q = R1*T1*P;
Q = eval(Q)
subplot(121)
ezmesh(P(1), P(2), P(3)); title('Original');
view(-13, 48);
subplot(122)
ezmesh(Q(1), Q(2), Q(3)); title('New');
view(-13, 48);
```

NOTE

`ezmesh`: creates a mesh for the function z = f (x, y)

8.3 BEZIER SURFACES

A Bezier surface is created from Bezier splines and is essentially a parametric surface (Hearn and Baker, 1996). However, Bezier surfaces are most often defined based on their associated control points. A Bezier curve using a degree 1 polynomial can be represented as: $f(t)=(1-t)\cdot P_0+t\cdot P_1=\begin{bmatrix} 1-t, & t \end{bmatrix}\begin{bmatrix} P_0 \\ P_1 \end{bmatrix}$.

A bi-linear Bezier surface is generated using two first degree Bezier splines along orthogonal directions and has the equation shown, below where P_{00} and P_{01} are the control points of the first spline and P_{10} and P_{11} are the control points of the second spline.

$$S(u, v)=\begin{bmatrix} 1-u, & u \end{bmatrix}\begin{bmatrix} P_{00} & P_{01} \\ P_{10} & P_{11} \end{bmatrix}\begin{bmatrix} 1-v \\ v \end{bmatrix} \tag{8.2}$$

Example 8.2

Find the equation of a bi-linear Bezier surface using the following control points: $\boldsymbol{P_{00} = (0, 0, 1), P_{01} = (1, 1, 1), P_{10} = (1, 0, 0), P_{11} = (0, 1, 0)}$

From Equation (8.2),

$$x(u, v)=\begin{bmatrix} 1-u, & u \end{bmatrix}\begin{bmatrix} 0 & 1 \\ 1 & 0 \end{bmatrix}\begin{bmatrix} 1-v \\ v \end{bmatrix}=u+v-2uv$$

$$y(u, v)=\begin{bmatrix} 1-u, & u \end{bmatrix}\begin{bmatrix} 0 & 1 \\ 0 & 1 \end{bmatrix}\begin{bmatrix} 1-v \\ v \end{bmatrix}=v$$

$$z(u, v)=\begin{bmatrix} 1-u, & u \end{bmatrix}\begin{bmatrix} 1 & 1 \\ 0 & 0 \end{bmatrix}\begin{bmatrix} 1-v \\ v \end{bmatrix}=1-u$$

Required surface equation: $S(u, v)=(u+v-2uv, v, 1-u)$ (Figure 8.5)

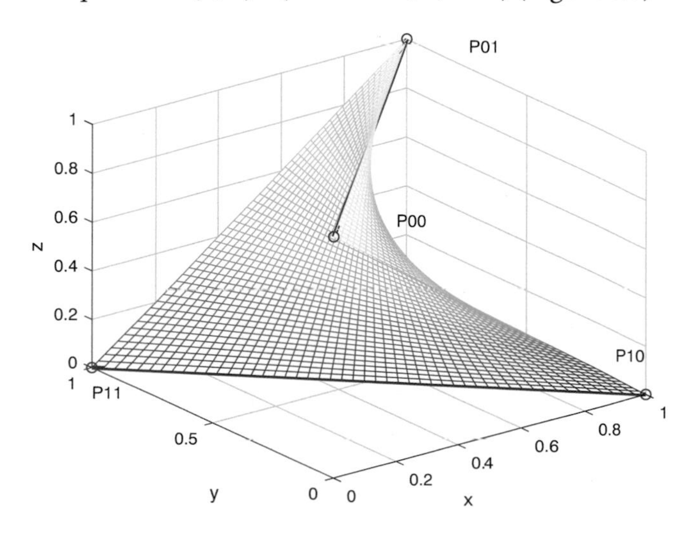

FIGURE 8.5 Plot for Example 8.2.

MATLAB Code 8.2

```
clear all; clc;
P00 = [0,0,1];
P01 = [1,1,1];
P10 = [1,0,0];
P11 = [0,1,0];

syms u v;

x00 = P00(1); x01 = P01(1); x10 = P10(1); x11 = P11(1);
y00 = P00(2); y01 = P01(2); y10 = P10(2); y11 = P11(2);
z00 = P00(3); z01 = P01(3); z10 = P10(3); z11 = P11(3);

x = [1-u, u]*[x00, x01 ; x10, x11]*[1-v ; v]; x = simplify(x)
y = [1-u, u]*[y00, y01 ; y10, y11]*[1-v ; v]; y = simplify(y)
z = [1-u, u]*[z00, z01 ; z10, z11]*[1-v ; v]; z = simplify(z)

ezmesh(x, y, z, [0,1,0,1]); hold on;
plot3([P00(1) P01(1)], [P00(2) P01(2)], [P00(3) P01(3)], 'b-', 'LineWidth', 2);
plot3([P00(1) P01(1)], [P00(2) P01(2)], [P00(3) P01(3)], 'bo');
plot3([P10(1) P11(1)], [P10(2) P11(2)], [P10(3) P11(3)], 'b-', 'LineWidth', 2);
plot3([P10(1) P11(1)], [P10(2) P11(2)], [P10(3) P11(3)], 'bo');

text(x00+0.2, y00, z00, 'P00');
text(x01+0.2, y01, z01, 'P01');
text(x10-0.1, y10, z10+0.2, 'P10');
text(x11, y11, z11-0.1, 'P11');
hold off;
```

A degree 2 Bezier curve is given by the equation: $\begin{bmatrix} (1-t)^2, & 2t(1-t), & t^2 \end{bmatrix} \begin{bmatrix} P_0 \\ P_1 \\ P_2 \end{bmatrix}$.

A bi-quadratic Bezier surface is, therefore, given by the following where a grid of 3 by 3 or 9 control points need to be specified.

$$S(u, v) = \begin{bmatrix} (1-u)^2, & 2u(1-u), & u^2 \end{bmatrix} \begin{bmatrix} P_{00} & P_{01} & P_{02} \\ P_{10} & P_{11} & P_{12} \\ P_{20} & P_{21} & P_{22} \end{bmatrix} \begin{bmatrix} (1-v)^2 \\ 2v(1-v) \\ v^2 \end{bmatrix} \quad (8.3)$$

Similarly, a cubic Bezier surface is associated with a grid of 4 by 4 or 16 control points.

Example 8.3

A bi-quadratic Bezier surface has the following control points: P_{00} = (1, −1, 0); P_{01} = (4, 3, 0); P_{02} = (5, −2, 0); P_{10} = (1, 1, 3); P_{11} = (3, 2, 3); P_{12} = (5, 1, 3); P_{20} = (1, −1, 5); P_{21} = (4, 3, 5); P_{22} = (5, −2, 5). Find its equation.

From Equation (8.3),

$$x(u, v) = \begin{bmatrix} (1-u)^2, & 2u(1-u), & u^2 \end{bmatrix} \begin{bmatrix} 1 & 4 & 5 \\ 1 & 3 & 5 \\ 1 & 4 & 5 \end{bmatrix} \begin{bmatrix} (1 \quad v)^2 \\ 2v(1-v) \\ v^2 \end{bmatrix}$$

$$= -4u^2v^2 + 4u^2v + 4uv^2 - 4uv - 2v^2 + 6v + 1$$

$$y(u, v) = \begin{bmatrix} (1-u)^2, & 2u(1-u), & u^2 \end{bmatrix} \begin{bmatrix} -1 & 3 & -2 \\ 1 & 2 & 1 \\ -1 & 3 & -2 \end{bmatrix} \begin{bmatrix} (1-v)^2 \\ 2v(1-v) \\ v^2 \end{bmatrix}$$

$$= -14u^2v^2 + 12u^2v - 4u^2 + 14uv^2 - 12uv + 4u - 9v^2 + 8v - 1$$

$$z(u, v) = \begin{bmatrix} (1-u)^2, & 2u(1-u), & u^2 \end{bmatrix} \begin{bmatrix} 0 & 0 & 0 \\ 3 & 3 & 3 \\ 5 & 5 & 5 \end{bmatrix} \begin{bmatrix} (1-v)^2 \\ 2v(1-v) \\ v^2 \end{bmatrix} = -u(u-6)$$

The required surface equation is $S(u, v) = \{x(u, v), y(u, v), z(u, v)\}$ (Figure 8.6).

MATLAB Code 8.3

```
clear all; clc;
P00 = [1, -1, 0];  P01 = [4 3, 0]; P02 = [5, -2, 0];
P10 = [1, 1, 3];  P11 = [3 2, 3]; P12 = [5, 1, 3];
P20 = [1, -1, 5];  P21 = [4 3, 5]; P22 = [5, -2, 5];

syms u v;

x00 = P00(1); x01 = P01(1); x02 = P02(1);
x10 = P10(1); x11 = P11(1); x12 = P12(1);
x20 = P20(1); x21 = P21(1); x22 = P22(1);
y00 = P00(2); y01 = P01(2); y02 = P02(2);
y10 = P10(2); y11 = P11(2); y12 = P12(2);
y20 = P20(2); y21 = P21(2); y22 = P22(2);
z00 = P00(3); z01 = P01(3); z02 = P02(3);
z10 = P10(3); z11 = P11(3); z12 = P12(3);
z20 = P20(3); z21 = P21(3); z22 = P22(3);

x = [(1 - u)^2, 2*u*(1 - u), u^2]*[x00 x01 x02 ; x10 x11 x12; x20 x21 x22]...
*[(1 - v)^2 ; 2*v*(1 - v) ; v^2] ; x = simplify (x)

y = [(1 - u)^2, 2*u*(1 - u), u^2]*[y00 y01 y02 ; y10 y11 y12; y20 y21 y22]...
*[(1 - v)^2 ; 2*v*(1 - v) ; v^2] ; y = simplify (y)

z = [(1 - u)^2, 2*u*(1 - u), u^2]*[z00 z01 z02 ; z10 z11 z12; z20 z21 z22]...
*[(1 - v)^2 ; 2*v*(1 - v) ; v^2] ; z = simplify (z)
```

```
ezmesh(x, y, z, [0,1,0,1]); hold on;
%plotting control points
plot3([P00(1) P01(1) P02(1)], [P00(2) P01(2) P02(2)], [P00(3) P01(3) P02(3)], 'b-')
plot3([P00(1) P01(1) P02(1)], [P00(2) P01(2) P02(2)], [P00(3) P01(3) P02(3)], 'bo')
plot3([P10(1) P11(1) P12(1)], [P10(2) P11(2) P12(2)], [P10(3) P11(3) P12(3)], 'b-')
plot3([P10(1) P11(1) P12(1)], [P10(2) P11(2) P12(2)], [P10(3) P11(3) P12(3)], 'bo')
plot3([P20(1) P21(1) P22(1)], [P20(2) P21(2) P22(2)], [P20(3) P21(3) P22(3)], 'b-')
plot3([P20(1) P21(1) P22(1)], [P20(2) P21(2) P22(2)], [P20(3) P21(3) P22(3)], 'bo')
plot3([P00(1) P10(1) P20(1)], [P00(2) P10(2) P20(2)], [P00(3) P10(3) P20(3)], 'b-')
plot3([P02(1) P12(1) P22(1)], [P02(2) P12(2) P22(2)], [P02(3) P12(3) P22(3)], 'b-')
plot3([P01(1) P11(1) P21(1)], [P01(2) P11(2) P21(2)], [P01(3) P11(3) P21(3)], 'b-')
view(157, -14);
hold off;
```

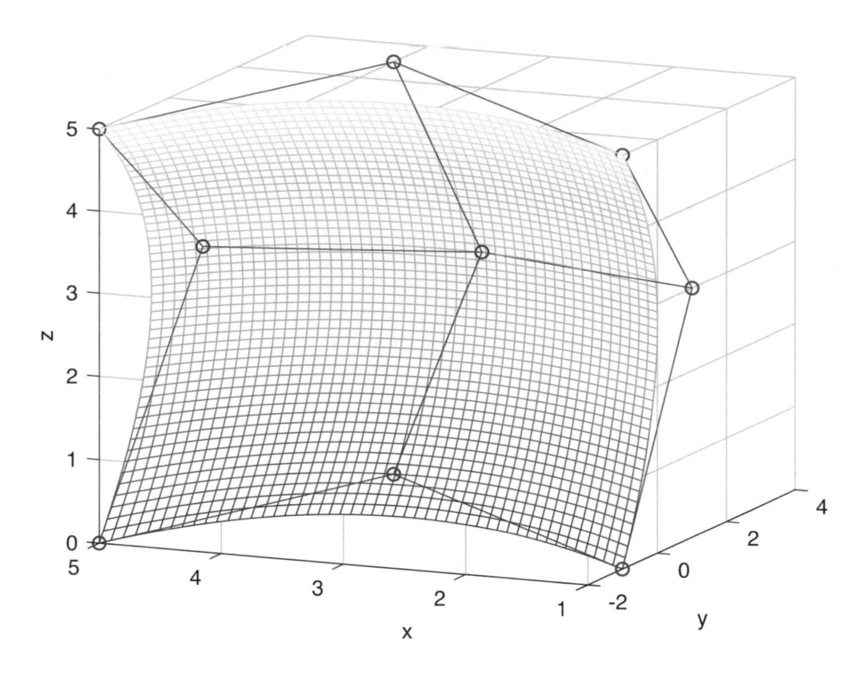

FIGURE 8.6 Plot for Example 8.3.

8.4 IMPLICIT SURFACES

Apart from parametric surfaces, the other type of surfaces most frequently encountered in graphics is called implicit surfaces and their equations are of the form $f(x, y, z)=0$. The second-degree implicit equations are also called quadric surfaces (Hearn and Baker, 1996), (Rovenski, 2010) and have a general form shown below, where A to J are constants (Figures 8.7–8.11):

$$Ax^2 + By^2 + Cz^2 + Dxy + Eyz + Fzx + Gx + Hy + Iz + J = 0 \tag{8.4}$$

Some commonly used quadric surfaces are listed below:

$$\text{Ellipsoid: } ax^2 + by^2 + cz^2 = k \tag{8.5}$$

$$\text{Elliptic cone: } ax^2 + by^2 - cz^2 = 0 \tag{8.6}$$

$$\text{Hyperboloid (1 sheet): } ax^2 + by^2 - cz^2 = k \tag{8.7}$$

$$\text{Hyperboloid (2 sheets): } ax^2 + by^2 - cz^2 = -k \tag{8.8}$$

$$\text{Elliptic paraboloid: } ax^2 + by^2 - z = 0 \tag{8.9}$$

$$\text{Hyperbolic paraboloid: } ax^2 - by^2 - z = 0 \tag{8.10}$$

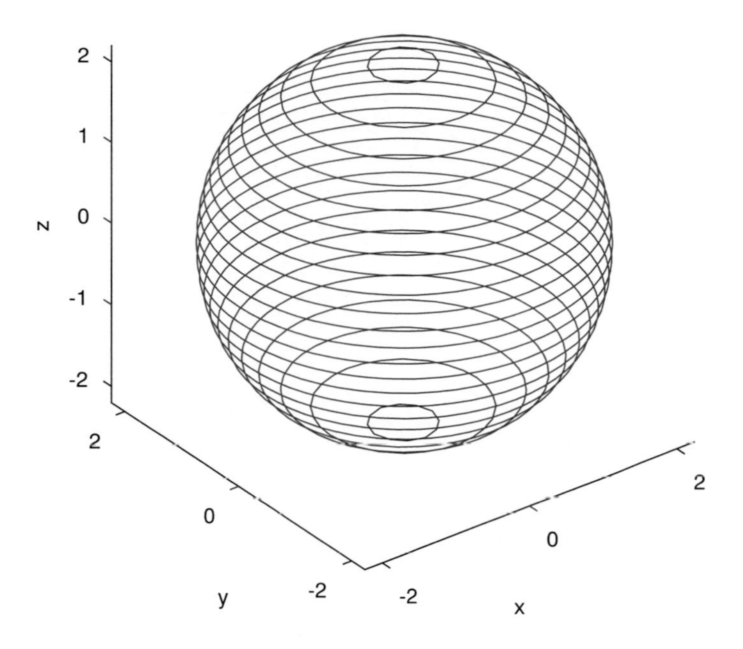

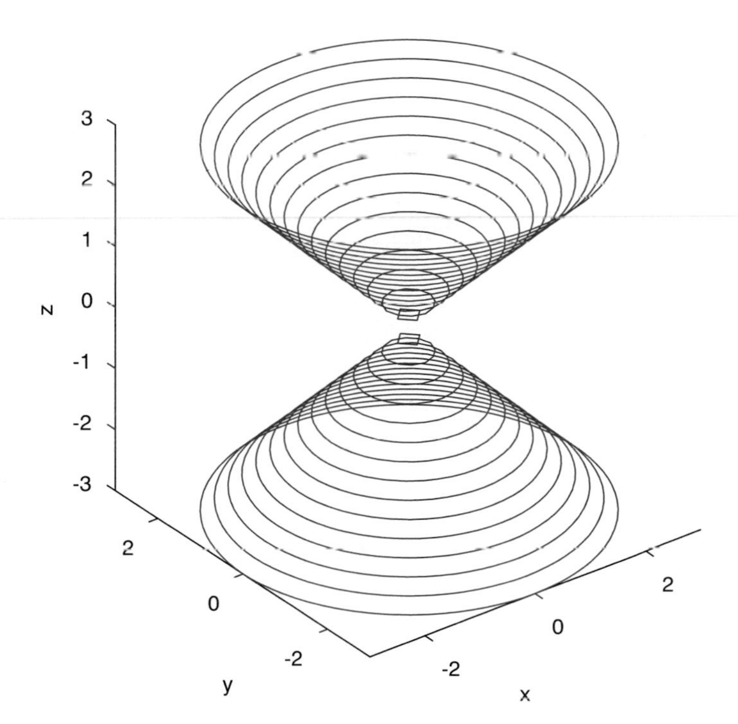

FIGURE 8.7 Quadric surfaces: ellipsoid and elliptic cone.

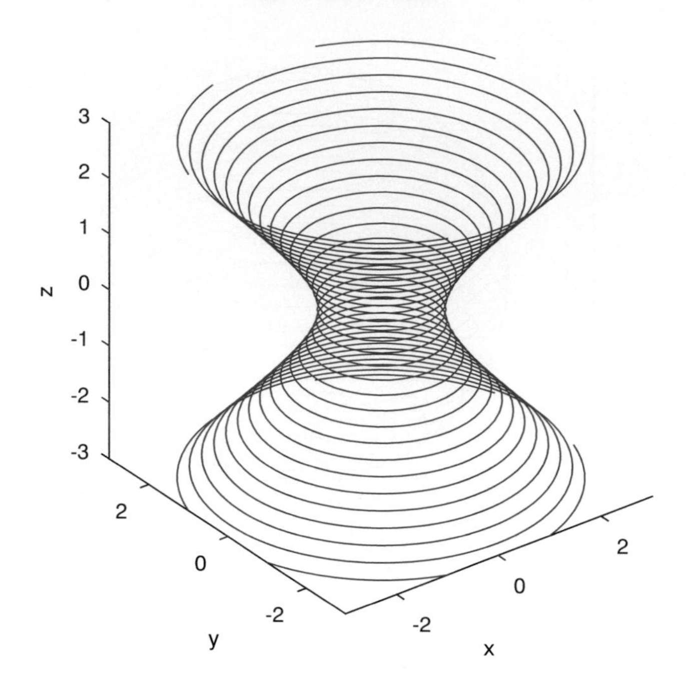

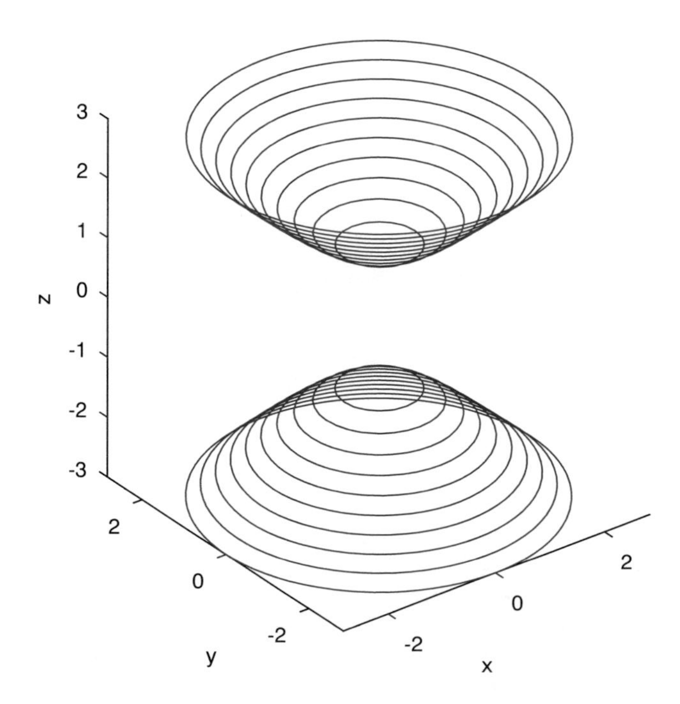

FIGURE 8.8 Quadric surfaces: hyperboloid (1 sheet) and hyperboloid (2 sheets).

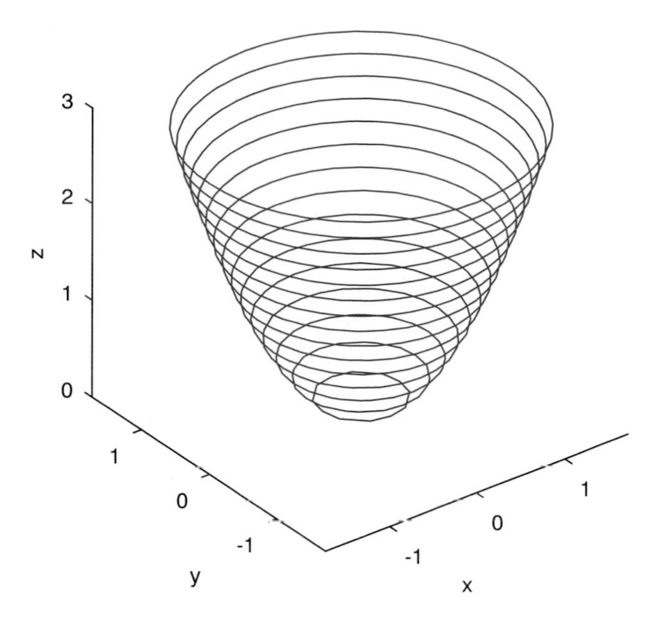

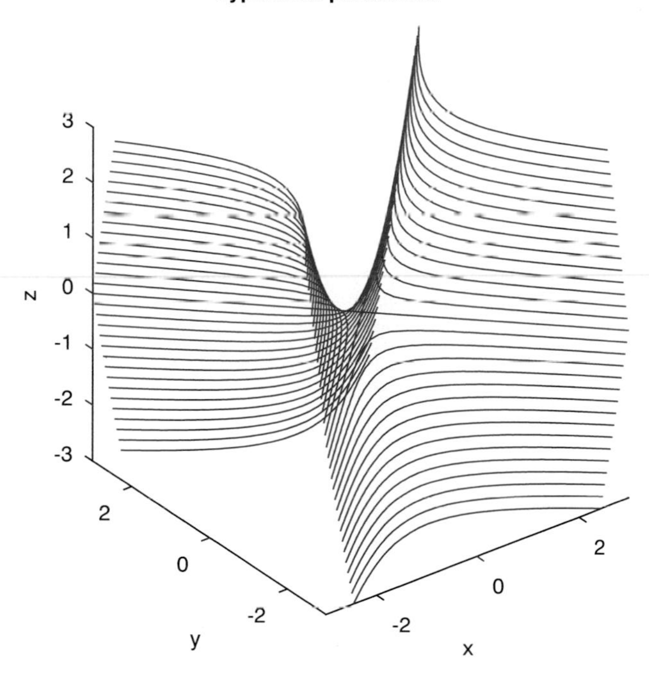

FIGURE 8.9 Quadric surfaces: elliptic paraboloid and hyperbolic paraboloid.

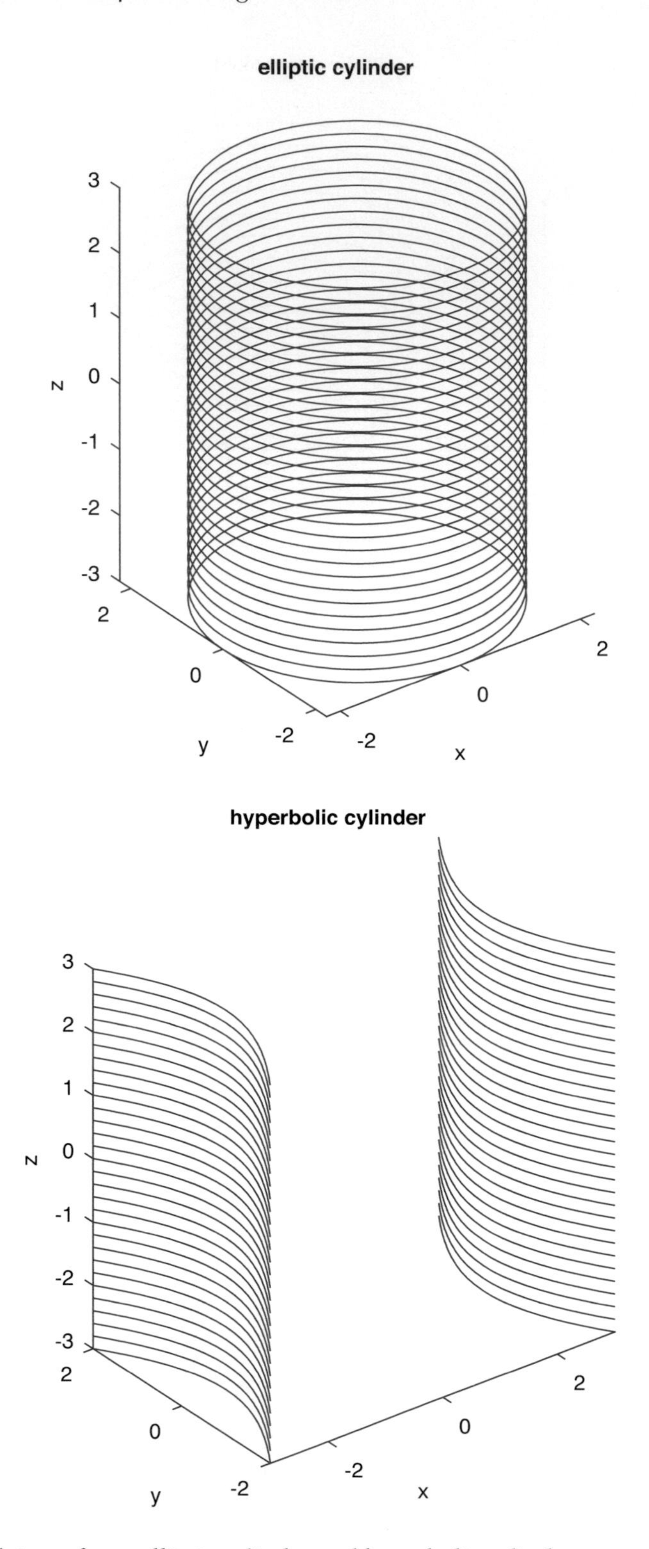

FIGURE 8.10 Quadric surfaces: elliptic cylinder and hyperbolic cylinder.

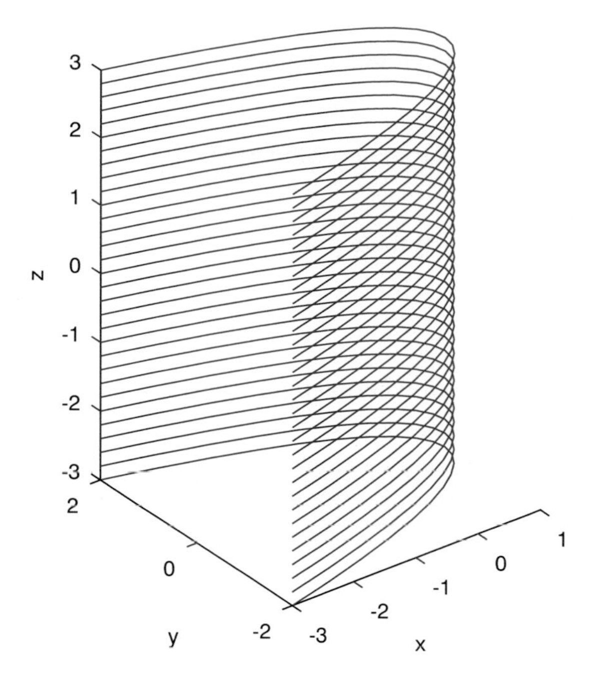

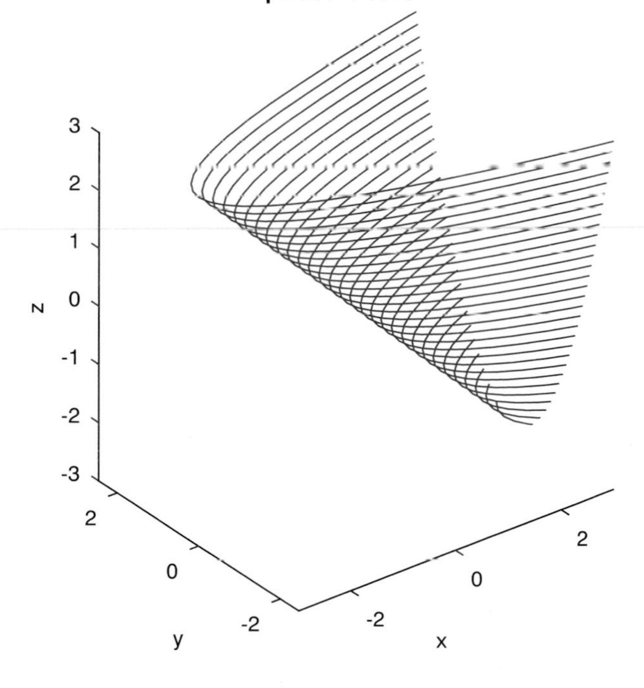

FIGURE 8.11 Quadric surfaces: parabolic cylinder and parabolic cone.

$$\text{Elliptic cylinder: } ax^2 + by^2 = k \tag{8.11}$$

$$\text{Hyperbolic cylinder: } ax^2 - by^2 = k \tag{8.12}$$

$$\text{Parabolic cylinder: } ax + by^2 = k \tag{8.13}$$

$$\text{Parabolic cone: } ax - by^2 + z = 0 \tag{8.14}$$

Example 8.4

Find the point(s) of intersection between the hyperboloid $2x^2 + 3y^2 - z^2 = 16$ and the line $(t-4,\ t-5,\ t+10)$

For any point on the line: $x = t-4,\ y = t-5,\ z = t+10$

At point of intersection this also should satisfy the equation of the surface

Substituting: $2(t-4)^2 + 3(t-5)^2 - (t+10)^2 = 16$

Simplifying: $4t^2 - 66t - 9 = 0$ whose solutions are $t = -0.1353,\ 16.6353$

Substituting, points of intersection are (−4.135255, −5.135255, 9.864745) and (12.635255, 11.635255, 26.635255)

Verification (Figure 8.12):

$$2(-4.135255)^2 + 3(-5.135255)^2 - (9.864745)^2 = 16$$

$$2(12.635255)^2 + 3(11.635255)^2 - (26.635255)^2 = 16$$

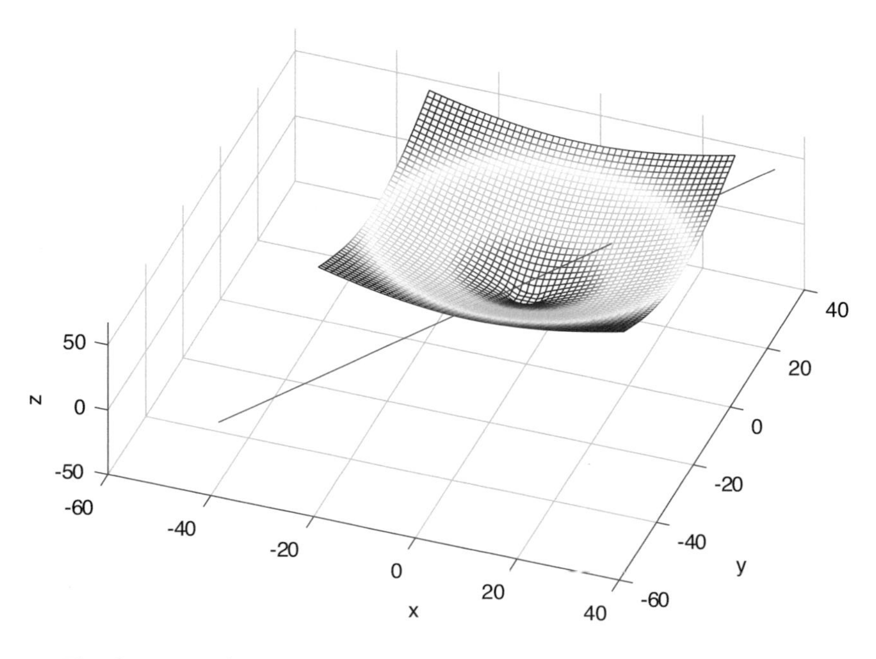

FIGURE 8.12 Plot for Example 8.4.

MATLAB Code 8.4

```
clear; clc;
syms t;
x = t - 4; y = t - 5; z = t + 10;
P = 2*x^2 + 3*y^2 - z^2 - 16;
R = solve(P);
R1 = R(1); eval(R1);
R2 = R(2); eval(R2);
X1 = subs(x, 't', R1); X1 = eval(X1);
Y1 = subs(y, 't', R1); Y1 = eval(Y1);
Z1 = subs(z, 't', R1); Z1 = eval(Z1);
X2 = subs(x, 't', R2); X2 = eval(X2);
Y2 = subs(y, 't', R2); Y2 = eval(Y2);
Z2 = subs(z, 't', R2); Z2 = eval(Z2);
fprintf('Point of intersection 1 : \n a = (%f, %f, %f)\n', X1, Y1, Z1);
fprintf('Point of intersection 2 : \n b = (%f, %f, %f)\n', X2, Y2, Z2);
%verification
vrf1 = subs(P, [x, y, z], [X1, Y1, Z1]); eval(vrf1)
vrf2 = subs(P, [x, y, z], [X2, Y2, Z2]); eval(vrf2)
```

```
%plotting
syms u v;
x = u;
y = v;
z = sqrt(2*u^2 + 3*v^2 - 16);
k = 30; ezmesh(x,y,z, [-k,k,-k,k,-k,k]);
hold on;
ezplot3('t-4', 't-5', 't + 10', [-40, 40]);
view(20, 64); hold off;
```

NOTE

`ezplot3`: plots symbolic variables directly in 3D environments

8.5 EXTRUDED SURFACES

Extruded surfaces are created when a 2D plane curve is moved along a straight line in a direction perpendicular to the plane (O'Rourke., 2003). If the generating curve is represented as $C(u) = \{x(u), y(u)\}$ then the resulting surface is expressed in parametric form:

$$S(u, v) = \{x(u), y(u), v\} \tag{8.15}$$

Here, $0 \le u \le 1, 0 \le v \le h$, where u is along the plane containing the curve and v is perpendicular to the plane, and h is the maximum distance by which the curve is moved.

Example 8.5

Find the extruded surface produced from the generating curves: (a) $x = \sin(u)$, $y = \cos(u)$ and (b) $x = \sin(2u)$, $y = \sin(u)$

(a) $C = \{\sin(u), \cos(u)\}$
$S(u, v) = \{\sin(u), \cos(u), v\}$
(b) $C = \{\sin(2u), \sin(u)\}$
$S(u, v) = \{\sin(2u), \sin(u), v\}$
(Figure 8.13)

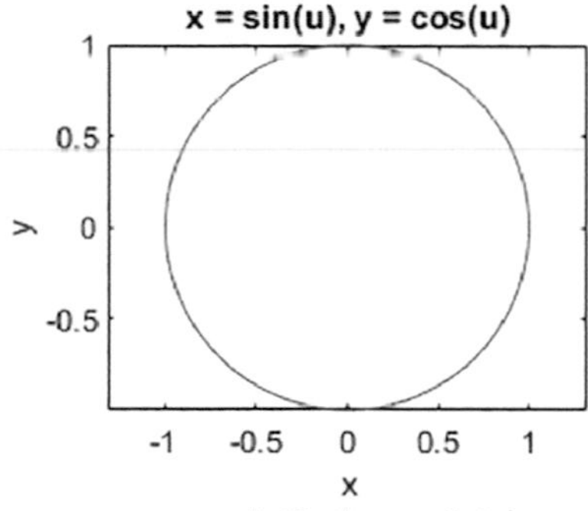

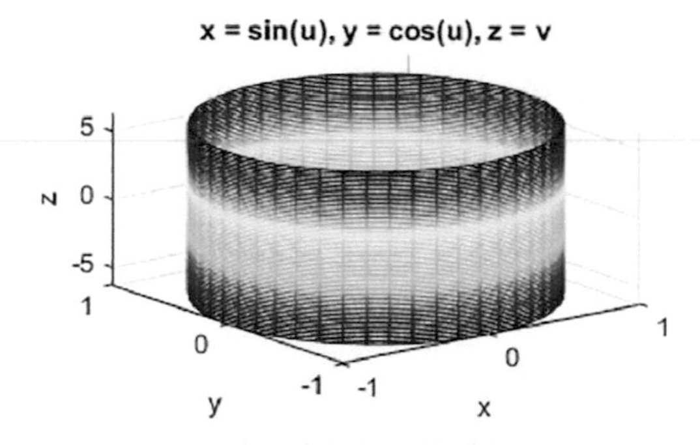

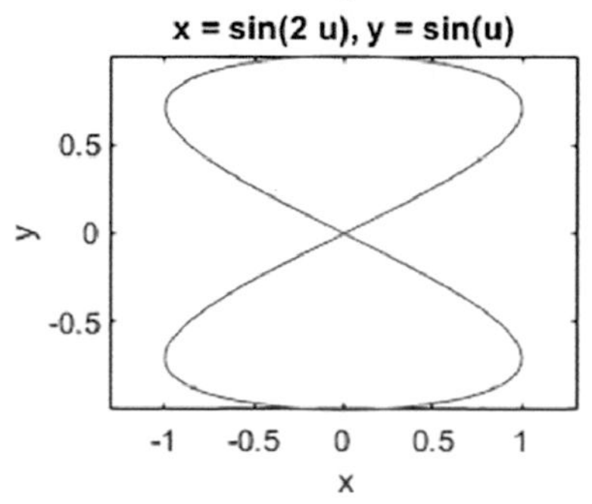

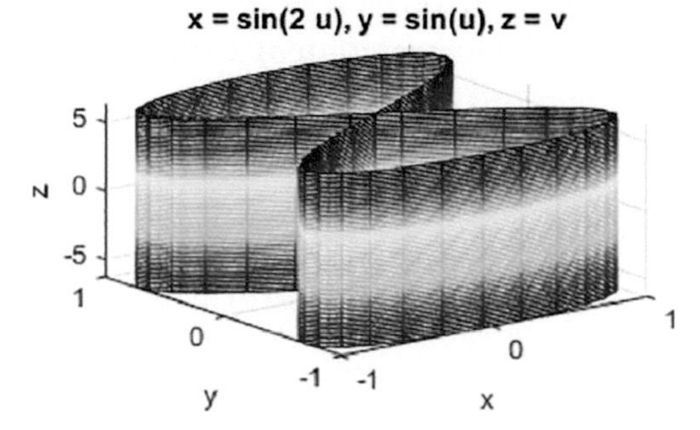

FIGURE 8.13 Plot for Example 8.5.

MATLAB Code 8.5

```
clear all; clc;
syms u v;

x1 = sin(u); y1 = cos(u); z1 = v;
C1 = [x1, y1]
S1 = [x1, y1, z1]
x2 = sin(2*u); y2 = sin(u); z2 = v;
C2 = [x2, y2]
S2 = [x2, y2, z2]

subplot(221), ezplot(C1(1), C1(2));
subplot(222), ezmesh(S1(1), S1(2), S1(3));
subplot(223), ezplot(C2(1), C2(2));
subplot(224), ezmesh(S2(1), S2(2), S2(3));
```

8.6 SURFACES OF REVOLUTION

Surfaces of revolution are created when a 2D plane curve is rotated about an axis (O'Rourke., 2003), (Rovenski, 2010). If the generating curve is represented as $C(u)=\{x(u), y(u)\}$, then the resulting surface is expressed in parametric form as:

$$S(u, v)=\{x(u)\cdot\cos v, x(u)\cdot\sin v, y(u)\} \tag{8.16}$$

Here, $0\le u\le 1$, $0\le v\le 2\pi$. If $v=2\pi$, the surface is closed, if $v<2\pi$ the surface is open. Note that in 3D space the 2D curve is generated on the X–Z plane and rotates on a circle on the X–Y plane. If the curve is required to be drawn on the X–Y plane and rotate on the X–Z plane then Equation (8.16) should be re-written as below:

$$S(u, v)=\{x(u)\cdot\cos v, y(u), x(u)\cdot\sin v\} \tag{8.17}$$

Example 8.6

Show that the revolution of a straight line can be used to produce a cylinder and cone, and the revolution of a circle can be used to produce a sphere and torus. Find the equations of resulting surfaces. Also generate a surface from a rotating sinusoid.

Cylinder:
Curve: $x=2$, $y=u$, $C(u)=\{x(u), y(u)\}$
Surface: $S(u, v)=\{2\cdot\cos(v), 2\cdot\sin(v), u\}$

Cone:
Curve: $x=u$, $y=u$, $C(u)=\{x(u), y(u)\}$
Surface: $S(u, v)=\{u\cdot\cos(v), u\cdot\sin(v), u\}$

Sphere:
Curve: $x = \cos(u),\ y = \sin(u),\ C(u) = \{x(u),\ y(u)\}$
Surface: $S(u,\ v) = \{\cos(u) \times \cos(v),\ \cos(u) \cdot \sin(v),\ \sin(u)\}$

Torus:
Curve: $x = 2 + \cos(u),\ y = 2 + \sin(u),\ \ C(u) = \{x(u),\ y(u)\}$
Surface: $S(u,\ v) = \{(2 + \cos(u)) \cdot \cos(v),\ (2 + \cos(u)) \cdot \sin(v),\ (2 + \sin(u))\}$

Sinusoid
Curve: $x = 2 + \sin(u),\ y = u,\ C(u) = \{x(u),\ y(u)\}$
Surface: $S(u,\ v) = \{(2 + \sin(u)) \cdot \cos(v),\ (2 + \sin(u)) \cdot \sin(v),\ u\}$
Curve: $x = 2 + \cos(u),\ y = u,\ C(u) = \{x(u),\ y(u)\}$
Surface: $S(u,\ v) = \{(2 + \cos(u)) \cdot \cos(v),\ (2 + \cos(u)) \cdot \sin(v),\ u\}$
(Figures 8.14–8.16)

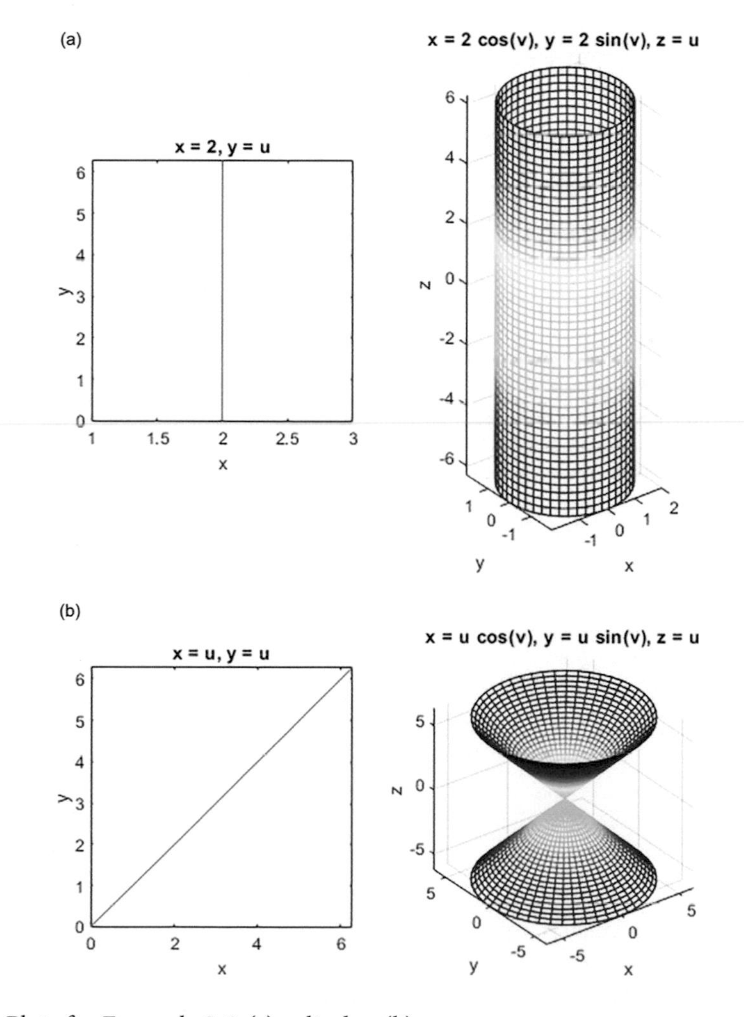

FIGURE 8.14 Plots for Example 8.6: (a) cylinder; (b) cone.

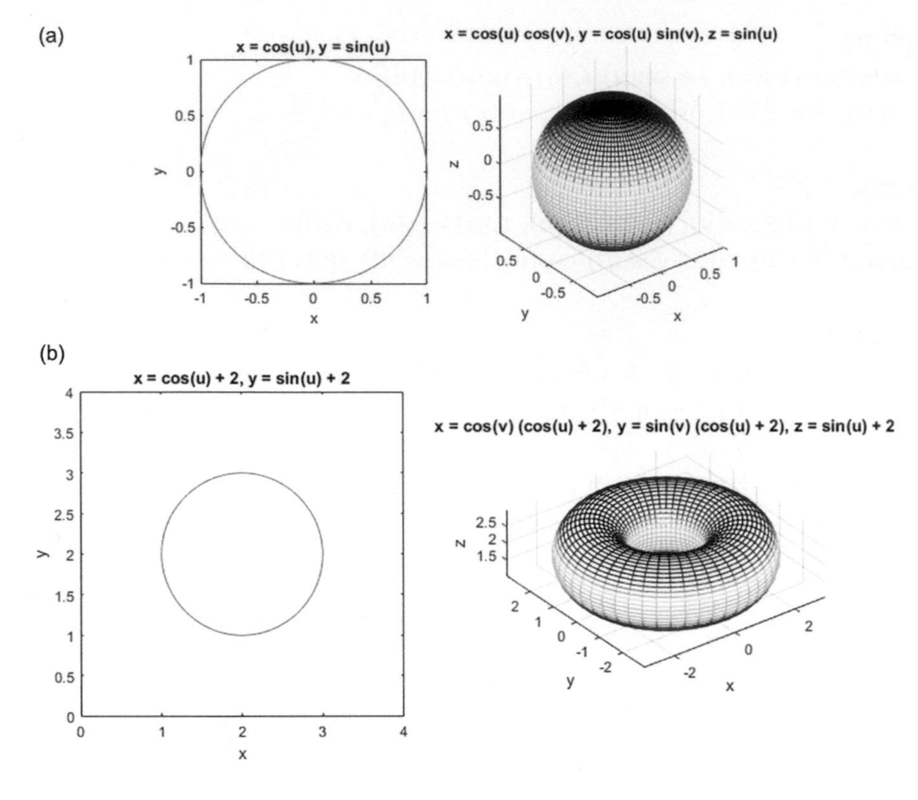

FIGURE 8.15 Plots for Example 8.6: (a) sphere; (b) torus

MATLAB Code 8.6

```
clear all; clc;
syms u v;

% cylinder
x = 2; y = u;
C = [x, y];
S = [x*cos(v), x*sin(v), y];
figure,
subplot(121), ezplot(C(1), C(2)); axis square;
subplot(122), ezmesh(S(1), S(2), S(3)); axis equal;
colormap(jet);

% cone
x = u; y = u;
C = [x, y];
S = [x*cos(v), x*sin(v), y];
figure,
subplot(121), ezplot(C(1), C(2)); axis square;
```

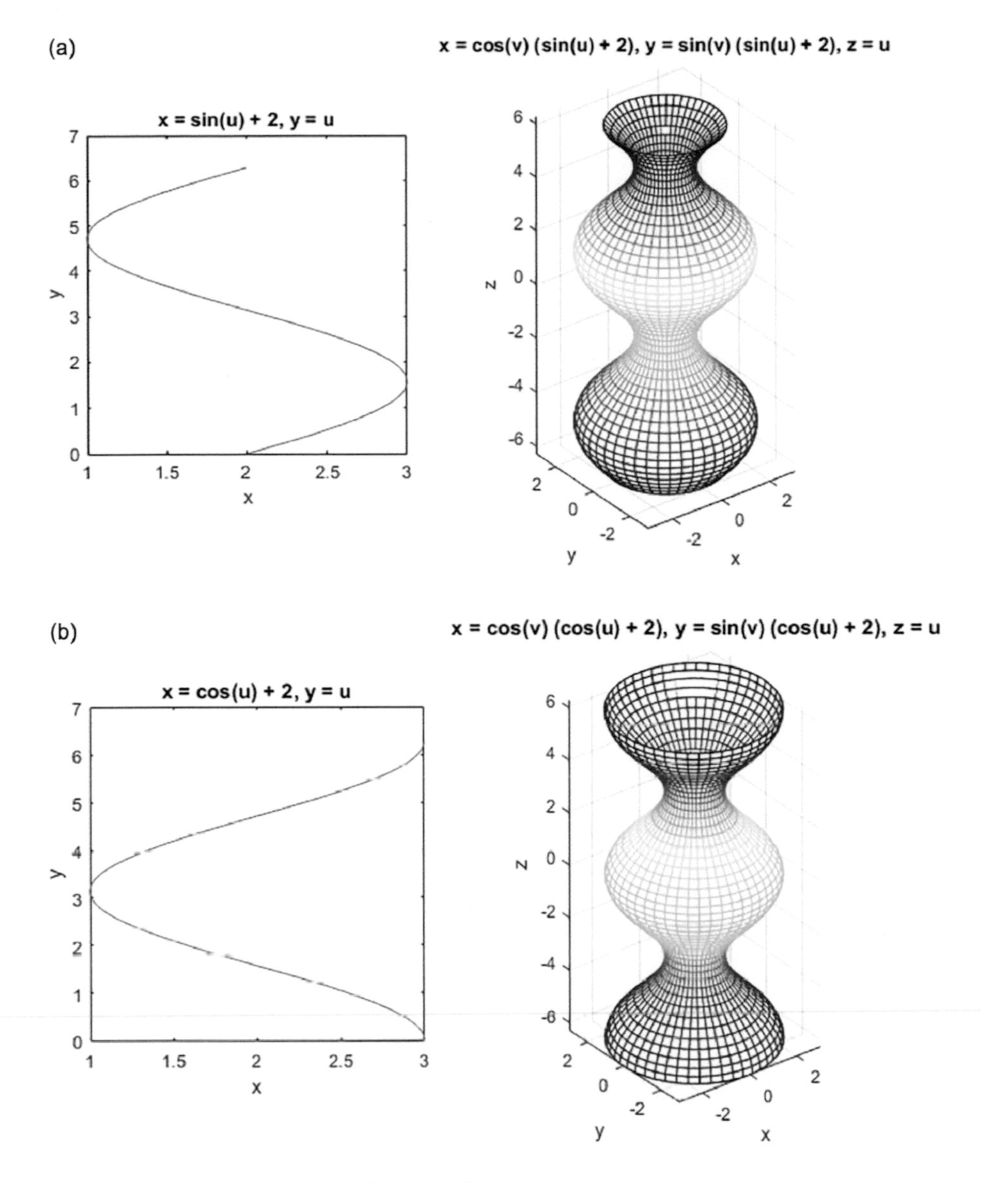

FIGURE 8.16 Plots for Example 8.6: (a) sine; (b) cosine.

```
subplot(122), ezmesh(S(1), S(2), S(3)); axis equal;
colormap(jet);

% sphere
x = cos(u); y = sin(u);
C = [x, y];
S = [x*cos(v), x*sin(v), y];
figure,
subplot(121), ezplot(C(1), C(2)); axis square;
```

```
subplot(122), ezmesh(S(1), S(2), S(3)); axis equal;
colormap(jet);

% torus
x = 2 + cos(u); y = 2 + sin(u);
C = [x, y];
S = [x*cos(v), x*sin(v), y];
figure,
subplot(121), ezplot(C(1), C(2)); axis ([0 4 0 4]); axis square;
subplot(122), ezmesh(S(1), S(2), S(3)); axis equal;
colormap(jet);

% sinusoid
x = 2 + sin(u); y = u;
C = [x, y];
S = [x*cos(v), x*sin(v), y];
figure,
subplot(121), ezplot(C(1), C(2)); axis square;
subplot(122), ezmesh(S(1), S(2), S(3)); axis equal;
colormap(jet);

x = 2 + cos(u); y = u;
C = [x, y];
S = [x*cos(v), x*sin(v), y];
figure,
subplot(121), ezplot(C(1), C(2)); axis square;
subplot(122), ezmesh(S(1), S(2), S(3)); axis equal;
colormap(jet);
```

8.7 NORMAL VECTOR AND TANGENT PLANE

This section takes a look at how the normal vector and the tangent plane can be calculated for a given surface, both for implicit equations and parametric equations.

For an implicit equation of the form $f(x, y, z)=0$, the normal to the surface at point $p(a, b, c)$ is given by the vector obtained by partial derivatives of the function f since these are proportional to the coefficients of the Cartesian equation of a plane perpendicular to the normal. See Chapter 6 to see why this so.

$$N(x, y, z)=\left(\frac{\partial f}{\partial x}, \frac{\partial f}{\partial y}, \frac{\partial f}{\partial z}\right) \tag{8.18}$$

The normal to this point p is, therefore, given by:

$$N_p = N(a, b, c)=\left(\frac{\partial f}{\partial x}, \frac{\partial f}{\partial y}, \frac{\partial f}{\partial z}\right)\Bigg|_{(a,b,c)} = \left(\frac{\partial f_p}{\partial x}, \frac{\partial f_p}{\partial y}, \frac{\partial f_p}{\partial z}\right) \tag{8.19}$$

If this plane needs to be the tangent plane then the plane should also pass through the point p on the surface. The equation of the tangent plane through p can therefore be written as (Rovenski, 2010):

$$T_p : \frac{\partial f_p}{\partial x} \cdot (x-a) + \frac{\partial f_p}{\partial y} \cdot (y-b) + \frac{\partial f_p}{\partial z} \cdot (z-c) = 0 \tag{8.20}$$

Example 8.7

Find the normal vector and tangent plane to the surface $f = x^3 + 3xy^2 - z = 0$ at point $p(1, 2, 13)$.

Here, $f = x^3 + 3xy^2 - z$

From Equation (8.18), normal N is given by

$$N(x, y, z) = \left(\frac{\partial f}{\partial x}, \frac{\partial f}{\partial y}, \frac{\partial f}{\partial z} \right) = \left(3x^2 + 3y^2, 6xy, -1 \right)$$

At point $p(1, 2, 13)$, normal vector $N_p = (15, 12, -1)$.

Equation of the tangent plane at point p: $N_p = 15(x-1) + 12(y-2) - 1(z-13) = 0$, which on simplification becomes $15x + 12y - z = 26$. (Figure 8.17)

NOTE

The z- coefficient for the vector N_p is negative, which means that the vector is pointing toward the negative Z-direction. Sometimes, if we are asked to calculate the upward pointing normal then we should simply choose the opposite facing vector toward the positive Z-direction i.e. $(-15, -12, 1)$

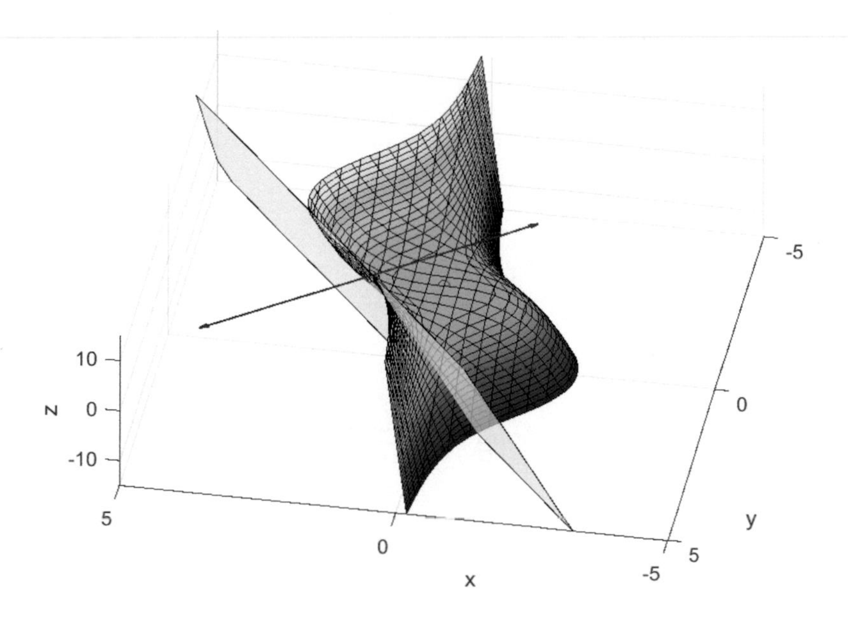

FIGURE 8.17 Plot for Example 8.7.

MATLAB Code 8.7

```
clear all; clc;
syms x y z;
p = [1, 2, 13];
f = x^3 + 3*x*y^2 - z;
dfx = diff(f, x);
dfy = diff(f, y);
dfz = diff(f, z);
dfxp = subs(dfx, [x, y, z], [p(1), p(2), p(3)]);
dfyp = subs(dfy, [x, y, z], [p(1), p(2), p(3)]);
dfzp = subs(dfz, [x, y, z], [p(1), p(2), p(3)]);
N = [dfx, dfy, dfz];
fprintf('Normal vector : \n');
Np = [dfxp, dfyp, dfzp]
Np = Np/5;
fprintf('Tangent plane : \n');
T = dfxp*(x - p(1)) + dfyp*(y - p(2)) + dfzp*(z - p(3))

%plotting
fimplicit3(f); hold on;
axis([-5 5 -5 5 -15 15]);
fimplicit3(T, 'MeshDensity', 2, 'FaceColor', 'y', 'FaceAlpha',0.6);
plot3([0 1], [0 2], [0 13], 'ro');
quiver3(p(1), p(2), p(3), Np(1), Np(2), Np(3), 'LineWidth', 1, 'color', 'r', 'MarkerSize', 10)
quiver3(p(1), p(2), p(3), -Np(1), -Np(2), -Np(3), 'LineWidth', 1, 'color', 'b', 'MarkerSize', 10)
xlabel('x'); ylabel('y'); zlabel('z');
view(-170,64); hold off;
```

If the surface equation is in parametric form $S=\{x(u, v), y(u, v), z(u, v)\}$, to compute the normal we first need to find out the component curves of the surface i.e. the curves along u and v, which has resulted in the surface being generated. The component curves are computed using partial derivatives of the surface (Rovenski, 2010).

$$r_u(u, v)=\frac{\partial S}{\partial u}$$
$$r_v(u, v)=\frac{\partial S}{\partial v} \tag{8.21}$$

Since the component curves lie on the surface, the normal is obtained as a cross product of these.

$$N=r_u(u, v)\times r_v(u, v) \tag{8.22}$$

To compute the normal vector at a given point $p=(u_0, v_0)$ the corresponding Cartesian coordinates are obtained from $S_p=\{x_0, y_0, z_0\}$. The component curves at the given point are $r_u(u_0, v_0)$ and $r_v(u_0, v_0)$ and the normal vector at the given point is

$$N_p=r_u(u_0, v_0)\times r_v(u_0, v_0)=(a, b, c) \tag{8.23}$$

The equation of the tangent plane at the point is

$$T_p:\ a(x-x_0)+b(y-y_0)+c(z-z_0)=0 \tag{8.24}$$

Example 8.8

Find the normal vector and tangent plane to the surface $S=\{u\cdot\sin v, u^2, 2u\cdot\cos v\}$ at point $(u=1, v=\pi)$.

Given surface: $S=\{u\cdot\sin v, u^2, 2u\cdot\cos v\}$

At $u=1, v=\pi$ we get by substitution, point p on the surface $S_p=\{x_0, y_0, z_0\}=\{0, 1, -2\}$

The component curves at p: $r_u(1, \pi)=(0, 2, -2)$ and $r_v(1, \pi)=(-1, 0, 0)$

From Equation (8.22), normal vector at point p: $N_p=r_u(1, \pi)\times r_v(1, \pi)=(0, 2, 2)$ i.e. $2j+2k$

From Equation (8.24), tangent plane at p: T_p: $2(y-1)+2(z+2)=0$ i.e. $y+z=-1$ (Figure 8.18)

MATLAB Code 8.8

```
clear all; clc;
syms u v x y z;
S = [u*sin(v), u^2, 2*u*cos(v)];
ru = diff(S, u);
rv = diff(S, v);
N = cross(ru, rv);

up = 1; vp = pi;
p = subs(S, [u, v], [up, vp]);
rup = subs(ru, [u, v], [up, vp]);
rvp = subs(rv, [u, v], [up, vp]);
fprintf('Normal vector : \n');
Np = cross(rup, rvp)
fprintf('Tangent plane : \n');
Tp = Np(1)*(x - p(1)) + Np(2)*(y - p(2)) + Np(3)*(z - p(3))

%plotting
Np = 10*Np; %scaling for visualization
ezmesh(S(1), S(2), S(3)); hold on;
f = @(x,y,z) Np(1)*(x - p(1)) + Np(2)*(y - p(2)) + Np(3)*(z - p(3));
fimplicit3(f, 'MeshDensity', 2, 'FaceColor', 'y', 'FaceAlpha',0.6)
axis([-10 10 -10 40 -15 15]);
plot3(p(1), p(2), p(3), 'ro');
quiver3(p(1), p(2), p(3), Np(1), Np(2), Np(3), 1, 'color', 'r');
quiver3(p(1), p(2), p(3), -Np(1), -Np(2), -Np(3), 1, 'color', 'b');
xlabel('X'); ylabel('Y'); zlabel('Z');
view(50,34); hold off;
```

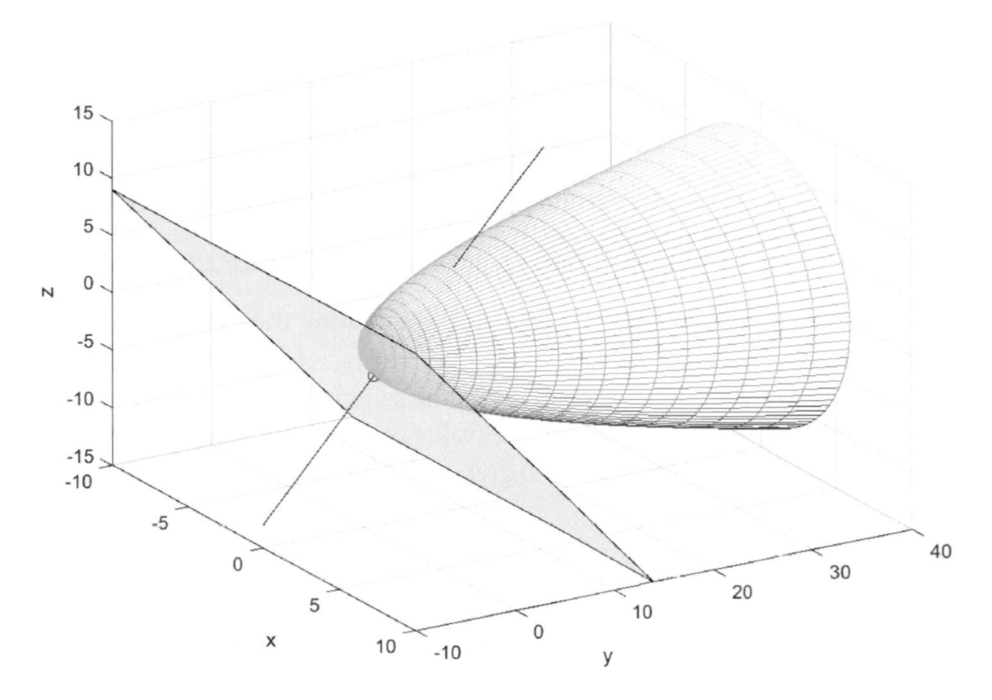

FIGURE 8.18 Plot for Example 8.8.

8.8 AREA AND VOLUME OF SURFACE OF REVOLUTION

This section discusses how to calculate the surface area and volume of the solid generated when a 2D curve is rotated about the X- or Y-axes perpendicular to the plane (Mathews, 2004). Consider a continuous function $y = f(x)$ in the interval $x \in [a, b]$ that needs to be rotated about the X-axis to generate a surface of revolution (see Figure 8.19). It is required to calculate the surface area and the volume of the surface.

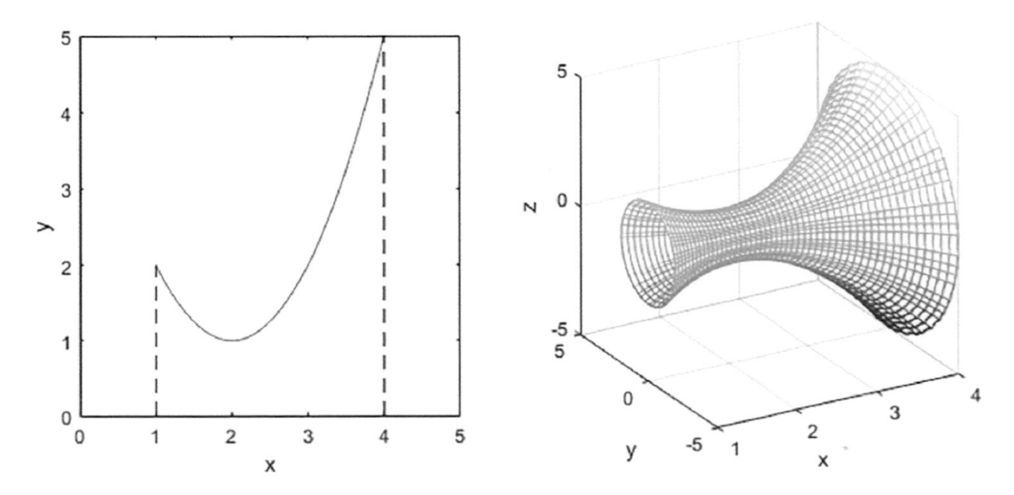

FIGURE 8.19 Curve and its surface of revolution.

The area of the surface generated is given by:

$$A = \int_a^b 2\pi y \cdot \sqrt{1 + \left(\frac{dy}{dx}\right)^2}\, dx \tag{8.25}$$

The above formula is derived by dividing the function into very thin slices each of arc length $\sqrt{1+\left(\frac{dy}{dx}\right)^2}$ and rotating each along a circle of circumference $2\pi y$ (See Section 5.4). The area of the whole surface is then obtained by integrating the individual areas within the specified limits.

The volume of the solid is derived by dividing the solid into very thin discs each of area πy^2 since the radius of each disc is simply the value of the function y measured from the X-axis. The total volume is obtained by integrating the areas of all discs over the specified limits and is given by:

$$V = \int_a^b \pi y^2\, dx \tag{8.26}$$

If the region is bounded by two curves y_1 and y_2 then the volume of the bounded region is given by: $V = \int_a^b \pi\left(y_1{}^2 - y_2{}^2\right) dx$

Example 8.9

Find the area and volume of the surface generated when part of the curve $y = \sqrt{9 - x^2}$ between $x = -2$ and $x = 2$ is rotated about the X-axis.

Given curve: $y = \sqrt{9 - x^2}$

Differentiating: $\frac{dy}{dx} = -x/\sqrt{9 - x^2}$

Thus: $\sqrt{1+\left(\frac{dy}{dx}\right)^2} = \frac{3}{\sqrt{9 - x^2}}$

From Equation (8.25), surface area: $A = \int_{-2}^{2} 2\pi\sqrt{9 - x^2} \cdot \frac{3}{\sqrt{9 - x^2}} = 24\pi$

From Equation (8.26), volume: $V = \int_{-2}^{2} \pi\left(9 - x^2\right) dx = \frac{92\pi}{3}$ (Figure 8.20)

MATLAB Code 8.9

```
clear all; clc;
syms x y u v;
y = sqrt(9 - x^2);
dy = diff(y, x);
a = sqrt(1 + dy^2);
fprintf('Area : \n');
A = int(2*pi*y*a, x, -2, 2)
```

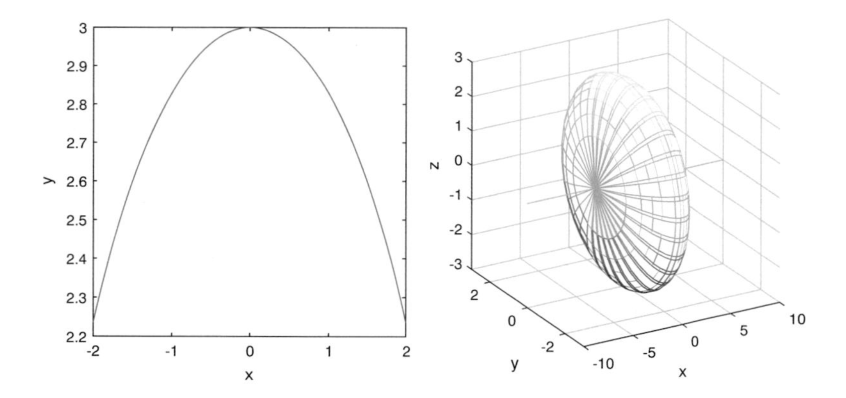

FIGURE 8.20 Plot for Example 8.9.

```
B = pi*y^2;
fprintf('Volume : \n');
V = int(B, x, -2, 2)

%plotting
C = [x, y];
figure,
subplot(121), ezplot(C(1), C(2), [-2 2]); axis square;
subplot(122), ezmesh(C(1), C(2)*cos(v), C(2)*sin(v), [-5 5 -10 10]);
axis square;
view(-30, 23);
```

If the given curve is represented as $x = f(y)$ in the interval $y \in [c, d]$ that needs to be rotated about the Y-axis to generate a surface of revolution, the area is derived by multiplying thin slices of arc length $\sqrt{1+\left(\frac{dx}{dy}\right)^2}$ by the circumference of $2\pi x$ and then integrating over the specified limits.

$$A = \int_c^d 2\pi x \cdot \sqrt{1+\left(\frac{dx}{dy}\right)^2}\, dy \tag{8.27}$$

The volume of the solid is derived by dividing the solid into very thin discs each of area πx^2 since the radius of each disc is simply the value of the function x measured from the Y-axis. The total volume is obtained by integrating the areas of all discs over the specified limits and is given by:

$$V = \int_c^d \pi x^2\, dy \tag{8.28}$$

If the region is bounded by two curves x_1 and x_2 then the volume of the bounded region is given by: $V = \int_c^d \pi\left(x_1{}^2 - x_2{}^2\right) dy$

Example 8.10

Find the area and volume of the surface generated when part of the curve $y = \sqrt[3]{x}$ between $y = 1$ and $y = 2$ is rotated about the Y-axis.

Given curve: $x = y^3$

Differentiating: $\dfrac{dx}{dy} = 3y^2$

Thus: $\sqrt{1 + \left(\dfrac{dx}{dy}\right)^2} = \sqrt{1 + 9y^4}$

From Equation (8.26), surface area: $A = \int_1^2 2\pi y^3 \cdot \sqrt{1 + 9y^4}$

Let $u = 1 + 9y^4$ so that when $y = 1$ then $u = 10$ and when $y = 2$ then $u = 145$

Also $\dfrac{du}{dy} = 36y^3$

From Equation (8.27), surface area: $A = \int_1^2 2\pi y^3 \cdot \sqrt{1 + 9y^4}$

$$dy = \int_{10}^{145} 2\pi \cdot \sqrt{u} \cdot \frac{du}{36} = 199.48$$

From Equation (8.28), volume: $V = \int_1^2 \pi y^6 \, dy = \dfrac{127\pi}{7} = 56.99$

MATLAB Code 8.10

```
clear all; clc;
syms y;
x = y^3;
d = diff(x);
e = sqrt(1 + d^2);
f = 2*pi*x*e;
fprintf('Area : \n');
A = int(f, 1, 2); A = eval(A)
fprintf('Volume : \n');
V = int(pi*x^2, 1, 2); V = eval(V)
```

8.9 TEXTURE MAPPING

Texture mapping is a process of applying an image, referred to as the texture, onto a surface, usually for the purpose of making the surface more realistic in appearance (Shirley, 2002) (see Figure 8.21).

The texture being an image is a rectangular 2D matrix of pixel values, usually represented using (u, v) coordinate values while the surface, generally in 3D space, can be of any arbitrary shape and is represented using (x, y, z) coordinate values. To determine which part of the image is applied to which part of the surface a mapping between the (u, v) and the (x, y, z) values is necessary to maintain homogeneous integrity (O'Rourke., 2003).

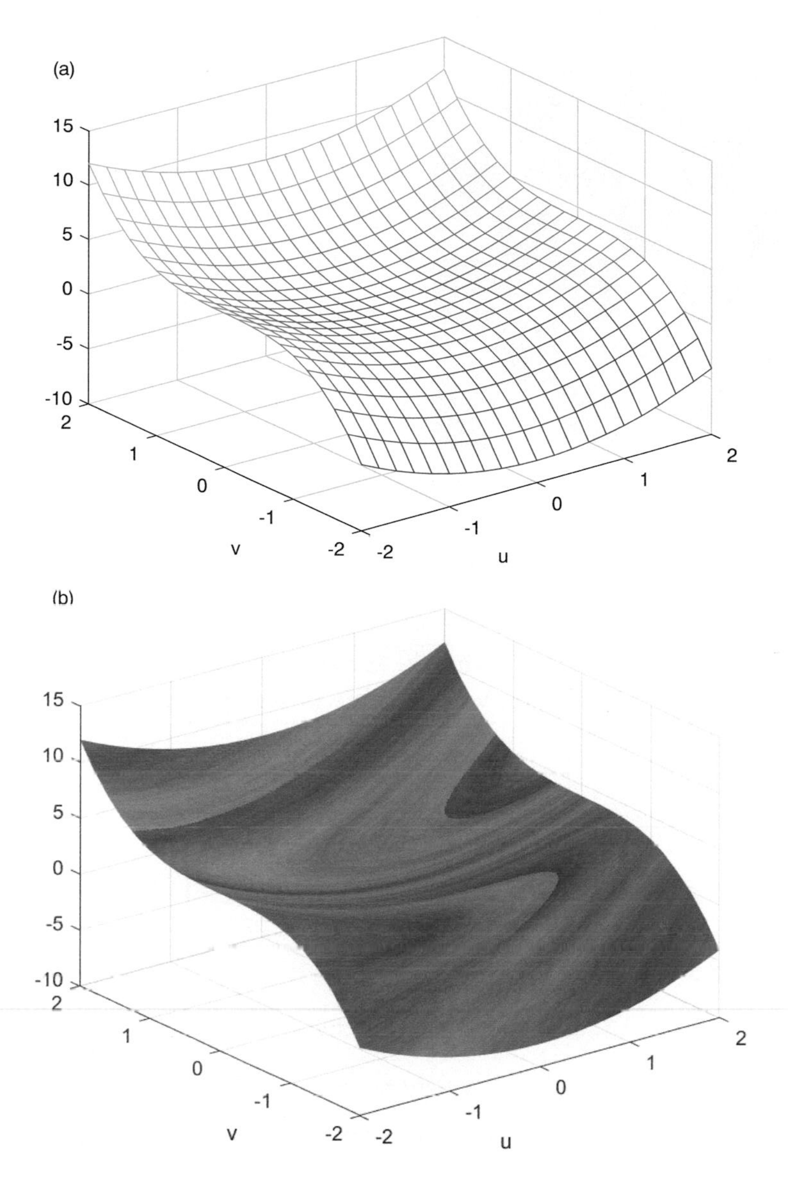

FIGURE 8.21 Surface (a) before and (b) after applying texture map.

We will study the mapping process and derive the mathematical transformations required for a 2D surface. The concepts could readily be extended for a 3D surface by the reader. There are broadly two types of mapping transformations: affine and perspective. An affine transformation can be generated when the mapping of the texture on the surface keeps parallel lines intact. Consider a texture described by normalized (u, v) coordinates in the range [0, 1] and the surface described by (x, y) coordinates (see Figure 8.22).

The corners of the rectangular texture image have (u, v) coordinates of (0, 0), (1, 0), (1, 1), (0, 1) and these points are mapped to the surface having (x, y) coordinates of $(x_0, y_0), (x_1, y_1), (x_2, y_2), (x_3, y_3)$. It is required to derive a mapping transformation for the operation. Let the shape of the surface has the following constraints: $(x_1 - x_0) = (x_2 - x_3)$

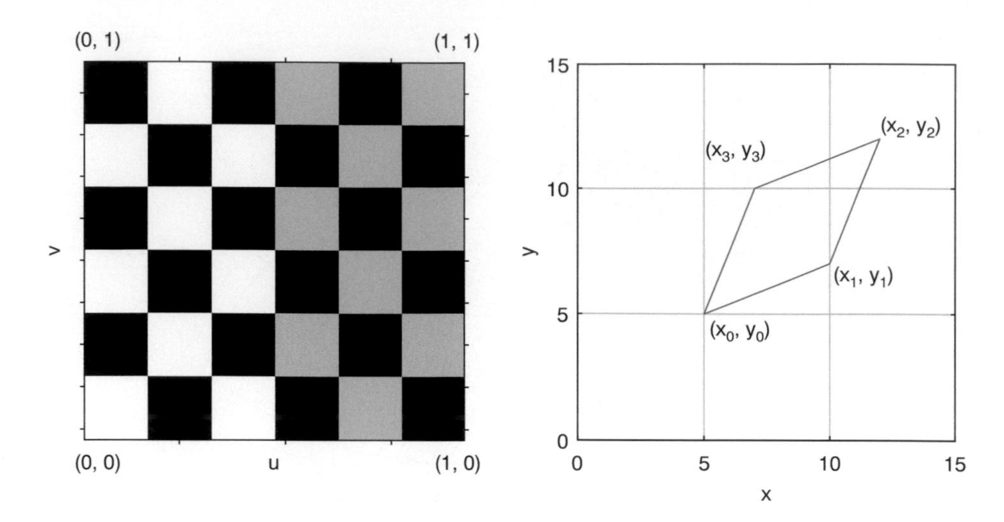

FIGURE 8.22 Texture with affine mapping.

and $(y_1 - y_0) = (y_2 - y_3)$. This implies that the surface has the shape of a parallelogram and hence the mapping transformation will be affine in nature. An affine transformation is given by the following:

$$\begin{bmatrix} x \\ y \\ 1 \end{bmatrix} = \begin{bmatrix} a & b & c \\ d & e & f \\ 0 & 0 & 1 \end{bmatrix} \begin{bmatrix} u \\ v \\ 1 \end{bmatrix} \tag{8.29}$$

From Equation (8.29) we have:

$$\begin{aligned} x &= au + bv + c \\ y &= du + ev + f \end{aligned} \tag{8.30}$$

Plugging in the given boundary conditions (BC), we can solve for the unknown coefficients a, b, c, d, e, f as follows:

BC-1: $u = 0, v = 0$ is mapped to $x = x_0, y = y_0$, which implies $c = x_0$ and $f = y_0$

BC-2: $u = 1, v = 0$ is mapped to $x = x_1, y = y_1$, which implies $a = x_1 - x_0$ and $d = y_1 - y_0$

BC-3: $u = 0, v = 1$ is mapped to $x = x_3, y = y_3$, which implies $b = x_3 - x_0$ and $e = y_3 - y_0$

Substituting the coefficient values in Equation (8.29):

$$\begin{bmatrix} x \\ y \\ 1 \end{bmatrix} = \begin{bmatrix} x_1 - x_0 & x_3 - x_0 & x_0 \\ y_1 - y_0 & y_3 - y_0 & y_0 \\ 0 & 0 & 1 \end{bmatrix} \begin{bmatrix} u \\ v \\ 1 \end{bmatrix} \tag{8.31}$$

Example 8.11

Derive a transformation for mapping a texture with (u, v) coordinates (0, 0), (1, 0), (1, 1), (0, 1) to a surface with (x, y) coordinates of (5, 5), (10, 7), (12, 12), (7, 10). Also for the texture point (u = 0.6, v = 0.7), find the corresponding point on the surface.

Surface coordinates: $(x_0, y_0) = (5, 5)$, $(x_1, y_1) = (10, 7)$, $(x_2, y_2) = (12, 12)$, $(x_3, y_3) = (7, 10)$

Now $(x_1 - x_0) = 5, (x_2 - x_3) = 5, (y_1 - y_0) = 2, (y_2 - y_3) = 2$

Since $(x_1 - x_0) = (x_2 - x_3)$ and $(y_1 - y_0) = (y_2 - y_3)$ the mapping transformation is affine in nature.

From Equation (8.31), transformation matrix $M = \begin{bmatrix} x_1 - x_0 & x_3 - x_0 & x_0 \\ y_1 - y_0 & y_3 - y_0 & y_0 \\ 0 & 0 & 1 \end{bmatrix}$

$$= \begin{bmatrix} 5 & 2 & 5 \\ 2 & 5 & 5 \\ 0 & 0 & 1 \end{bmatrix}$$

From Equation (8.30),

$$x = 5u + 2v + 5$$

$$y = 2u + 5v + 5$$

For $(u = 0.6, v = 0.7)$,

$$x = 5(0.6) + 2(0.7) + 5 = 9.4$$

$$y = 2(0.6) + 5(0.7) + 5 = 9.7$$

Verification:

$x(0,0) = 5$, $y(0,0) = 5$, $x(0,1) = 7$, $y(0,1) = 10$, $x(1,0) = 10$, $y(1,0) = 7$, $x(1,1) = 12$, $y(1,1) = 12$

(Figure 8.23):

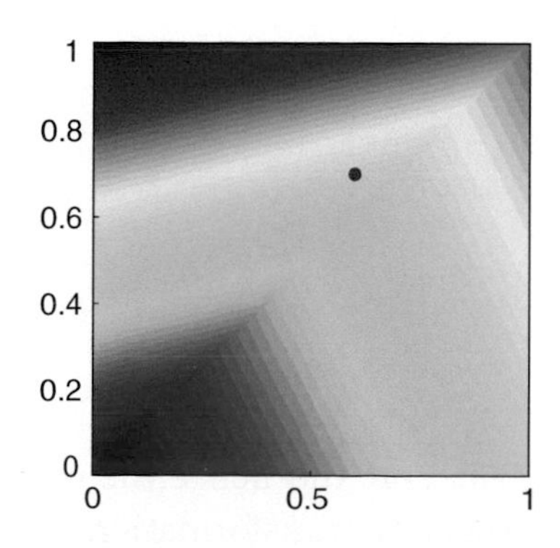

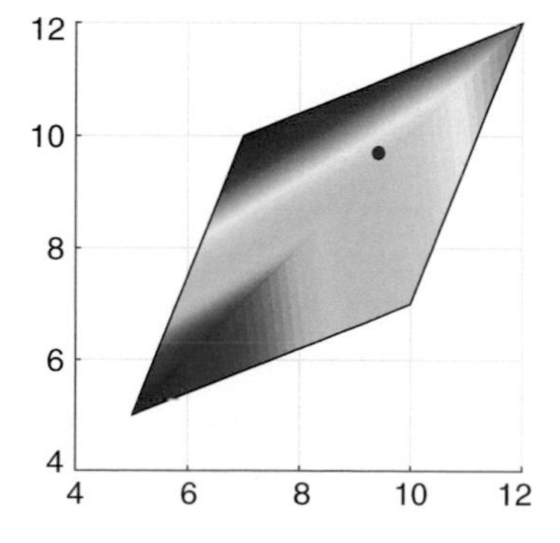

FIGURE 8.23 Plot for Example 8.11.

MATLAB Code 8.11

```
clear all; clc;

x0 = 5; y0 = 5;
x1 = 10; y1 = 7;
x2 = 12; y2 = 12;
x3 = 7; y3 = 10;

d1 = x1 - x0; d2 = x2 - x3;

if d1 == d2
    fprintf('Transformation is affine\n');
else
    fprintf('Transformation is perspective\n');
end

U = [0 1 1 ; 0 0 1 ; 1 1 1];
X = [x0 x1 x2 ; y0 y1 y2 ; 1 1 1];
fprintf('Transformation matrix : \n');
M = X*inv(U)

X1 = M*[0.6 ; 0.7 ; 1];
fprintf('For u=0.6, v=0.7, (x,y) = (%.2f, %.2f) \n', X1(1),X1(2) );

subplot(121)
x = [0 1 1 0];
y = [0 0 1 1];
c = [0 4 6 8];
colormap(jet);
patch(x,y,c); hold on;
scatter(0.6, 0.7, 20, 'r', 'filled');
axis square; hold off;

subplot(122)
x = [5 10 12 7];
y = [5 7 12 10];
c = [0 4 6 8];
colormap(jet);
patch(x,y,c); hold on;
scatter(X1(1), X1(2), 20, 'r', 'filled');
axis square; grid; hold off;
```

If the shape of the surface be such that: $(x_1 - x_0) \neq (x_2 - x_3)$ and $(y_1 - y_0) \neq (y_2 - y_3)$, it implies that the surface is an arbitrary quadrilateral and hence the mapping transformation will be perspective in nature. A perspective transformation is given by the following:

$$\begin{bmatrix} x' \\ y' \\ w \end{bmatrix} = \begin{bmatrix} a & b & c \\ d & e & f \\ g & h & 1 \end{bmatrix} \begin{bmatrix} u \\ v \\ 1 \end{bmatrix} \tag{8.32}$$

where x' and y' are in homogeneous coordinates. The Cartesian coordinates are $x = x'/w$ and $y = y'/w$. From Equation (8.32) we get:

$$x = \frac{x'}{w} = \frac{au + bv + c}{gu + hv + 1}$$
$$y = \frac{y'}{w} = \frac{du + ev + f}{gu + hv + 1} \tag{8.33}$$

Plugging in the given BC, we can solve for the unknown coefficients a, b, c, d, e, f, g, h as follows:

BC-1: $u = 0, v = 0$ is mapped to $x = x_0, y = y_0$, which implies $c = x_0$ and $f = y_0$

BC-2: $u = 1, v = 0$ is mapped to $x = x_1, y = y_1$, which implies $x_1 = (a + x_0)/(g + 1)$ and $y_1 = (d + y_0)/(g + 1)$

BC-3: $u = 0, v = 1$ is mapped to $x = x_3, y = y_3$, which implies $x_3 = (b + x_0)/(h + 1)$ and $y_3 = (e + y_0)/(h + 1)$

BC-4: $u = 1, v = 1$ is mapped to $x = x_2, y = y_2$, which implies $x_2 = (a + b + x_0)/(g + h + 1)$ and $y_2 = (d + e + y_0)/(g + h + 1)$

The above eight equations can be solved to find out the values of the eight unknown coefficients. The solution of the above equations is given by the following:

$$\Delta x_1 = x_1 - x_2$$

$$\Delta x_2 = x_3 - x_2$$

$$\Delta x_3 = x_0 - x_1 + x_2 - x_3$$

$$\Delta y_1 = y_1 - y_2$$

$$\Delta y_2 = y_3 - y_2$$

$$\Delta y_3 = y_0 - y_1 + y_2 - y_3$$

$$g = \frac{\det\begin{bmatrix} \Delta x_3 & \Delta x_2 \\ \Delta y_3 & \Delta y_2 \end{bmatrix}}{\det\begin{bmatrix} \Delta x_1 & \Delta x_2 \\ \Delta y_1 & \Delta y_2 \end{bmatrix}}$$

$$h = \frac{\det\begin{bmatrix} \Delta x_1 & \Delta x_3 \\ \Delta y_1 & \Delta y_3 \end{bmatrix}}{\det\begin{bmatrix} \Delta x_1 & \Delta x_2 \\ \Delta y_1 & \Delta y_2 \end{bmatrix}}$$

$$a = x_1 - x_0 + g \cdot x_1$$

$$b = x_3 - x_0 + h \cdot x_3$$

$$c = x_0$$

$$d = y_1 - y_0 + g \cdot y_1$$

$$e = y_3 - y_0 + h \cdot y_3$$

$$f = y_0$$

Example 8.12

Derive a transformation for mapping a texture with (u, v) coordinates $(0, 0)$, $(1, 0)$, $(1, 1)$, $(0, 1)$ to a surface with (x, y) coordinates of $(5, 5)$, $(10, 7)$, $(10, 14)$, $(7, 10)$. Specify mapping relations and hence for the texture point $(u = 0.6, v = 0.7)$, find the corresponding point on the surface.

Surface coordinates: $(x_0, y_0) = (5, 5)$, $(x_1, y_1) = (10, 7)$, $(x_2, y_2) = (10, 14)$, $(x_3, y_3) = (7, 10)$

Now $(x_1 - x_0) = 5$, $(x_2 - x_3) = 3$, $(y_1 - y_0) = 2$, $(y_2 - y_3) = 7$

Since $(x_1 - x_0) \neq (x_2 - x_3)$ and $(y_1 - y_0) \neq (y_2 - y_3)$ the mapping transformation is perspective in nature.

Here,

$$\Delta x_1 = x_1 - x_2 = 0$$

$$\Delta x_2 = x_3 - x_2 = -3$$

$$\Delta x_3 = x_0 - x_1 + x_2 - x_3 = -2$$

$$\Delta y_1 = y_1 - y_2 = -7$$

$$\Delta y_2 = y_3 - y_2 = -4$$

$$\Delta y_3 = y_0 - y_1 + y_2 - y_3 = 2$$

$$g = \frac{\det\begin{bmatrix} \Delta x_3 & \Delta x_2 \\ \Delta y_3 & \Delta y_2 \end{bmatrix}}{\det\begin{bmatrix} \Delta x_1 & \Delta x_2 \\ \Delta y_1 & \Delta y_2 \end{bmatrix}} = -0.6667$$

$$h = \frac{\det\begin{bmatrix} \Delta x_1 & \Delta x_3 \\ \Delta y_1 & \Delta y_3 \end{bmatrix}}{\det\begin{bmatrix} \Delta x_1 & \Delta x_2 \\ \Delta y_1 & \Delta y_2 \end{bmatrix}} = 0.6667$$

$$a = x_1 - x_0 + g \cdot x_1 = -1.6667$$

$$b = x_3 - x_0 + h \cdot x_3 = 6.6667$$

$$c = x_0 = 5$$

$$d = y_1 - y_0 + g \cdot y_1 = -2.6667$$

$$e = y_3 - y_0 + h \cdot y_3 = 11.6667$$

$$f = y_0 - 5$$

From Equation (8.32), transformation matrix $M = \begin{bmatrix} a & b & c \\ d & e & f \\ g & h & 1 \end{bmatrix}$

$$= \begin{bmatrix} -1.6667 & 6.6667 & 5 \\ -2.6667 & 11.6667 & 5 \\ -0.6667 & 0.6667 & 1 \end{bmatrix}$$

From Equation (8.33), mapping relations are:

$$x = \frac{au + bv + c}{gu + hv + 1} = \frac{(-1.6667)u + (6.6667)v + 5}{(-0.6667)u + (0.6667)v + 1}$$

$$y = \frac{du + ev + f}{gu + hv + 1} = \frac{(-2.6667)u + (11.6667)v + 5}{(-0.6667)u + (0.6667)v + 1}$$

For $(u = 0.6, v = 0.7)$,

$$x(0.6, 0.7) = 8.6667$$

$$y(0.6, 0.7) = 11.5667$$

Verification:

$$x(0,0)=5, y(0,0)=5, x(0,1)=7, y(0,1)=10, x(1,0)=10, y(1,0)=7, x(1,1)=10, y(1,1)=14$$

MATLAB Code 8.12

```
clear all; clc;

x0 = 5; y0 = 5;
x1 = 10; y1 = 7;
x2 = 10; y2 = 14;
x3 = 7; y3 = 10;

d1 = x1 - x0; d2 = x2 - x3;

if d1 == d2
    fprintf('Transformation is affine\n');
else
    fprintf('Transformation is perspective\n');
end

dx1 = x1 - x2;
dx2 = x3 - x2;
dx3 = x0 - x1 + x2 - x3;
dy1 = y1 - y2;
dy2 = y3 - y2;
dy3 = y0 - y1 + y2 - y3;

g = det([dx3 dx2 ; dy3 dy2 ])/det([dx1 dx2 ; dy1 dy2]);
h = det([dx1 dx3 ; dy1 dy3 ])/det([ dx1 dx2 ; dy1 dy2]);
a = x1 - x0 + g*x1;
b = x3 - x0 + h*x3;
c = x0;
d = y1 - y0 + g*y1;
e = y3 - y0 + h*y3;
f = y0;

fprintf('Transformation matrix : \n');
M = [a b c ; d e f ; g h 1]

X1 = M*[0.6 ; 0.7 ; 1];
fprintf('For u=0.6, v=0.7, (x,y) = (%.2f, %.2f) \n', X1(1),X1(2) );
```

```
% Verification
fprintf('\n Verification : \n');
u=0; v=0; x = (a*u + b*v + c)/(g*u + h*v + 1);
fprintf('u = %d, v = %d, x = %d \n', u, v, x)

u=1; v=0; x = (a*u + b*v + c)/(g*u + h*v + 1);
fprintf('u = %d, v = %d, x = %d \n', u, v, x)

u=1; v=1; x = (a*u + b*v + c)/(g*u + h*v + 1);
fprintf('u = %d, v = %d, x = %d \n', u, v, x)

u=0; v=1; x = (a*u + b*v + c)/(g*u + h*v + 1);
fprintf('u = %d, v = %d, x = %d \n', u, v, x)

u=0; v=0; y = (d*u + e*v + f)/(g*u + h*v + 1);
fprintf('u = %d, v = %d, y = %d \n', u, v, y)

u=1; v=0; y = (d*u + e*v + f)/(g*u + h*v + 1);
fprintf('u = %d, v = %d, y = %d \n', u, v, y)

u=1; v=1; y = (d*u + e*v + f)/(g*u + h*v + 1);
fprintf('u = %d, v = %d, y = %d \n', u, v, y)

u=0; v=1; y = (d*u + e*v + f)/(g*u + h*v + 1);
fprintf('u = %d, v = %d, y = %d \n', u, v, y)
```

8.10 SURFACE ILLUMINATION

Surface illumination determines the brightness of a surface given the parameters regarding the light sources such as the light intensity, angle of incidence light, angle of the observer looking at the surface, and also the reflectance properties of the surface. An illumination model takes these parameters as input and produces an output regarding the brightness of the surface. Since the brightness depends on the angle of reflected light and the viewpoint of the observer in some cases, we need to derive a mathematical model for this purpose (Shirley, 2002).

In Figure 8.24, let PA indicate the direction of an incident light ray striking the surface at P making an angle θ with the normal along PC and let PD indicate the direction of the reflected ray such that $PA = PD$. Let AC be parallel to PD and CD be parallel to PA intersecting at C. Project PA onto N at B and extend PB to C.

Writing in terms of vectors, let L and R represent the unit vectors for incidence and reflected rays and N is the unit vector along the normal. The laws of reflection dictate that angles of incidence and reflection have to be equal and also the incident ray, reflected ray and surface normal should lie on the same plane. According to the triangle rule for vectors, $R = PC + CD$. The projection of L onto N is given by $(L \cdot \cos\theta)\, N = PB$. Hence $PC = 2PB = 2(|L| \cdot \cos\theta)\, N$. Also $CD = -L$. Combining these expressions, we get the following, remembering that vectors L, N, and R are the unit vectors and have magnitude 1:

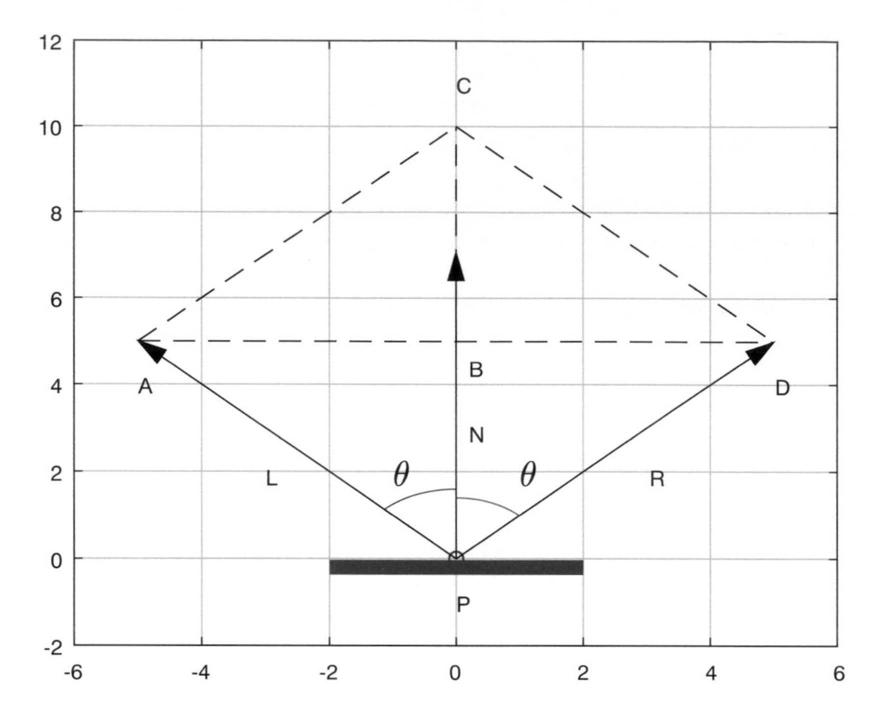

FIGURE 8.24 Relation between incident ray, reflected ray, and normal to a surface.

$$R = PC + CD = 2PB + CD = 2(|L| \cdot \cos\theta)N - L = 2(|L||N| \cdot \cos\theta)N - L = 2(L \cdot N)N - L \tag{8.34}$$

Example 8.13

Light is incident along L = − i + 2j − k on a surface with normal N = j. Calculate the reflected ray and angle of incidence. Verify that angles of incidence and reflection are equal.

Here, $L = -i + 2j - k$ and $N = j$

$$L_u = \frac{L}{|L|} = (-0.4082 \quad 0.8165 \quad -0.4082)$$

$$N_u = \frac{N}{|N|} = (0,1,0)$$

From Equation (8.34), reflected ray: $R_u = 2(L_u \bullet N_u)N_u - L_u = (0.4082 \quad 0.8165 \quad 0.4082)$

Angle of incidence: $\cos\theta_i = (L_u \bullet N_u) = 2/\sqrt{6} = 0.8165$ i.e. $\theta_i = 35.26°$

Angle of reflection: $\cos\theta_r = (N_u \bullet R_u) = 2/(\sqrt{6}) = 0.8165$ i.e. $\theta_r = 35.26°$ (Figure 8.25)

MATLAB Code 8.13

```
clear all; clc;

L = [-1, 2, -1];     Lu = L/norm(L); L = Lu;
N = [0, 1, 0];       Nu = N/norm(N); N = Nu;
Ru = 2*dot(Lu,Nu)*Nu - Lu;           R = Ru;
fprintf('Reflected ray : (%.2f)i + (%.2f)j +(%.2f)k \n', R(1), R(2), R(3));

ci = dot(L, N); ai = acosd(ci);
cr = dot(R, N); ar = acosd(cr);
fprintf('Angle of incidence : %.2f deg\n', ai);
fprintf('Angle of reflection : %.2f deg\n', ar);

%plotting
quiver3(0, 0, 0, L(1), L(2), L(3)); hold on;
quiver3(0, 0, 0, R(1), R(2), R(3));
quiver3(0, 0, 0, N(1), N(2), N(3));
plot3(0, 0, 0, 'bo'); grid on;
view(17, 53);
xlabel('i'); ylabel('j'); zlabel('k');
text(L(1), L(2), L(3), 'L');
text(N(1), N(2), N(3), 'N');
text(R(1), R(2), R(3), 'R');
hold off;
```

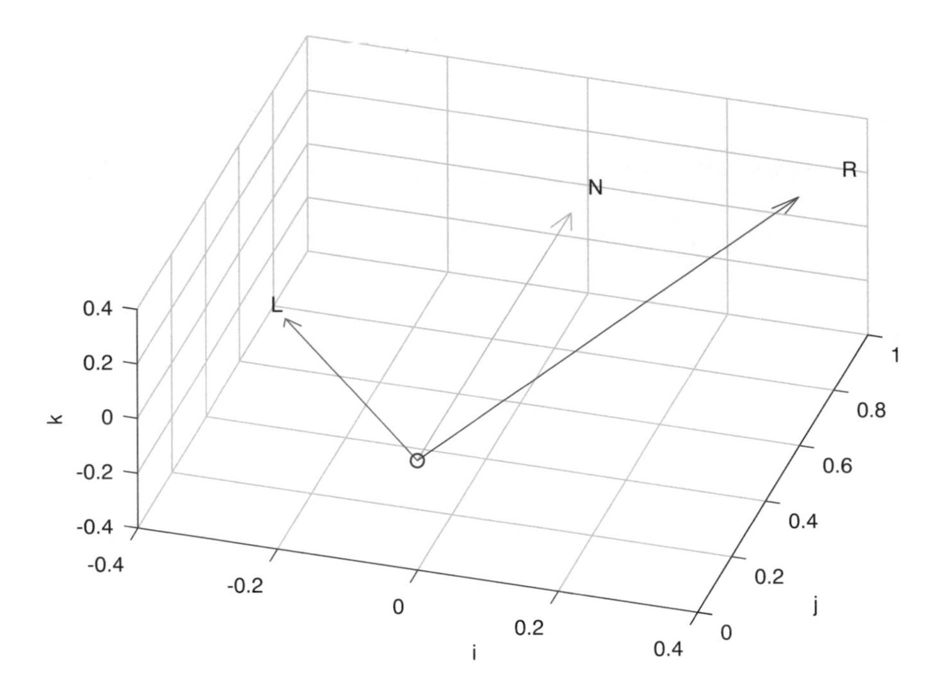

FIGURE 8.25 Plot for Example 8.13.

Intensity at any point on the surface depends on three types of reflections: ambient reflection, diffuse reflection, and specular reflection (Hearn and Baker, 1996). Ambient reflection assumes that all objects receive the same amount of light from all directions and simulates a constant background illumination. Intensity I_a at a point P due to ambient light is given by the following, where L_a is the intensity of ambient light, and k_a is the ambient reflection coefficient of a surface i.e. what percentage of the incident ambient light the surface reflects $0 \le k_a \le 1$ (Foley et al., 1995)

$$I_a = L_a \cdot k_a \tag{8.35}$$

Diffuse reflection assumes the surface is perfectly diffusing i.e. light is scattered equally in all directions as it strikes a surface. In this case, the intensity of a point would depend on the angle of reflection θ. If the angle is very small, then most of the incident light would bounce back along the same path and the observed brightness would be high. However as the angle increase, light would be scattered in other directions so that the percentage of reflected light coming to the observer would be smaller. This would reduce the observed intensity of the surface. Intensity I_d at a point P due to diffused reflection is given by the following, where L_p is the intensity of the light source P, θ is the angle of reflection, and k_d is the diffuse reflection coefficient of a surface i.e. what percentage of the incident diffused light the surface reflects $0 \le k_d \le 1$, and remembering that L and N are the unit vectors (Foley et al., 1995).

$$I_d = L_p \cdot k_d \cdot \cos\theta = L_p \cdot k_d \cdot (L \bullet N) \tag{8.36}$$

Specular reflection occurs when light is reflected at a certain angle from a shiny surface. A shiny surface behaves like an imperfect mirror and produces specular highlights, which is actually the image of the light source itself reflected from the surface. Unlike diffuse reflection, specular reflection is dependent on position of viewer. Let V be the unit vector along the viewing direction and φ be the angle between V and R (see Figure 8.26). Observed intensity would be highest when the observer views the surface exactly along the reflected ray i.e. $\varphi = 0$ but intensity would reduce as φ increases and the observer moves further away from the reflected light, intensity I_s at a point P due to specular reflection is given by the following, where L_p is the intensity of the light source P, φ is the angle between R and V, m is a positive number dependent on the material of the surface, and k_s is the specular reflection coefficient of a surface, $0 \le k_s \le 1$, and remembering that R and V are the unit vectors (Foley et al., 1995).

$$I_s = L_p \cdot k_s \cdot (\cos\varphi)^m = L_p \cdot k_s \cdot (R \bullet V)^m \tag{8.37}$$

Hence, total intensity of reflected light at a point on a surface is the combined effect of all the above factors:

$$I = I_a + I_d + I_s = L_a \cdot k_a + L_p \cdot k_d \cdot \cos\theta + L_p \cdot k_s \cdot (\cos\varphi)^m \tag{8.38}$$

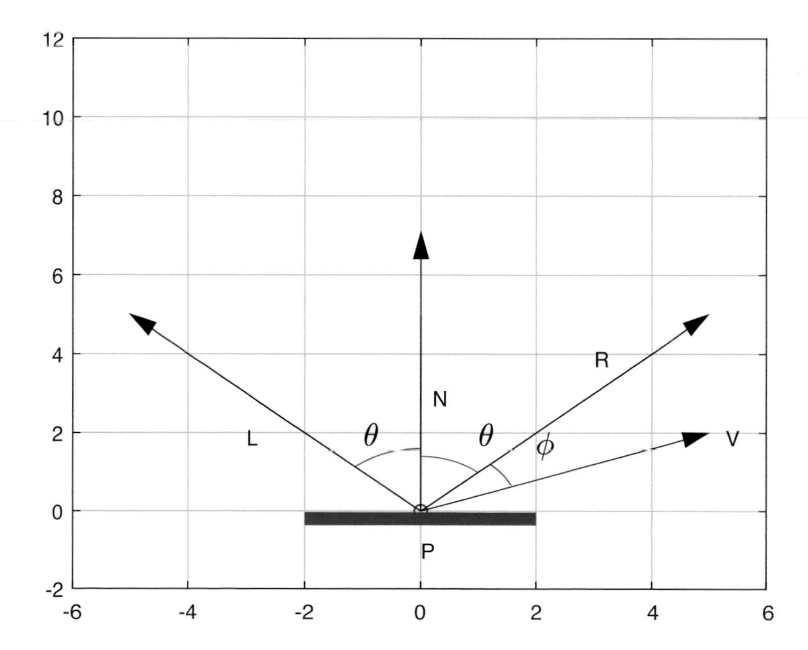

FIGURE 8.26 Relation between incident ray, reflected ray, normal, and viewing direction.

Example 8.14

Light is incident along $L = -i + 2j - k$ at a point on a surface with normal $N = j$. Calculate the intensity at the point if the source light is 10 times more intense than the ambient light. Assume viewing direction as $V = i + 1.5j + 0.5\,k$ and $k_a = 0.15, k_d = 0.4, k_s = 0.8, m = 5$.

Incident ray: $L = -i + 2j - k$
Normal: $N = j$
Reflected ray: $R = i + 2j + k$ (see previous example)
Viewing direction: $V = i + 1.5j + 0.5\,k$

$$L_u = \frac{(-i + 2j - k)}{\sqrt{6}} = (-0.4082 \quad 0.8165 \quad -0.4082)$$

$$N_u = \frac{j}{\sqrt{1}} = (0 \quad 1 \quad 0)$$

$$R_u = \frac{(i + 2j + k)}{\sqrt{6}} = (0.4082 \quad 0.8165 \quad 0.4082)$$

$$V_u = \frac{i + 1.5j + 0.5\,k}{\sqrt{1.8708}} = (0.5345 \quad 0.8018 \quad 0.2673)$$

$$L_u \bullet N_u = 0.8165$$

$$R_u \bullet V_u = 0.9820$$

$$I_a = L_a \cdot k_a = 1 \times 0.15 = 0.15$$

$$I_d = L_p \cdot k_d \cdot (L_u \bullet N_u) = 10 \times 0.25 \times 0.8165 = 2.0412$$

$$I_s = L_p \cdot k_s \cdot (R_u \bullet V_u)^5 = 10 \times 0.5 \times 0.9820^5 = 4.5655$$

$$I = I_a + I_d + I_s = 6.7567$$

Thus the intensity on the surface is approximately 67.5% of the source intensity.

MATLAB Code 8.14

```
clear all; clc;
L = [-1, 2, -1];                    Lu = L/norm(L);
N = [0, 1, 0];                      Nu = N/norm(N);
V = [1, 1.5, 0.5];                  Vu = V/norm(V);
Ru = 2*dot(Lu,Nu)*Nu - Lu;

La = 1;
Lp = 10;
```

```
ka = 0.15;
kd = 0.25;
ks = 0.5;
m = 5;

fprintf('Ambient intensity : \n');
Ia = La*ka
fprintf('Diffused intensity : \n');
Id = Lp*kd*dot(Lu,Nu)
fprintf('Specular intensity : \n');
Is = Lp*ks*(dot(Ru,Vu))^m
fprintf('Total intensity : \n');
I =  Ia + Id + Is
```

8.11 NOTES ON 3D PLOTTING FUNCTIONS

This section summarizes MATLAB 3D plotting functions used and some additional ones (Marchand, 2002). The reader is encouraged to explore further details about these functions from MATLAB documentations (Figures 8.27–8.42).

(a) `ezplot3 & fplot3`: Used to plot functions using parametric variables:

```
ezplot3('cos(t)','t*sin(t)','sqrt(t)')

xt = @(t) cos(t); yt = @(t) t.*sin(t); zt = @(t) sqrt(t);
fplot3(xt,yt,zt)
```

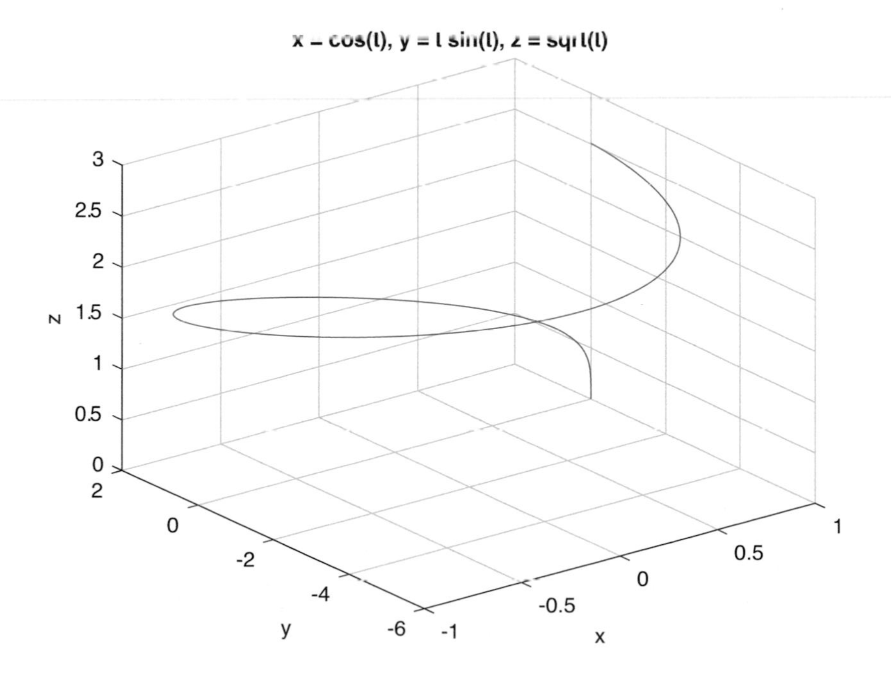

FIGURE 8.27 Plotting with `ezplot & fplot3`.

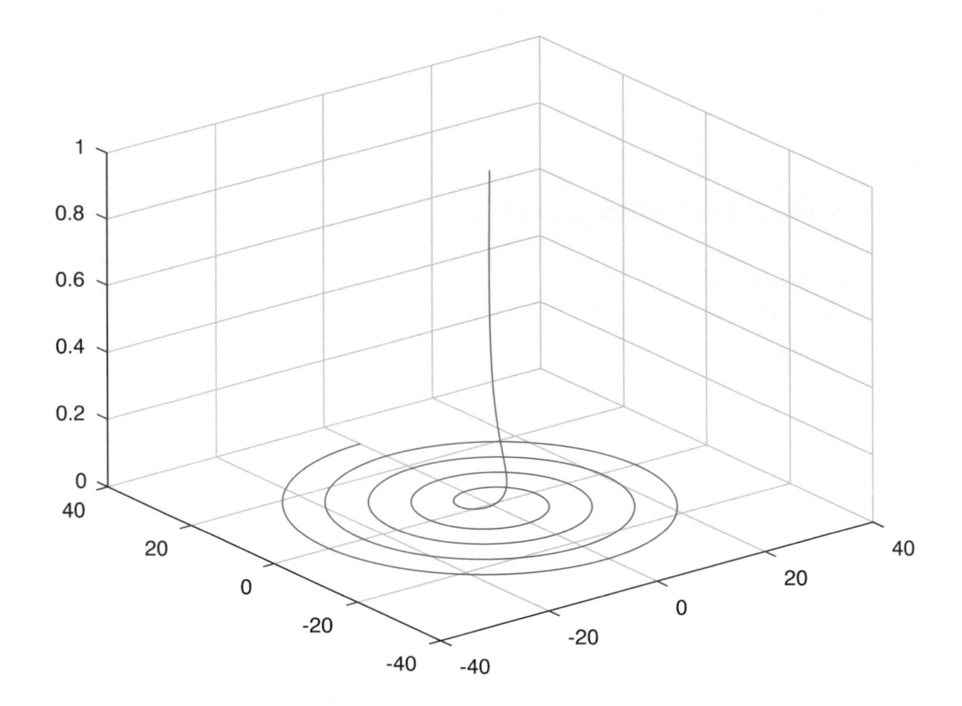

FIGURE 8.28 Plotting with `plot3`.

(b) `plot3`: Used to plot functions using vector of values:

```
t = 0:pi/50:10*pi; plot3(t.*sin(t), t.*cos(t), exp(-t));
```

(c) `ezmesh` & `ezsurf`: Generates a mesh plot and surface plot using symbolic variables

```
ezmesh('x.*y.*exp(-x.^2 - y.^2)')
ezsurf('x.*y.*exp(-x.^2 - y.^2)')
```

(d) `ezcontour`: Generates a contour plot and filled contour plot using symbolic variables

```
ezcontour('x.*y.*exp(-x.^2 - y.^2)')
ezcontourf('x.*y.*exp(-x.^2 - y.^2)')
```

(e) `ezsurfc`: Combines a surface plot with a contour plot

```
ezsurfc('x.*y.*exp(-x.^2 - y.^2)')
```

Views on graphs can be changed by specifying the view function, which takes in two arguments: the first for the horizontal rotation angle (azimuth) and the second for the vertical rotation angle (elevation). The colormap function can be used to change the color scheme.

```
ezsurfc('x.*y.*exp(-x.^2 - y.^2)'); colormap(summer); view(18, 26)
```

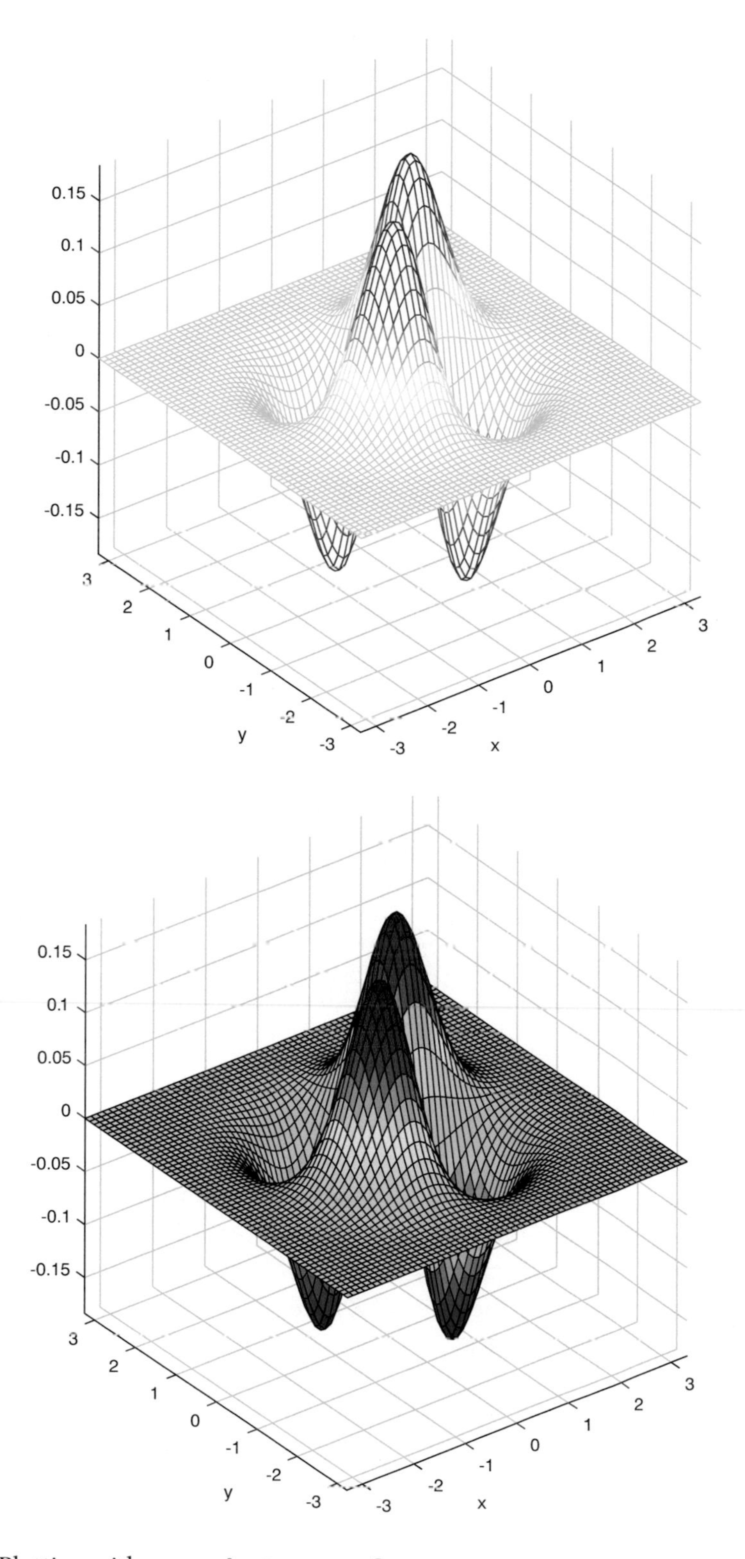

FIGURE 8.29 Plotting with `ezmesh` & `ezsurf`.

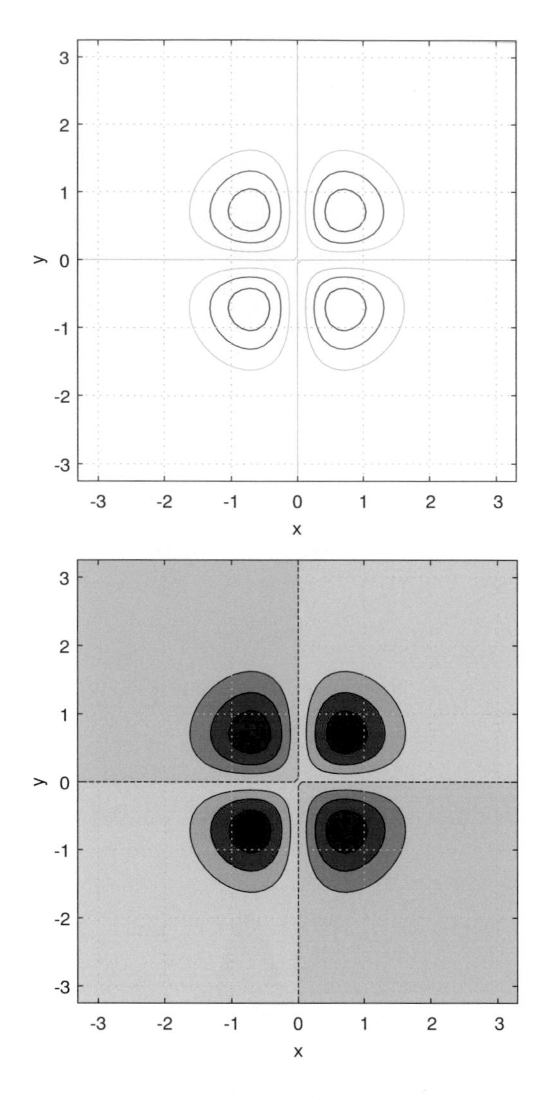

FIGURE 8.30 Plotting with `ezcontour & ezcontourf`.

(f) `mesh & surf`: Generates a surface using a matrix of values. In the first step, a grid of points using function "meshgrid" is created on the *X–Y* plane. For each point a *Z* value is defined using a specific function. The "mesh" or "surf" function is used to create a surface by plotting the value of *Z* for each point on the *X–Y* grid and joining the values of *Z* by colored lines.

```
[X,Y] = meshgrid(-2:.2:2, -2:.2:2);
Z = X .*Y.* exp(-X.^2 - Y.^2);
mesh(X,Y,Z);

[X,Y] = meshgrid(-2:.2:2, -2:.2:2);
Z = X .*Y.* exp(-X.^2 - Y.^2);
surf(X,Y,Z);
```

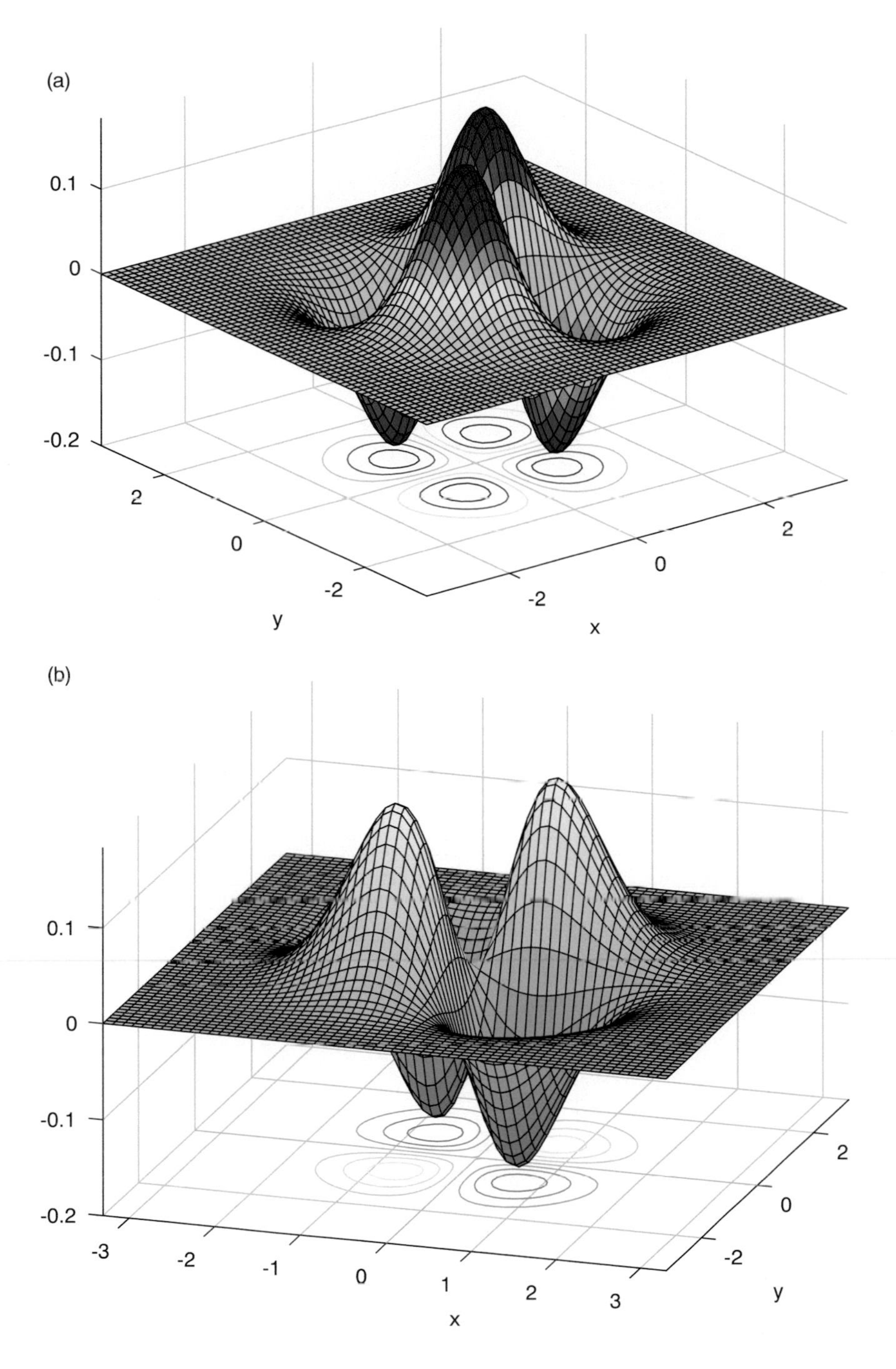

FIGURE 8.31 (a) Plotting with ezsurfc; (b) using view to change orientation.

(g) `patch`: Creates filled polygons given the vertices and colors

```
x = [4 6 11 9];
y = [2 7 9 4];
z = [3 5 10 6];
c = [0 4 6 8];
colormap(jet);
patch(x,y,z, c);
colorbar;
hold; grid;
v1 = [2 4 5; 2 12 6; 8 4 1];
patch('Vertices', v1, 'FaceColor', 'red', 'FaceAlpha', 0.3, 'EdgeColor', 'red');
axis([0 12 0 12 0 12]);
view(-164,-56);
xlabel('x'); ylabel('y'); zlabel('z');
```

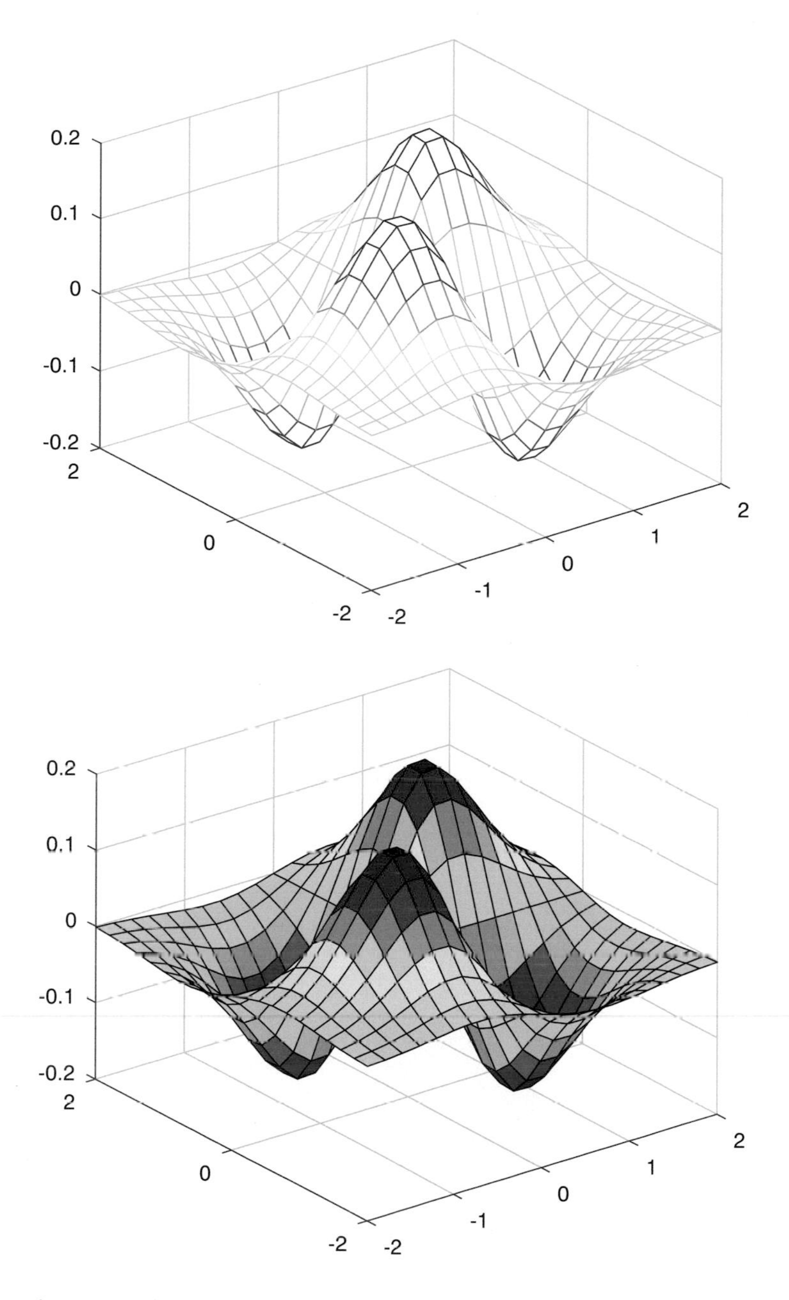

FIGURE 8.32 Plotting with `mesh` & `surf`.

(h) `isosurface`: computes a surface geometry for the specified n-dimensional grid and function f.

```
[y,x,z] = ndgrid(linspace(-5,5,64));
f = (x.^2 + y.^2 + z.^2 - 5);
isosurface(x,y,z,f,.01);
axis equal; grid;
```

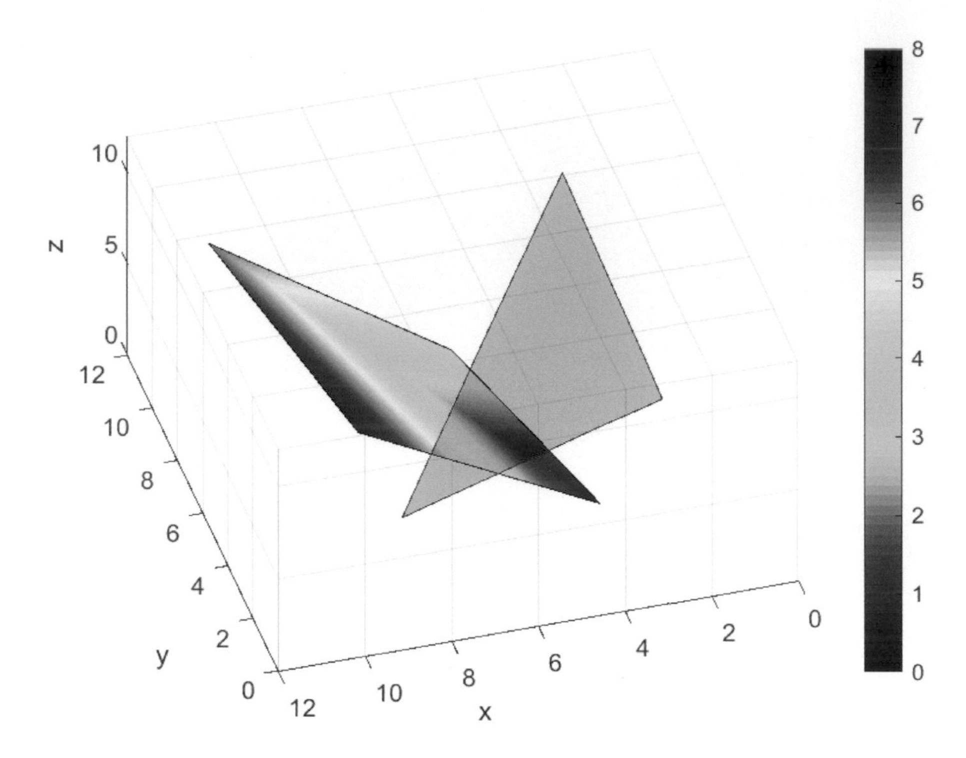

FIGURE 8.33 Plotting with `patch`.

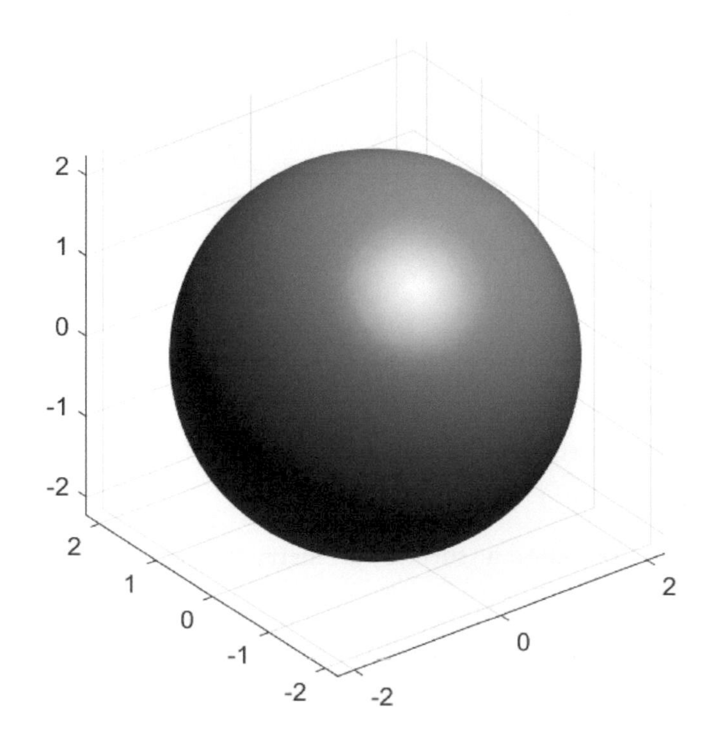

FIGURE 8.34 Plotting with `isosurface`.

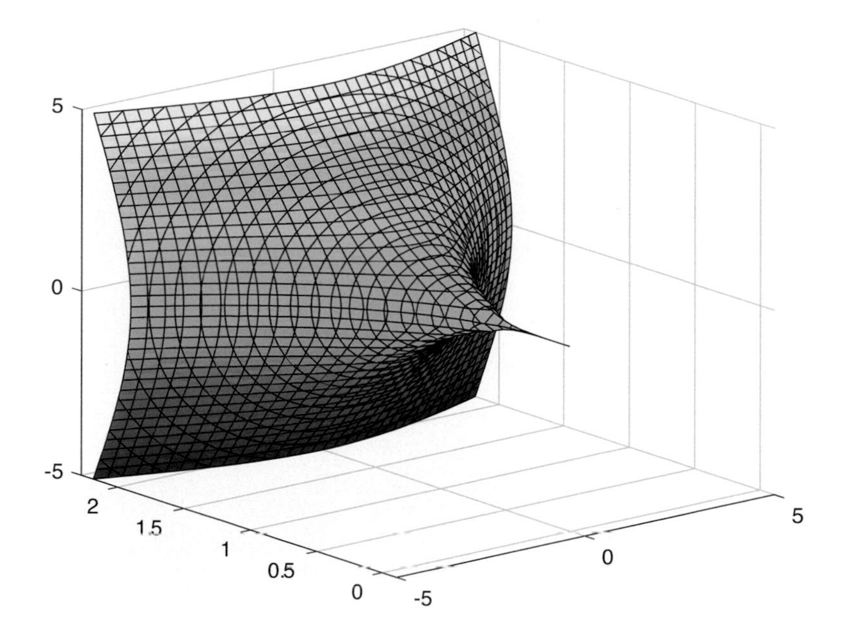

FIGURE 8.35 Plotting with fimplicit3.

(i) fimplicit3: This function has been introduced from MATLAB version 2016 and takes as an argument the implicit function

```
f = @(x,y,z) x.^2 - y.^5 + z.^2; fimplicit3(f);
```

(j) lightangle: Specifies lighting parameters on a surface

```
h = ezsurf('sin(sqrt(x^2+y^2))/sqrt(x^2+y^2)', [-4*pi,4*pi]);
view(0,75);
figure
h = ezsurf('sin(sqrt(x^2+y^2))/sqrt(x^2+y^2)', [-4*pi,4*pi]);
view(0,75);
lightangle(-45,30);
h.AmbientStrength = 0.3;
h.DiffuseStrength = 0.8;
h.SpecularStrength = 0.9;
h.SpecularExponent = 25;
```

(k) warp: Used to map a texture image over a surface with a known equation

```
I = imread('peppers.png');
[X,Y] = meshgrid(-10:10,-10:10);
Z = -sqrt(X.^2 + Y.^2 + 10);
surf(X,Y,Z)
figure; warp(X,Y,Z,I);
```

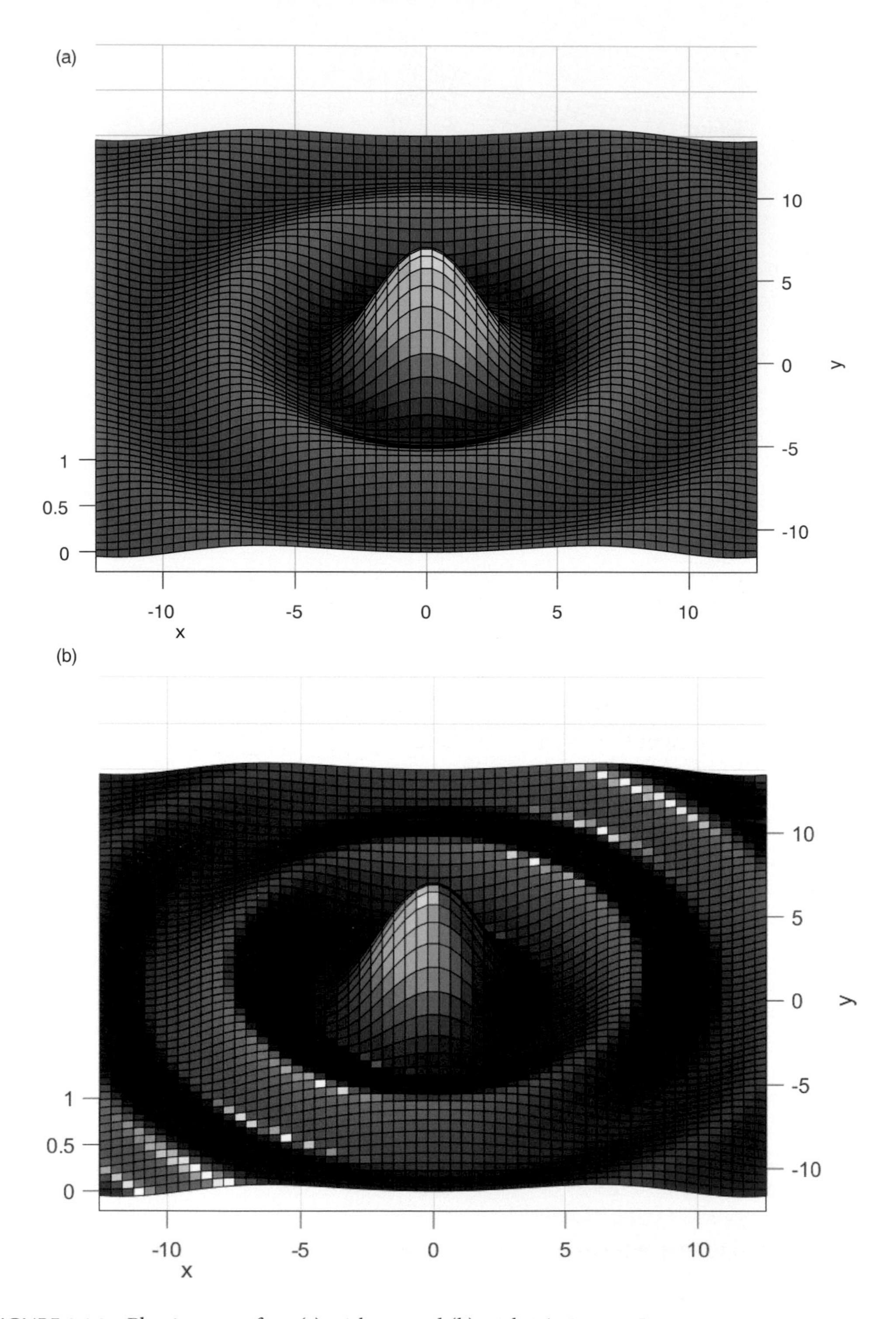

FIGURE 8.36 Plotting a surface (a) without and (b) with `lightangle`.

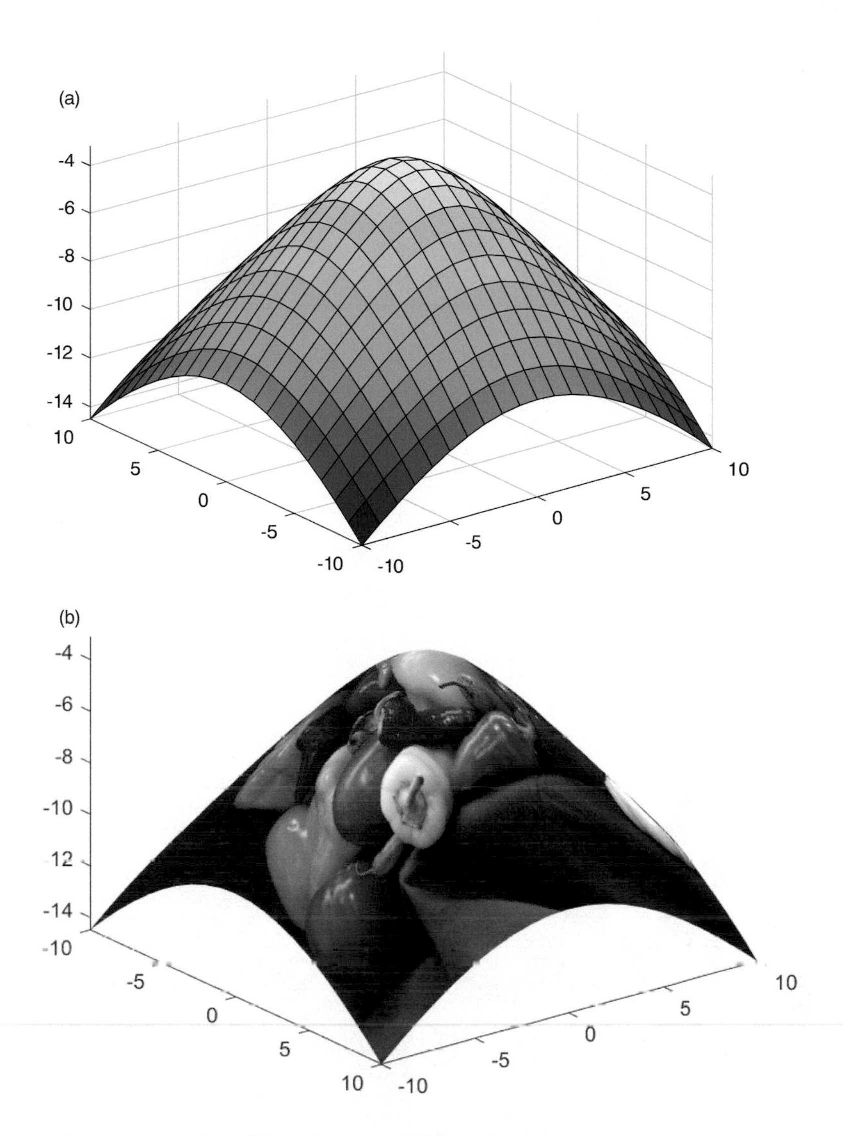

FIGURE 8.37 Plotting a surface (a) without and (b) with texture mapping using `warp`.

(l) `set`: Used to set object properties, which can change over time and thereby generate animations

```
[x, y, z] = ellipsoid(0, 0, 0, 10, 10, 10);
h = surf(x,y,-z);
im = imread('world-map.jpg');
set(h, 'CData', im, 'FaceColor', 'texturemap', 'edgecolor', 'none');
el = 24;
for az=0:360
    view(az, el);
    pause(0.1);
end
```

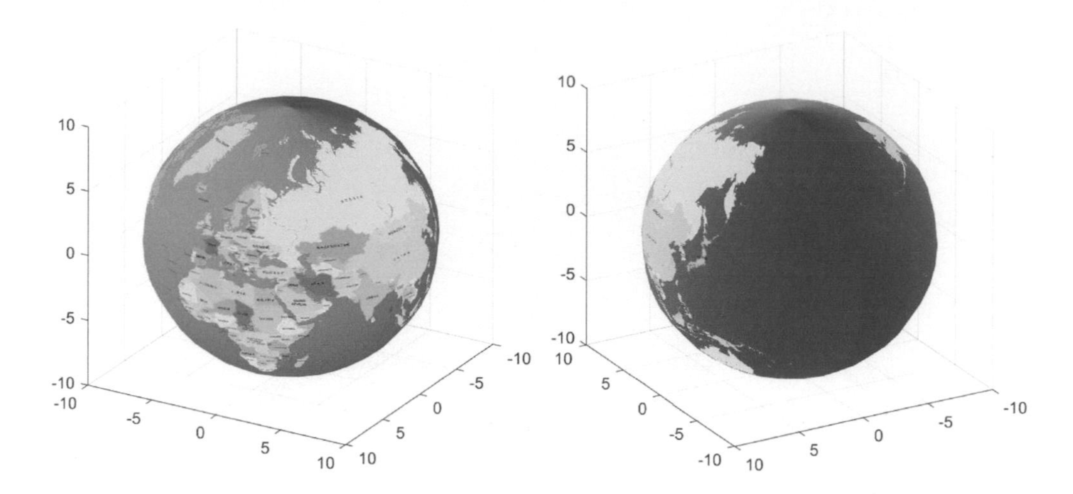

FIGURE 8.38 Animations with `set`.

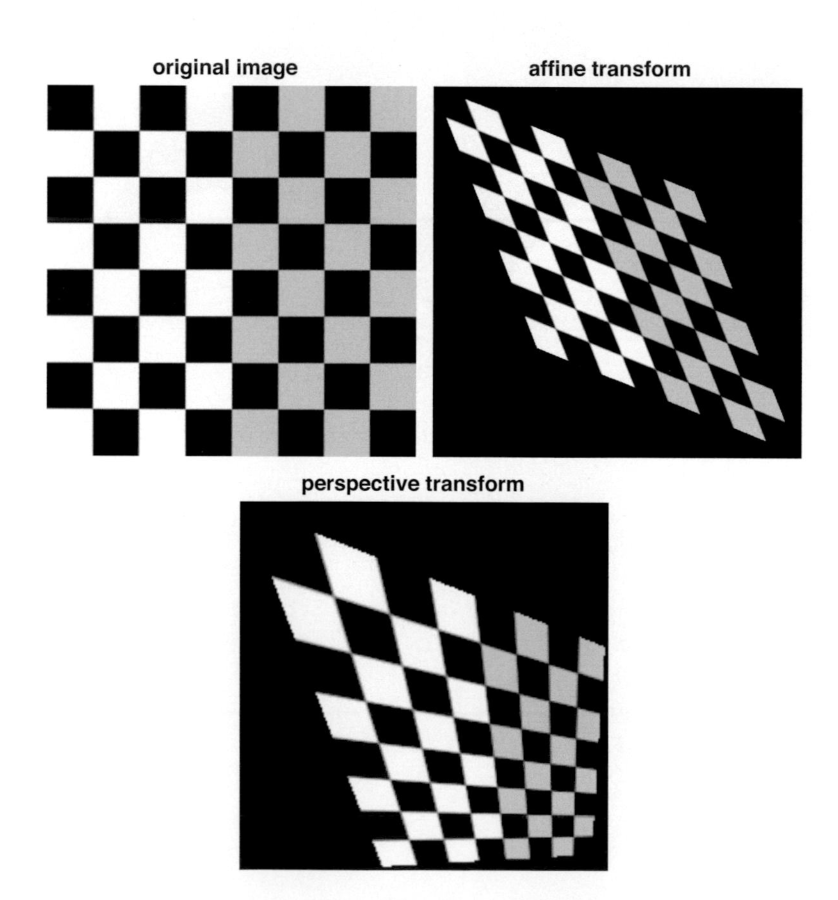

FIGURE 8.39 Transformations of images with `affine2d` and `projective2d`.

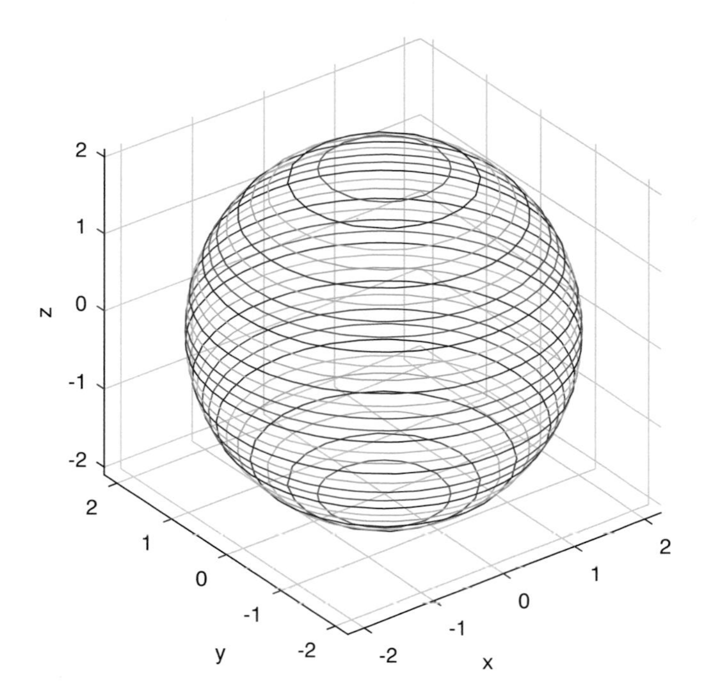

FIGURE 8.40 Plotting with `impl`.

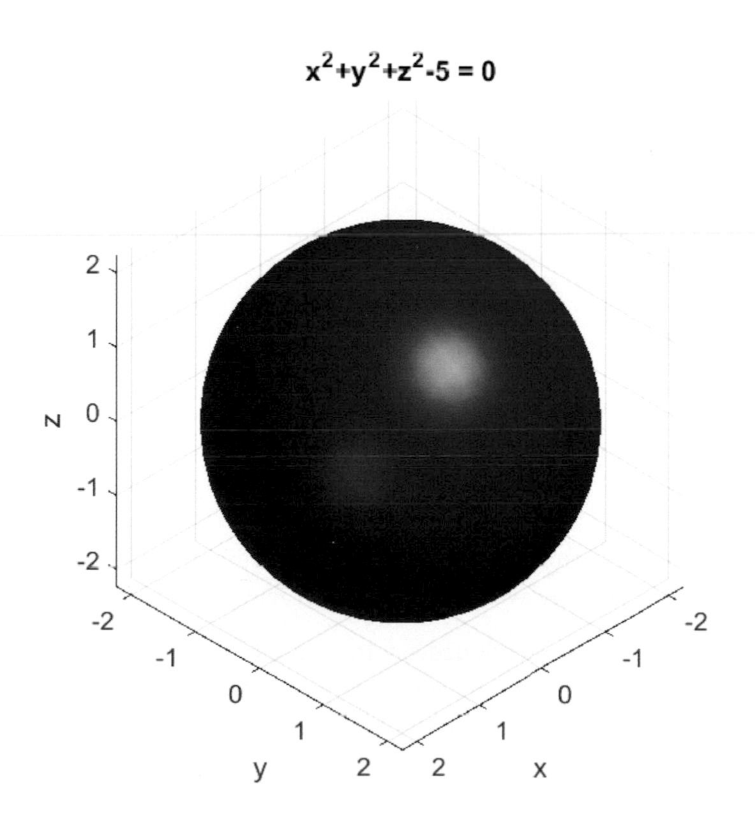

FIGURE 8.41 Plotting with `ezimplot3`.

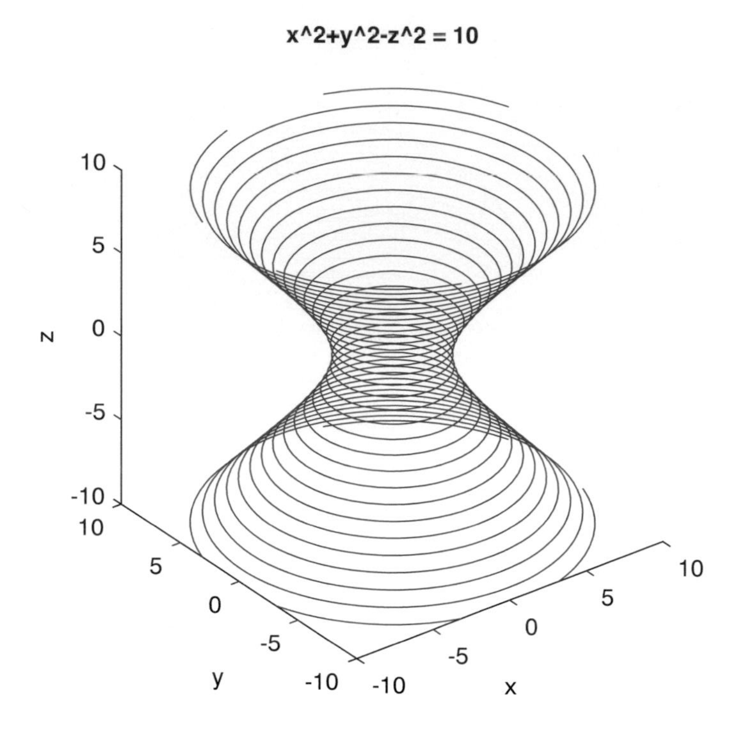

FIGURE 8.42 Plotting with `implicitplot3d`.

(m) `affine2d` and `projective2d`: Used to create affine and perspective (or projective) transformations of images

```
A = checkerboard(10);
M = [5 2 5 ; 2 5 5 ; 0 0 1];
tform = affine2d(M');
B = imwarp(A, tform);
figure, imshow(B); title('affine transform');
M = [5 2 5 ; 2 5 5 ; 0.01 0.01 1];
tform = projective2d(M');
C = imwarp(A, tform);
figure, imshow(C); title('perspective transform');
```

In addition to in-built functions listed above, a number of user-created functions have been uploaded in publicly available websites. A few of them are mentioned below.

(n) `impl`: This function is mentioned on the website of Jeffery Cooper, University of Maryland (http://www.math.umd.edu/~jcooper/matcomp/matcompmfiles/). Generates a surface from an implicit equation of the form $f(x, y, z) = 0$.

```
syms x y z; f = inline('x.^2 + y.^2 + z.^2 - 5', 'x', 'y', 'z');
impl(f, [-3, 3, -3, 3, -3, 3], 0), axis equal; grid;
```

(o) `ezimplot3`: This function is mentioned on the MathWorks File Exchange web site (https://in.mathworks.com/matlabcentral/fileexchange/300-implot-m). Generates a surface from an implicit equation of the form $f(x, y, z) = 0$.

```
f = 'x^2+y^2+z^2-5' ; ezimplot3(f);
```

(p) `implicitplot3d`: This function is mentioned on the web site of Jonathan M. Rosenberg, University of Maryland (www2.math.umd.edu/~jmr). Generates a surface from an implicit equation of the form $f(x, y, z) = 0$.

```
implicitplot3d('x^2+y^2-z^2', 10, -10, 10, -10, 10, -10, 10, 30);
```

NOTE

`imshow`: displays an image in a figure window
`imwarp`: applies geometric transformation to an image for mapping it to a surface

8.12 CHAPTER SUMMARY

The following points summarize the topics discussed in this chapter:

- Surfaces can be created due to a combination of two splines along orthogonal directions.
- Based on mathematical representations, surfaces are categorized as parametric form or implicit form.
- Parametric Bezier surfaces can be modified by adjusting a grid of control points.
- Implicit surfaces can be used to represent ellipsoids and are known as quadric surfaces.
- Based on creation methods surfaces are categorized based on extrusion and revolution.
- The normal vector to a surface can be computed based on its partial derivatives.
- The tangent plane is the plane perpendicular to the normal and passing through a given point.
- The area and volume of the solid are obtained by rotating a curve about a primary axis.
- Texture mapping is a process of applying an image to a surface to improve its appearance.
- The texture mapping transformation can be represented as either affine or perspective.

- The surface illumination is determined based on light sources and surface reflection properties.
- Surface reflection can be either ambient or diffuse or specular.

8.13 REVIEW QUESTIONS

1. What is the difference between parametric surfaces and implicit surfaces?
2. How is a Bezier surface generated from two Bezier splines?
3. Specify how commonly used quadric surfaces are represented using implicit equations?
4. How are surfaces categorized based on creation methods?
5. What is the difference between extruded surface and surfaces of revolution?
6. How is the normal vector and tangent plane of a surface calculated?
7. How is the area and volume of a surface of revolution computed?
8. How is an affine texture mapping transformation computed?
9. How is a perspective texture mapping transformation computed?
10. What is an illumination model? How can it used to calculate brightness at a surface point?

8.14 PRACTICE PROBLEMS

1. A parametric surface $\left(s^2, 2s, t\right)$ is translated using $T(1, 0, -1)$ and then scaled using $S(-1, 0, 1)$. Find the parametric representation of the resulting surface.
2. Find the parametric representation of the elliptic paraboloid $x = 5y^2 + 2z^2 - 10$ that is in front of the YZ-plane.
3. A bi-quadratic Bezier surface has the following control points: $P_{00}(0, 0, 0)$, $P_{01}\left(0, \frac{1}{2}, 0\right)$, $P_{02}\left(0, 1, \frac{1}{b}\right)$, $P_{10}\left(\frac{1}{2}, 0, 0\right)$, $P_{11}\left(\frac{1}{2}, \frac{1}{2}, 0\right)$, $P_{12}\left(\frac{1}{2}, 1, \frac{1}{b}\right)$, $P_{20}\left(1, 0, \frac{1}{a}\right)$, $P_{21}\left(1, \frac{1}{2}, \frac{1}{a}\right)$, $P_{22}\left(1, 1, \frac{1}{a} + \frac{1}{b}\right)$ Find the surface equation.
4. Find the equation of the tangent plane of the cone $x^2 + y^2 = z^2$ at point (0.6, 0.8, 1).
5. Find the normal to the surface $S(u, v) = \left(2u, v, u^2 + v^2\right)$ at point (4, 3, 13).
6. Find the volume of the solid obtained by rotating about the x-axis the region bounded by the curve $y = x^2 - 2x + 3$, $x = 0$, $x = 3$.

7. Find the volume of the solid obtained by rotating about the y-axis the portion of the region bounded by the curves $y = x^{1/3}$ and $y = x/4$, in the first quadrant between $x = 0$, $x = 2$.

8. A light source positioned at $P(0, 10, 20)$ is shining on the surface $S(u, v) = \left(u, v, -u^2 - v^2\right)$ for $0 \leq u, v \leq 1$. Determine the incident ray, the reflected ray at point $Q = (½, ½, −½)$ on the surface and angle of reflection.

9. Light along direction $L = 2i + j + 3k$ falls on a surface with normal $N = i - 2j + k$. Calculate the reflected ray and incidence angle. Also verify that angle of incidence is equal to angle of reflection.

10. Light is striking the plane P: $-2x - 8y + 10z - 10 = 0$ along the direction $L = k$, and the observer viewpoint is along $V = (i + j + k)$. Assuming the ambient intensity to be 1/6th the source intensity and $m = 2$, $k_a = 0.2$, $k_d = 0.3$, $k_s = 0.4$, determine intensity of reflected light at a point on the plane as a percentage of the source intensity.

CHAPTER 9

Projection

9.1 INTRODUCTION

Projection is used to map a higher-dimensional object to a lower-dimensional view i.e. from 2D to 1D or from 3D to 2D. The lower-dimensional entity is called a viewline or viewplane, respectively. In this chapter, we will mostly discuss about projection of 3D objects on to a 2D viewplane but we will introduce the concepts using a 2D projection on a 1D viewline and then extend the concepts for the 3D case.

Projection can be of two types: parallel and perspective. In parallel projection, projection lines are parallel to each other. This type of projection produces unrealistic views in the sense that it is not what a viewer in the physical world would see as the apparent size of an object would not depend on its distance from the viewer, and all things near and far would appear at their true sizes. However, parallel projection is useful as it keeps intact the true sizes and angles of objects. In perspective projection, the projection lines appear to converge to a point called projection reference point (PRP). This is actually how we see in the real world as parallel lines appear to converge at a distance to our eyes. This is known as the perspective effect and happens because the size of objects depend on their distances from the observer—as objects move farther their apparent sizes reduce. Although perspective projection produces realistic views of scenes; however, it distorts the true lengths and angles of lines and surfaces. For 2D projection, points are projected on a line called the viewline, while for 3D projection points are projected on a plane called the viewplane. Parallel projection can again be of two types: orthographic and oblique. In parallel orthographic projection, the projection lines are perpendicular to the view plane. In parallel oblique projection, the projection lines can be oriented at any arbitrary angles to the view plane (Hearn and Baker, 1996) (see Figure 9.1).

Usually for 3D projection, parallel orthographic projection can also be sub-divided into two types: multi-view and axonometric. In multi-view projection, the projection occurs on the primary planes i.e. *XY*-, *YZ*-, or *XZ*-planes. Such views are called top, side, and front views, and display only one face of the object. These are used in engineering and architectural drawings as length and angles can be accurately measured. In axonometric projection, the projection occurs on an arbitrary plane that does not coincide with any of

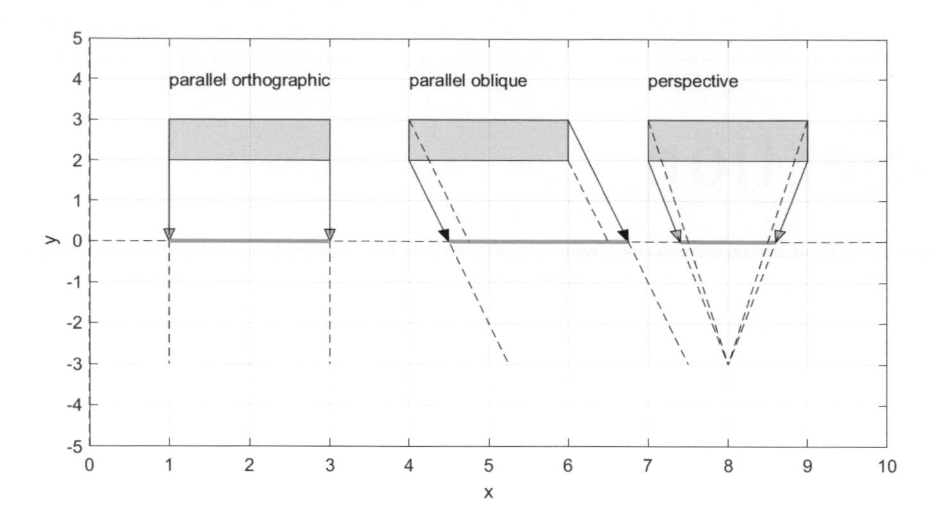

FIGURE 9.1 Projection types.

the primary planes. In this case, more than one face of the object can be viewed. The ratios of the actual lengths of the object along the three axes to their projected lengths are called foreshortening factors.

9.2 2D PROJECTION

For studying 2D projection, we use vector equations in homogeneous coordinates, as explained in Section 6.9. The main results are summarized here for the benefit of the reader.

Cartesian equation of a line is: $ax+by+c=0$ and is expressed in vector form as $\ell=(a, b, c)$.

$P(X, Y, W)$ are homogeneous coordinates of point (x, y) i.e. $x=X/W, y=Y/W$.

For line $\ell=(a, b, c)$ passing through $P(X, Y, W)$ we must have: $\ell \bullet P=0$.

For line ℓ through two given points $P_1(X_1, Y_1, W_1)$ and $P_2(X_2, Y_2, W_2)$ we have: $\ell=P_1 \times P_2$.

The intersection point P of two lines ℓ_1 and ℓ_2 is given by: $P=\ell_1 \times \ell_2$

Let P be a point on an object, which is projected on the viewline L along direction VP at Q, where V is the PRP. Let the projection line through P and V be K. Since, line K passes through two points P and V we have (Figure 9.2a):

$$K=V \times P \tag{9.1}$$

Also point of intersection Q between lines L and K is given by:

$$Q=L \times K=L \times (V \times P) \tag{9.2}$$

Now using the vector identity $A \times (B \times C)=(C \bullet A)B-(A \bullet B)C=\left[B^T \cdot A-\left(BA^T\right) \cdot I\right] \cdot C$ we get:

$$Q=L \times (V \times P)=\left[V^T \cdot L-\left(VL^T\right) \cdot I\right] \cdot P=M \cdot P \tag{9.3}$$

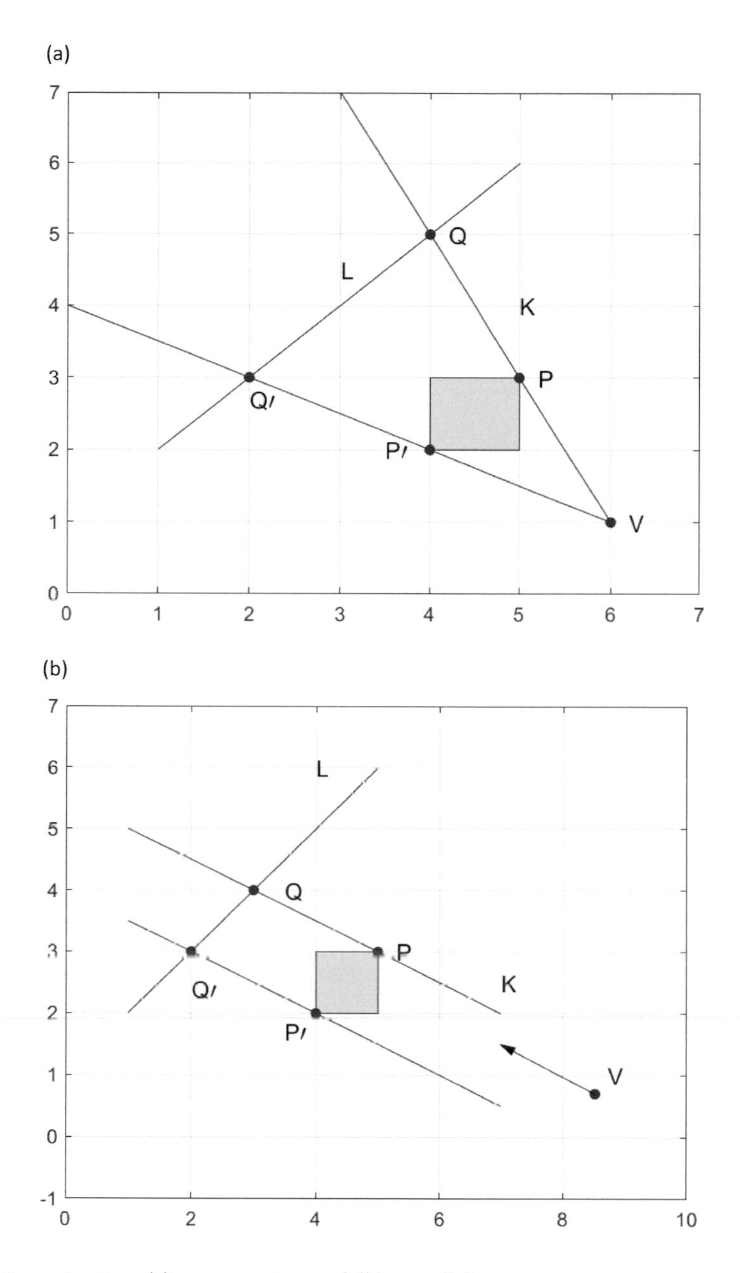

FIGURE 9.2 2D projection (a) perspective and (b) parallel.

Here, I is the identity matrix and M is the perspective projection matrix of point P onto viewline L along direction V (Marsh, 2005).

As shown above the points and lines are represented using homogeneous vector notations of the form $L=(a, b, c)$ and $V=(v_1, v_2, 1)$. For parallel projection V can be thought to be at infinity so that projection lines are parallel (Figure 9.2b). To represent a point at infinity in the direction of V we use the notation $V=(v_1, v_2, 0)$. The vector from V towards the viewline L along the direction of projection is called the projection vector.

Example 9.1

Consider a viewline $L(x, y): -3x+12y-5=0$ and a point P(–8, –6). Determine the projection matrix and projected coordinates in each case: (a) Perspective projection of P with viewpoint (–3, 11) (b) Parallel oblique projection of P in the direction (3, –2) (c) Parallel orthographic projection of P at right angles to the viewline

(a)

$V = [-3, 11, 1]$, $L = [-3, 12, -5]$, $P = [-8, 6, 1]$

From Equation (9.3)

$$\text{Projection matrix: } M = \left[V^T \cdot L - \left(VL^T\right) \cdot I\right] = \begin{bmatrix} -3 \\ 11 \\ 1 \end{bmatrix} \begin{bmatrix} -3 & 12 & -5 \end{bmatrix}$$

$$- \begin{bmatrix} -3 & 11 & 1 \end{bmatrix} \begin{bmatrix} -3 \\ 12 \\ -5 \end{bmatrix} \begin{bmatrix} 1 & 0 & 0 \\ 0 & 1 & 0 \\ 0 & 0 & 1 \end{bmatrix}$$

$$\text{Simplifying, } M = \begin{bmatrix} -127 & -36 & 15 \\ -33 & -4 & -55 \\ -3 & 12 & -141 \end{bmatrix} \text{(in homogeneous coordinates)}$$

$$\text{Original coordinates of point } P = \begin{bmatrix} -8 \\ 6 \\ 1 \end{bmatrix}$$

$$\text{Projected coordinates: } Q = M \cdot P = \begin{bmatrix} 815 \\ 185 \\ -45 \end{bmatrix} \text{(in homogeneous coordinates)}$$

$$\text{Projected coordinates: } Q' = \begin{bmatrix} -18.11 \\ -4.11 \\ 1 \end{bmatrix} \text{(in Cartesian coordinates)}$$

Verification: The projected point must lie on the view line: $L(-18.11, -4.11) = 0$ (Figure 9.3a).

(b)

$V = [3, -2, 0]$, $L = [-3, 12, -5]$, $P = [-8, 6, 1]$

From Equation (9.3)

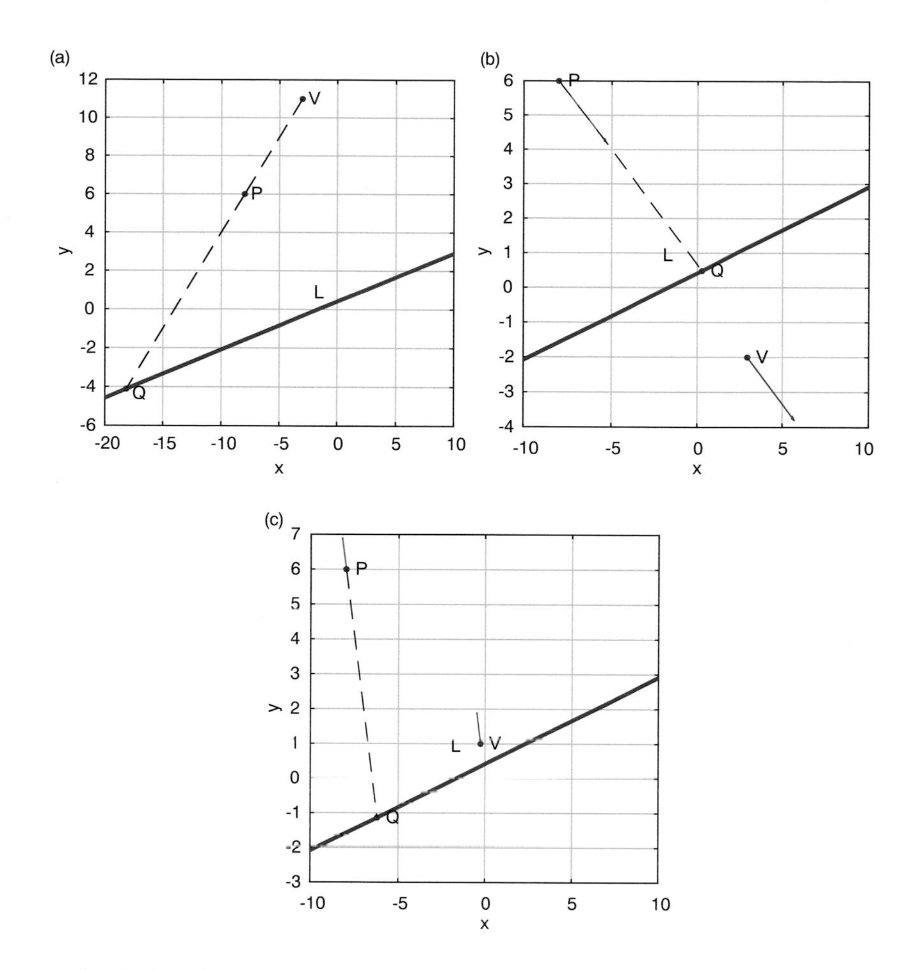

FIGURE 9.3 (a–c) Plots for Example 9.1.

$$\text{Projection matrix: } M = \left[V^T \cdot L - \left(VL^T\right) \cdot I\right] = \begin{bmatrix} -3 \\ 2 \\ 0 \end{bmatrix} \begin{bmatrix} -3 & 12 & -5 \end{bmatrix}$$

$$- \begin{bmatrix} -3 & 2 & 0 \end{bmatrix} \begin{bmatrix} -3 \\ 12 \\ -5 \end{bmatrix} \begin{bmatrix} 1 & 0 & 0 \\ 0 & 1 & 0 \\ 0 & 0 & 1 \end{bmatrix}$$

$$\text{Simplifying, } M = \begin{bmatrix} 24 & 36 & -15 \\ 6 & 9 & 10 \\ 0 & 0 & 33 \end{bmatrix} \text{(in homogeneous coordinates)}$$

$$\text{Original coordinates of point } P = \begin{bmatrix} -8 \\ 6 \\ 1 \end{bmatrix}$$

$$\text{Projected coordinates: } Q = M \cdot P = \begin{bmatrix} 9 \\ 16 \\ 33 \end{bmatrix} \text{(in homogeneous coordinates)}$$

$$\text{Projected coordinates: } Q' = \begin{bmatrix} 0.2727 \\ 0.4848 \\ 1 \end{bmatrix} \text{(in Cartesian coordinates)}$$

Verification: The projected point must lie on the view line: $L(0.2727, 0.4848) = 0$ (Figure 9.3b).

(c)

Slope of line $ax + by + c = 0$ is given by $-\frac{a}{b} = -\frac{-3}{12} = 0.25$
Tangent vector: $t = [1, 0.25, 1]$
Normal vector: $n = R(90) * t = [-0.25, 1, 1]$
Parallel orthographic projection should be along the direction of the normal vector
Hence, $V = [-0.25, 1, 0]$, $L = [-3, 12, -5]$, $P = [-8, 6, 1]$
From Equation (9.3)

$$\text{Projection matrix: } M = \left[V^T \cdot L - \left(VL^T\right) \cdot I\right] = \begin{bmatrix} -0.25 \\ 1 \\ 0 \end{bmatrix} \begin{bmatrix} -3 & 12 & -5 \end{bmatrix}$$

$$-\begin{bmatrix} -0.25 & 1 & 0 \end{bmatrix} \begin{bmatrix} -3 \\ 12 \\ -5 \end{bmatrix} \begin{bmatrix} 1 & 0 & 0 \\ 0 & 1 & 0 \\ 0 & 0 & 1 \end{bmatrix}$$

$$\text{Simplifying, } M = \begin{bmatrix} -12 & -3 & 1.25 \\ -3 & -0.75 & -5 \\ 0 & 0 & -12.75 \end{bmatrix} \text{(in homogeneous coordinates)}$$

$$\text{Projected coordinates: } Q = M \cdot P = \begin{bmatrix} 79.25 \\ 14.5 \\ -12.75 \end{bmatrix} \text{(in homogeneous coordinates)}$$

$$\text{Projected coordinates: } Q' = \begin{bmatrix} -6.21 \\ -1.14 \\ 1 \end{bmatrix} \text{(in Cartesian coordinates)}$$

Verification: The projected point must lie on the view line: $L(-6.21, -1.14) = 0$ (Figure 9.3c).

MATLAB® Code 9.1

```
clear all; clc; format compact;
P = [-8, 6, 1];
L = [-3, 12, -5];
syms x y;
f = -3*x + 12*y - 5;

% (a) perspective projection

V = [-3, 11, 1];
M = V'*L - V*L'*eye(3);
Q = M*P';
Qc = Q/Q(3)
%plotting
y1 = (-L(1)*x - L(3))/L(2);
xx = -20:10;
yy = subs(y1, x, xx);
plot(xx, yy, 'b-', 'LineWidth', 1.5);
hold on; grid;
scatter(P(1), P(2), 20, 'r', 'filled');
scatter(V(1), V(2), 20, 'r', 'filled');
scatter(Qc(1), Qc(2), 20, 'r', 'filled');
plot([V(1), P(1)], [V(2), P(2)], 'k--');
plot([P(1), Qc(1)], [P(2), Qc(2)], 'k--');
xlabel('x'); ylabel('y'); axis square;
text(-2, 1, 'L', 'FontSize', 15);
text(P(1)+0.5, P(2), 'P');
text(Qc(1)+0.5, Qc(2), 'Q');
text(V(1)+0.5, V(2), 'V');
%verification
vrf1 = subs(f, [x, y], [Qc(1), Qc(2)])

% (b) oblique projection

V = [3, -2, 0];
M = V'*L - V*L'*eye(3);
Q = M*P';
Qc = Q/Q(3)
%plotting
figure
syms x y;
y1 = (-L(1)*x - L(3))/L(2);
xx = -10:10;
yy = subs(y1, x, xx);
plot(xx, yy, 'b-', 'LineWidth', 1.5);
```

```
hold on; grid;
quiver(P(1), P(2), V(1), V(2));
quiver(V(1), V(2), V(1), V(2));
scatter(P(1), P(2), 20, 'r', 'filled');
scatter(V(1), V(2), 20, 'r', 'filled');
scatter(Qc(1), Qc(2), 20, 'r', 'filled');
plot([P(1), Qc(1)], [P(2), Qc(2)], 'k--');
xlabel('x'); ylabel('y'); axis square;
text(P(1)+0.5, P(2), 'P');
text(V(1)+0.5, V(2), 'V');
text(Qc(1)+0.5, Qc(2), 'Q');
text(-2, 1, 'L', 'FontSize', 15);
%verification
vrf2 = subs(f, [x, y], [Qc(1), Qc(2)])

% (c) orthographic projection

m = -L(1)/L(2);  % slope
t = [1, m, 1];  % tangent
R90 = [cosd(90) -sind(90) 0 ; sind(90) cosd(90) 0 ; 0 0 1];
n = R90*t';        % normal
V = [n(1), n(2), 0];
M = V'*L - V*L'*eye(3);
Q = M*P';
Qc = Q/Q(3)
%plotting
figure
syms x y;
y1 = (-L(1)*x - L(3))/L(2);
xx = -10:10;
yy = subs(y1, x, xx);
plot(xx, yy, 'b-', 'LineWidth', 1.5);
hold on; grid;
quiver(P(1), P(2), V(1), V(2));
quiver(V(1), V(2), V(1), V(2));
scatter(P(1), P(2), 20, 'r', 'filled');
scatter(V(1), V(2), 20, 'r', 'filled');
scatter(Qc(1), Qc(2), 20, 'r', 'filled');
plot([P(1), Qc(1)], [P(2), Qc(2)], 'k--');
text(Qc(1)+0.5, Qc(2), 'Q');
text(V(1)+0.5, V(2), 'V');
text(P(1)+0.5, P(2), 'P');
text(-2, 1, 'L', 'FontSize', 15);
xlabel('x'); ylabel('y'); axis square;
```

```
%verification
vrf3 = subs(f, [x, y], [Qc(1), Qc(2)]);
hold off;
```

NOTE

`eye`: generates an identity matrix of specified size

9.3 3D PROJECTION

Similar to the discussions of the previous section, it can be shown that the projected coordinates Q of the point P in 3D space onto a viewplane with normal N and viewpoint V is given by:

$$Q = L \times (V \times P) = \left[V^T \cdot N - \left(VN^T\right) \cdot I\right] \cdot P = M \cdot P \tag{9.4}$$

Here, I is the identity matrix and M is the perspective transformation matrix (Marsh, 2005).

Example 9.2

Consider a viewplane $F(x, y, z): -x + 3y + 2z - 4 = 0$ and a point P(−4, 2, 2). Determine the projection matrix and projected coordinates in each case: (a) Perspective projection of P with viewpoint (2, −1, 1) (b) Parallel oblique projection of P in the direction (1, 2, 1) (c) Parallel orthographic projection of P at right angles to the viewplane

(a)

Point $P = [-4, 2, 2, 1]$
Normal $N = [-1, 3, 2, -4]$
Viewpoint $V = [2, -1, 1, 1]$
From Equation (9.4)

$$\text{Projection matrix } M = \left[V^T \cdot N - \left(VN^T\right) \cdot I\right] = \begin{bmatrix} 2 \\ -1 \\ 1 \\ 1 \end{bmatrix} \begin{bmatrix} -1 & 3 & 2 & -4 \end{bmatrix}$$

$$-\begin{bmatrix} 2 & -1 & 1 & 1 \end{bmatrix} \begin{bmatrix} -1 \\ 3 \\ 2 \\ -4 \end{bmatrix} \begin{bmatrix} 1 & 0 & 0 & 0 \\ 0 & 1 & 0 & 0 \\ 0 & 0 & 1 & 0 \\ 0 & 0 & 0 & 1 \end{bmatrix}$$

Simplifying, $M = \begin{bmatrix} 5 & 6 & 4 & -8 \\ 1 & 4 & -2 & 4 \\ -1 & 3 & 9 & -4 \\ -1 & 3 & 2 & 3 \end{bmatrix}$

Projected coordinates: $Q_h = M \cdot P = \begin{bmatrix} -8 \\ 4 \\ 24 \\ 17 \end{bmatrix}$ (homogeneous coordinates)

Projected coordinates: $Q = Q_h/17 = \begin{bmatrix} -0.47 \\ 0.23 \\ 1.41 \\ 1 \end{bmatrix}$ (Cartesian coordinates)

Verification: $F(-0.47, 0.23, 1.41) = 0$ (Figure 9.4a)

(b)

Point $P = [-4, 2, 2, 1]$
Normal $N = [-1, 3, 2, -4]$
Viewpoint $V = [1, 2, 1, 1]$
From Equation (9.4)

Projection matrix $M = \left[V^T \cdot N - \left(VN^T\right) \cdot I\right] = \begin{bmatrix} 1 \\ 2 \\ 1 \\ 1 \end{bmatrix} \begin{bmatrix} -1 & 3 & 2 & -4 \end{bmatrix}$

$-\begin{bmatrix} 1 & 2 & 1 & 1 \end{bmatrix} \begin{bmatrix} -1 \\ 3 \\ 2 \\ -4 \end{bmatrix} \begin{bmatrix} 1 & 0 & 0 & 0 \\ 0 & 1 & 0 & 0 \\ 0 & 0 & 1 & 0 \\ 0 & 0 & 0 & 1 \end{bmatrix}$

Simplifying, $M = \begin{bmatrix} -8 & 3 & 2 & -4 \\ -2 & -1 & 4 & -8 \\ -1 & 3 & -5 & -4 \\ 0 & 0 & 0 & -7 \end{bmatrix}$

Projected coordinates: $Q_h = M \cdot P = \begin{bmatrix} 38 \\ 6 \\ -4 \\ -7 \end{bmatrix}$ (homogeneous coordinates)

$$\text{Projected coordinates: } Q = Q_h/(-7) = \begin{bmatrix} -5.43 \\ -0.85 \\ 0.57 \\ 1 \end{bmatrix} \text{(Cartesian coordinates)}$$

Verification: $F(-5.43, -0.85, 0.57) = 0$ (Figure 9.4b)

(c)

Point $P = [-4, 2, 2, 1]$

Normal $N = [-1, 3, 2, -4]$

Viewpoint $V = N = [-1, 3, 2, -4]$

From Equation (9.4)

$$\text{Projection matrix } M = \left[V^T \cdot N - \left(VN^T\right) \cdot I\right] = \begin{bmatrix} -1 \\ 3 \\ 2 \\ -4 \end{bmatrix} \begin{bmatrix} -1 & 3 & 2 & -4 \end{bmatrix}$$

$$- \begin{bmatrix} -1 & 3 & 2 & -4 \end{bmatrix} \begin{bmatrix} -1 \\ 3 \\ 2 \\ -4 \end{bmatrix} \begin{bmatrix} 1 & 0 & 0 & 0 \\ 0 & 1 & 0 & 0 \\ 0 & 0 & 1 & 0 \\ 0 & 0 & 0 & 1 \end{bmatrix}$$

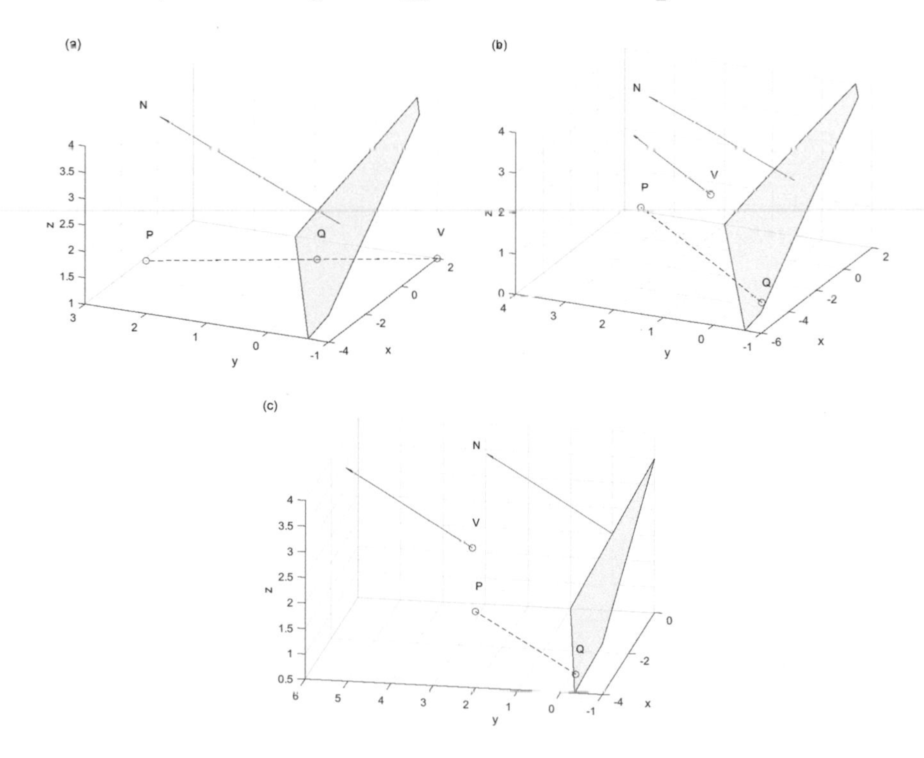

FIGURE 9.4 Plots for Example 9.2 (a) perspective projection (b) oblique projection (c) orthographic projection.

$$\text{Simplifying, } M = \begin{bmatrix} -13 & -3 & -2 & 4 \\ -3 & -5 & 6 & -12 \\ -2 & 6 & -10 & -8 \\ 0 & 0 & 0 & -14 \end{bmatrix}$$

$$\text{Projected coordinates: } Q_h = M \cdot P = \begin{bmatrix} 46 \\ 2 \\ -8 \\ -14 \end{bmatrix} \text{(homogeneous coordinates)}$$

$$\text{Projected coordinates: } Q = Q_h/(-14) = \begin{bmatrix} -3.28 \\ -0.14 \\ 0.57 \\ 1 \end{bmatrix} \text{(Cartesian coordinates)}$$

Verification: $F(-3.28, -0.14, 0.57) = 0$ (Figure 9.4c)

MATLAB Code 9.2

```
clear all; clc; format compact;
N = [-1, 3, 2, -4];
P = [-4, 2, 2, 1];
syms x y z;
f = -x + 3*y +2*z - 4;

% (a) perspective projection

V = [2, -1, 1, 1];
M = V'*N - V*N'*eye(4);
Qh = M*P';
Q = Qh/Qh(4)
figure
plot3(P(1), P(2), P(3), 'ro');
hold on; grid; view(-66, 30);
xlabel('x'); ylabel('y'); zlabel('z');
plot3(Q(1), Q(2), Q(3), 'ro');
plot3(V(1), V(2), V(3), 'ro');
plot3([V(1) P(1)], [V(2) P(2)], [V(3) P(3)], 'k--');
quiver3(0, 0, 2, N(1), N(2), N(3));
text(P(1), P(2), P(3)+0.5, 'P');
text(Q(1), Q(2), Q(3)+0.5, 'Q');
text(V(1), V(2), V(3)+0.5, 'V');
text(N(1), N(2), N(3)+2, 'N');
%verification
vrf1 = subs(f, [x, y, z], [Q(1), Q(2), Q(3)])
fimplicit3(f, 'MeshDensity', 2, 'FaceColor', 'y', 'FaceAlpha',0.3);
```

```
% (b) oblique projection

V = [1, 2, 1, 0];
M = V'*N - V*N'*eye(4);
Qh = M*P';
Q = Qh/Qh(4)
figure
plot3(P(1), P(2), P(3), 'ro');
hold on; grid; view(-66, 30);
quiver3(0, 0, 2, N(1), N(2), N(3));
quiver3(V(1), V(2), V(3), V(1), V(2), V(3));
plot3(Q(1), Q(2), Q(3), 'ro');
plot3(V(1), V(2), V(3), 'ro');
plot3([Q(1) P(1)], [Q(2) P(2)], [Q(3) P(3)], 'k--');
xlabel('x'); ylabel('y'); zlabel('z');
text(P(1), P(2), P(3)+0.5, 'P');
text(Q(1), Q(2), Q(3)+0.5, 'Q');
text(V(1), V(2), V(3)+0.5, 'V');
text(N(1), N(2), N(3)+2, 'N');
%verification
vrf2 = subs(f, [x, y, z], [Q(1), Q(2), Q(3)])
fimplicit3(f, 'MeshDensity', 2, 'FaceColor', 'y', 'FaceAlpha',0.3);

% (c) orthographic projection

V = [N(1), N(2), N(3), 0];
M = V'*N - V*N'*eye(4);
Qh = M*P';
Q = Qh/Qh(4)
figure
plot3(P(1), P(2), P(3), 'ro');
hold on; grid; view(-80, 25);
quiver3(0, 0, 2, N(1), N(2), N(3));
quiver3(V(1), V(2), V(3), V(1), V(2), V(3));
plot3(Q(1), Q(2), Q(3), 'ro');
plot3(V(1), V(2), V(3), 'ro');
plot3([Q(1) P(1)], [Q(2) P(2)], [Q(3) P(3)], 'k--');
xlabel('x'); ylabel('y'); zlabel('z');
text(P(1), P(2), P(3)+0.5, 'P');
text(Q(1), Q(2), Q(3)+0.5, 'Q');
text(V(1), V(2), V(3)+0.5, 'V');
text(N(1), N(2), N(3)+2, 'N')
%verification
vrf3 = subs(f, [x, y, z], [Q(1), Q(2), Q(3)])
fimplicit3(f, 'MeshDensity', 2, 'FaceColor', 'y', 'FaceAlpha',0.3);
hold off;
```

9.4 MULTI-VIEW PROJECTION

In parallel orthographic multi-view projection, the projection lines are parallel to each other and perpendicular to the primary planes. Such views are called top, side, and front views, and display only one face of the object (Foley et al., 1995). These are used in engineering and architectural drawings as length and angles can be accurately measured. If view plane coincides with primary planes then projection matrices are as shown below, where the subscripts indicate projection on the *XY*-, *XZ*-, and *YZ*-planes.

$$P_{xy} = \begin{bmatrix} 1 & 0 & 0 & 0 \\ 0 & 1 & 0 & 0 \\ 0 & 0 & 0 & 0 \\ 0 & 0 & 0 & 1 \end{bmatrix} \quad (9.5)$$

$$P_{xz} = \begin{bmatrix} 1 & 0 & 0 & 0 \\ 0 & 0 & 0 & 0 \\ 0 & 0 & 1 & 0 \\ 0 & 0 & 0 & 1 \end{bmatrix} \quad (9.6)$$

$$P_{yz} = \begin{bmatrix} 0 & 0 & 0 & 0 \\ 0 & 1 & 0 & 0 \\ 0 & 0 & 1 & 0 \\ 0 & 0 & 0 & 1 \end{bmatrix} \quad (9.7)$$

If the projection planes are parallel to the primary planes say at $z = k$ then the following steps are performed to derive the transformation matrix:

- Translate the $z = k$ plane to the $z = 0$ plane (*XY*-plane): $T_1 = T(0, 0, -k)$
- Perform projection on the *XY*-plane: P_{xy}
- Reverse translate the plane to original location: $T_2 = T(0, 0, k)$
- Composite transformation matrix: $M = T(0, 0, k) \cdot P_{xy} \cdot T(0, 0, -k)$

Example 9.3

A cube with center at origin and vertices at (–1, 1, 1), (1, 1, 1), (1, –1, 1), (–1, –1, 1), (–1, 1, –1), (1, 1, –1), (1, –1, –1), and (–1, –1, –1). Derive a parallel projection of the cube onto the $z = 3$ plane in a direction parallel to the Z-axis

Here, normal to viewplane $N = [0, 0, 1, -3]$

The viewpoint is located at infinity along the *Z*-axis, hence $V = N$

From Equation (9.4)

Projection matrix $M = \left[V^T \cdot N - \left(VN^T\right) \cdot I\right] = \begin{bmatrix} 0 \\ 0 \\ 1 \\ -3 \end{bmatrix} \begin{bmatrix} 0 & 0 & 1 & -3 \end{bmatrix}$

$$-\begin{bmatrix} 0 & 0 & 1 & -3 \end{bmatrix} \begin{bmatrix} 0 \\ 0 \\ 1 \\ -3 \end{bmatrix} \begin{bmatrix} 1 & 0 & 0 & 0 \\ 0 & 1 & 0 & 0 \\ 0 & 0 & 1 & 0 \\ 0 & 0 & 0 & 1 \end{bmatrix}$$

Simplifying, $M = \begin{bmatrix} -1 & 0 & 0 & 0 \\ 0 & -1 & 0 & 0 \\ 0 & 0 & 0 & -3 \\ 0 & 0 & 0 & -1 \end{bmatrix}$

Original coordinate matrix: $C = \begin{bmatrix} -1 & 1 & 1 & -1 & -1 & 1 & 1 & -1 \\ 1 & 1 & -1 & -1 & 1 & 1 & -1 & -1 \\ 1 & 1 & 1 & 1 & -1 & -1 & -1 & -1 \\ 1 & 1 & 1 & 1 & 1 & 1 & 1 & 1 \end{bmatrix}$

New coordinate matrix:

$$D_h = M \cdot C = \begin{bmatrix} 1 & -1 & -1 & 1 & 1 & -1 & -1 & 1 \\ -1 & -1 & 1 & 1 & -1 & -1 & 1 & 1 \\ -3 & -3 & -3 & -3 & -3 & -3 & -3 & -3 \\ 1 & 1 & 1 & 1 & 1 & 1 & 1 & 1 \end{bmatrix} \text{(homogeneous coordinates)}$$

New coordinate matrix:

$$D = D_h/(-1) = \begin{bmatrix} -1 & 1 & 1 & -1 & -1 & 1 & 1 & -1 \\ 1 & 1 & -1 & -1 & 1 & 1 & -1 & -1 \\ 3 & 3 & 3 & 3 & 3 & 3 & 3 & 3 \\ 1 & 1 & 1 & 1 & 1 & 1 & 1 & 1 \end{bmatrix} \text{(Cartesian coordinates)}$$

New coordinates: (–1, 1, 3), (1, 1, 3), (1, –1, 3), (–1, –1, 3), (–1, 1, 3), (1, 1, 3), (1, –1, 3), and (–1, –1, 3) (Figure 9.5)

MATLAB Code 9.3

```
clear all; clc; format compact;
N = [0, 0, 1, -3];
p1 = [-1,1,1];
p2 = [1,1,1];
p3 = [1,-1,1];
p4 = [-1,-1,1];
```

```
p5 = [-1,1,-1];
p6 = [1,1,-1];
p7 = [1,-1,-1];
p8 = [-1,-1,-1];
C = [p1' p2' p3' p4' p5' p6' p7' p8' ;
    1 1 1 1 1 1 1 1 ];

% orthographic multi-view projection
V = [N(1), N(2), N(3), 0];
M = V'*N - V*N'*eye(4);
Dh = M*C;
D = Dh/Dh(4);

fprintf('New vertices : \n')
for i=1:8
    fprintf('(%.2f, %.2f, %.2f) \n',D(1,i), D(2,i), D(3,i));
end

figure
C = [p1' p2' p3' p4' p1' p5' p6' p7' p8' p5' p8' p4' p3' p7' p6' p2' ;
    1 1 1 1 1 1 1 1 1 1 1 1 1 1 1 1];
Dh = M*C; D = Dh/Dh(4);
plot3(C(1,:), C(2,:), C(3,:), 'b'); hold on;
plot3(D(1,:), D(2,:), D(3,:), 'r');
xlabel('x'); ylabel('y'); zlabel('z');
legend('original', 'new'); axis equal;
grid; hold off;
```

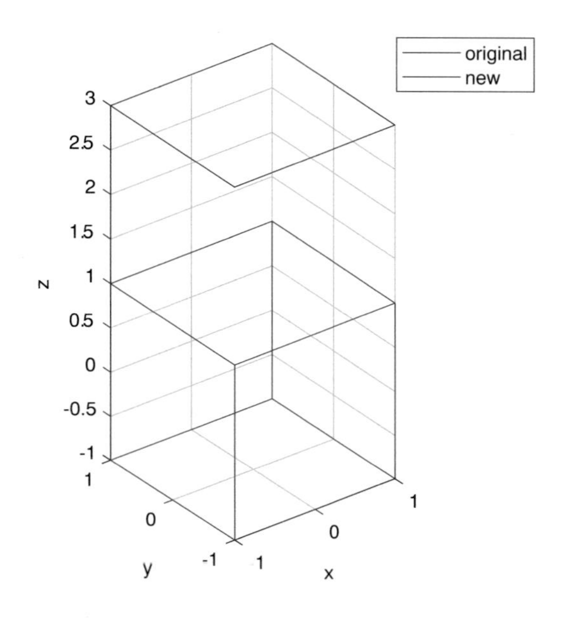

FIGURE 9.5 Plot for Example 9.3.

9.5 AXONOMETRIC PROJECTION

In parallel orthographic axonometric projection, the projection lines are parallel to each other and perpendicular to the viewplanes but the viewplanes do not coincide with the primary planes (Chakraborty, 2010). Unlike multi-view projection where only a single face of the object is visible, in this case multiple views of the object can be viewed. To derive the projection matrix the following steps are followed:

- Align the normal to the viewplane to coincide with one of the primary axes, say the Z-axis
- Perform projection on the corresponding primary plane i.e. XY-plane
- Reverse align the viewplane to original location

Example 9.4

A cube with center at origin and vertices at (–1, 1, 1), (1, 1, 1), (1, –1, 1), (–1, –1, 1), (–1, 1, –1), (1, 1, –1), (1, –1, –1), and (–1, –1, –1). Derive an axonometric projection of the cube onto the plane having normal vector along $i+j+k$ and passing through the point (0, 0, 5).

The viewplane has normal $i+j+k$ and passes through the point (0, 0, 5). Hence, equation of the viewplane is $(x-0)+(y-0)+(z-5)=0$. Thus $N=[1, 1, 1, -5]$.

Also since the projection is parallel and orthographic the projection lines are along the normal to the plane and the viewpoint is located at infinity. Thus $V=[1, 1, 1, 0]$.

From Equation (9.4)

$$\text{Projection matrix: } M=\left[V^T\cdot N-\left(VN^T\right)\cdot I\right]=\begin{bmatrix}1\\1\\1\\0\end{bmatrix}\begin{bmatrix}1 & 1 & 1 & -5\end{bmatrix}$$

$$-\begin{bmatrix}1 & 1 & 1 & 0\end{bmatrix}\begin{bmatrix}1\\1\\1\\-5\end{bmatrix}\begin{bmatrix}1 & 0 & 0 & 0\\0 & 1 & 0 & 0\\0 & 0 & 1 & 0\\0 & 0 & 0 & 1\end{bmatrix}$$

$$\text{Simplifying, } M=\begin{bmatrix}-2 & 1 & 1 & -5\\1 & -2 & 1 & -5\\1 & 1 & -2 & -5\\0 & 0 & 0 & -3\end{bmatrix}$$

$$\text{Original coordinate matrix: } C = \begin{bmatrix} -1 & 1 & 1 & -1 & -1 & 1 & 1 & -1 \\ 1 & 1 & -1 & -1 & 1 & 1 & -1 & -1 \\ 1 & 1 & 1 & 1 & -1 & -1 & -1 & -1 \\ 1 & 1 & 1 & 1 & 1 & 1 & 1 & 1 \end{bmatrix}$$

New coordinate matrix: $D_h = M \cdot C$

$$= \begin{bmatrix} 1 & -5 & -7 & -3 & -3 & -7 & -9 & -5 \\ -7 & -5 & -1 & -3 & -9 & -7 & -3 & -5 \\ -7 & -5 & -7 & -9 & -3 & -1 & -3 & -5 \\ -3 & -3 & -3 & -3 & -3 & -3 & -3 & -3 \end{bmatrix} \text{(homogeneous coordinates)}$$

New coordinate matrix:

$$D = D_h/(-3) = \begin{bmatrix} 0.33 & 1.67 & 2.33 & 1 & 1 & 2.33 & 3 & 1.67 \\ 2.33 & 1.67 & 0.33 & 1 & 3 & 2.33 & 1 & 1.67 \\ 2.33 & 1.67 & 2.33 & 3 & 1 & 0.33 & 1 & 1.67 \\ 1 & 1 & 1 & 1 & 1 & 1 & 1 & 1 \end{bmatrix} \text{(Cartesian coordinates)}$$

New coordinates: (0.33, 2.33, 2.33), (1.67, 1.67, 1.67), (2.33, 0.33, 2.33), (1, 1, 3), (1, 3, 1), (2.33, 2.33, 0.33), (3, 1, 1), and (1.67, 1.67, 1.67) (Figure 9.6)

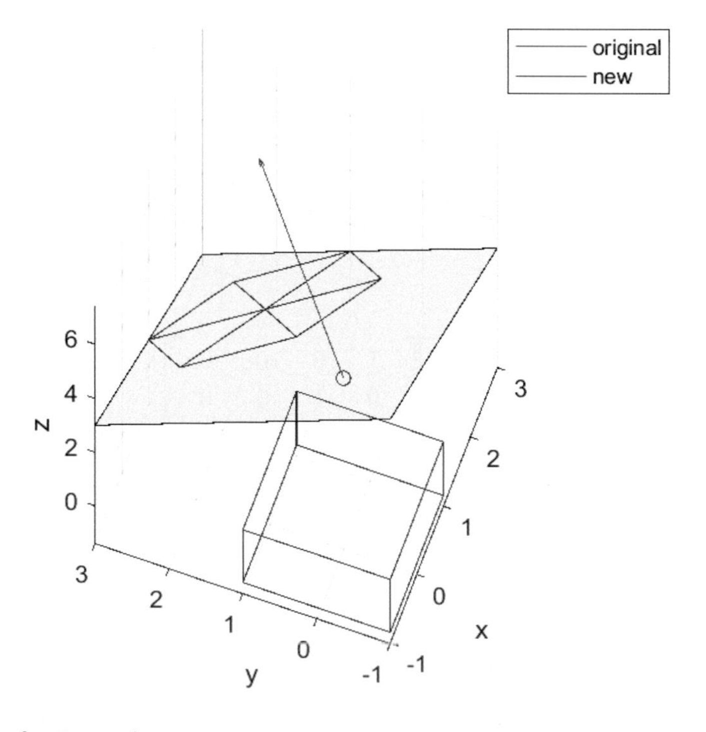

FIGURE 9.6 Plot for Example 9.4.

MATLAB Code 9.4

```
clear all; clc; format compact;
p1 = [-1,1,1];
p2 = [1,1,1];
p3 = [1,-1,1];
p4 = [-1,-1,1];
p5 = [-1,1,-1];
p6 = [1,1,-1];
p7 = [1,-1,-1];
p8 = [-1,-1,-1];
C = [p1' p2' p3' p4' p5' p6' p7' p8' ;
    1 1 1 1 1 1 1 1 ];
N = [1, 1, 1, -5];
V = [1, 1, 1, 0];
M = V'*N - V*N'*eye(4);
Dh = M*C;
D = Dh/Dh(4);

fprintf('New vertices : \n')
for i=1:8
    fprintf('(%.2f, %.2f, %.2f) \n',D(1,i), D(2,i), D(3,i));
end

figure
syms x y z;
f = x + y + z - 5;
C = [p1' p2' p3' p4' p1' p5' p6' p7' p8' p5' p8' p4' p3' p7' p6' p2' ;
    1 1 1 1 1 1 1 1 1 1 1 1 1 1 1 1];
Dh = M*C; D = Dh/Dh(4);
plot3(C(1,:), C(2,:), C(3,:), 'b'); hold on; grid;
plot3(D(1,:), D(2,:), D(3,:), 'r');
plot3(0, 0, 5, 'ro');
quiver3(0, 0, 5, 2, 2, 2);
fimplicit3(f, 'MeshDensity', 2, 'FaceColor', 'y', 'FaceAlpha',0.3);
xlabel('x'); ylabel('y'); zlabel('z');
legend('original', 'new'); axis equal;
view(-70, 70); hold off;
```

9.6 FORESHORTENING FACTORS

Foreshortening factors are ratios of the projected lengths to the original lengths of components of a vector along the three principle axes. Let a line PQ on a plane be projected on the XY-plane at pq. Then the ratio of length (pq) to the ratio of length (PQ) is referred to as the foreshortening scaling factor and defines how much the original line has been scaled due to projection (Marsh, 2005). It turns out that this ratio is same for all lines on the plane and does not depend on the actual coordinates of the line (Figure 9.7).

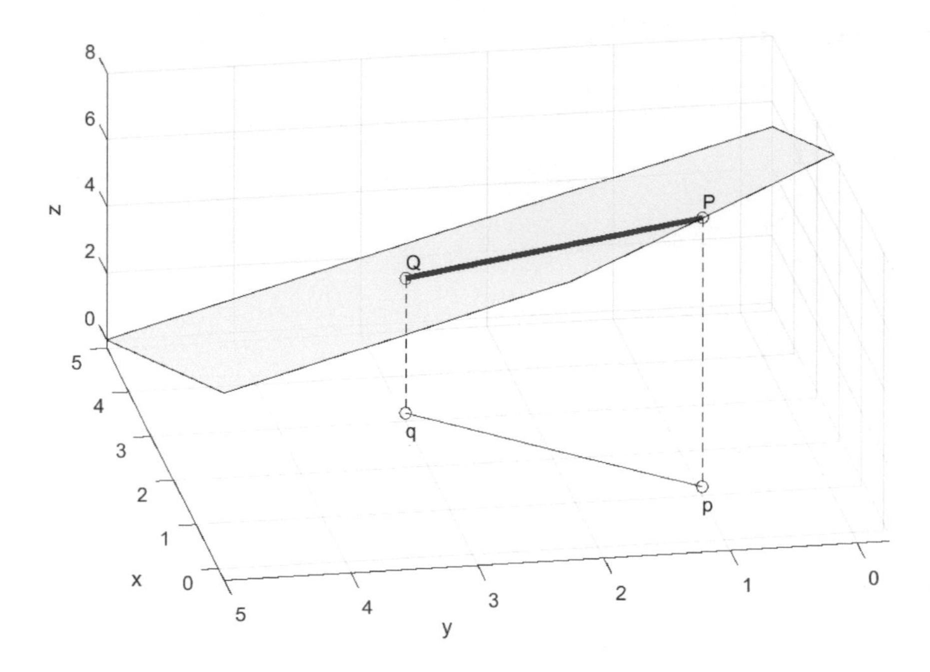

FIGURE 9.7 Projection of a line.

Consider projection of a plane whose normal is given by the vector $N = ai + bj + ck$. To calculate projection matrix for the plane, let the vector N be aligned along a principle axis, say Z-axis. As discussed in Chapter 7, Section 7.8, the processes for vector alignment are as follows:

- Rotate vector by angle α CCW around X-axis to place it on the XZ-plane: $R_x(\alpha)$.
- Rotate vector by angle $-\varphi$ CW around Y-axis to align it along Z-axis: $R_y(-\varphi)$.
- Project on XY-plane along Z-axis: P_{xy}.
- Calculate combined transformation matrix: $M = P_{xy} \cdot R_y(-\varphi) \cdot R_x(\alpha)$.

Expanding M,

$$M = \begin{bmatrix} \cos\varphi & -\sin\alpha \cdot \sin\varphi & -\cos\alpha \cdot \sin\varphi & 0 \\ 0 & \cos\alpha & -\sin\alpha & 0 \\ 0 & 0 & 0 & 0 \\ 0 & 0 & 0 & 1 \end{bmatrix} \tag{9.8}$$

Now consider a vector $P = Ai + Bj + Ck$. Its component vectors along the three principle axes are given by $P_x = (A, 0, 0)$, $P_y = (0, B, 0)$, $P_z = (0, 0, C)$. The lengths of the vectors are:

$$L(P_x) = A,\ L(P_y) = B,\ L(P_z) = C \tag{9.9}$$

The projected components of P onto the primary axes are calculated as follows:

$$\begin{aligned} Q_x &= M \cdot P_x = (A \cdot \cos\varphi, 0, 0) \\ Q_y &= M \cdot P_y = (-B \cdot \sin\alpha \cdot \sin\varphi, B \cdot \cos\alpha, 0) \\ Q_z &= M \cdot P_z = (-C \cdot \cos\alpha \cdot \sin\varphi, -C \cdot \sin\alpha, 0) \end{aligned} \tag{9.10}$$

New lengths of the vector components are calculated as follows:

$$\begin{aligned} L(Q_x) &= A \cdot \cos\varphi \\ L(Q_y) &= B \cdot \sqrt{(\cos\alpha)^2 + (\sin\alpha \cdot \sin\varphi)^2} \\ L(Q_z) &= C \cdot \sqrt{(\sin\alpha)^2 + (\cos\alpha \cdot \sin\varphi)^2} \end{aligned} \tag{9.11}$$

Foreshortening factors are calculated as ratio of modified lengths to original lengths:

$$\begin{aligned} ff_x &= L(Q_x)/L(P_x) = \cos\varphi \\ ff_y &= L(Q_y)/L(P_y) = \sqrt{(\cos\alpha)^2 + (\sin\alpha \cdot \sin\varphi)^2} \\ ff_z &= L(Q_z)/L(P_z) = \sqrt{(\sin\alpha)^2 + (\cos\alpha \cdot \sin\varphi)^2} \end{aligned} \tag{9.12}$$

Observation: Foreshortening factors are independent of vector components A, B, and C. They only depend on the angles of vertical rotation α and horizontal rotation φ.

The projection on XY-plane was an arbitrary choice. Let the projection be on another plane say YZ-plane. In that case the calculation of the transformation matrix is as follows:

- Rotate vector by angle α CCW around X-axis to place it on the XZ-plane: $R_x(\alpha)$
- Rotate vector by angle φ CCW around Y-axis to align it along X-axis: $R_y(\varphi)$.
- Project about YZ-plane along X-axis: P_{yz}.
- Calculate combined transformation matrix: $M = P_{yz} \cdot R_y(\varphi) \cdot R_x(\alpha)$.

Expanding M,

$$M = \begin{bmatrix} 0 & 0 & 0 & 0 \\ 0 & \cos\alpha & -\sin\alpha & 0 \\ -\sin\varphi & \sin\alpha \cdot \cos\varphi & \cos\alpha \cdot \cos\varphi & 0 \\ 0 & 0 & 0 & 1 \end{bmatrix} \tag{9.13}$$

The projected components of P onto the primary axes are calculated as follows:

$$\begin{aligned} Q_x &= M \cdot P_x = (0, 0, A \cdot \sin\varphi) \\ Q_y &= M \cdot P_y = (0, B \cdot \cos\alpha, B \cdot \sin\alpha, \cos\varphi) \\ Q_z &= M \cdot P_z = (0, -C \cdot \sin\alpha, C \cdot \cos\alpha \cdot \cos\varphi) \end{aligned} \quad (9.14)$$

New lengths of the vector components are calculated as follows:

$$\begin{aligned} L(Q_x) &= A \cdot \sin\varphi \\ L(Q_y) &= B \cdot \sqrt{(\cos\alpha)^2 + (\sin\alpha \cdot \cos\varphi)^2} \\ L(Q_z) &= C \cdot \sqrt{(\sin\alpha)^2 + (\cos\alpha \cdot \cos\varphi)^2} \end{aligned} \quad (9.15)$$

Foreshortening factors are calculated as ratio of modified lengths to original lengths:

$$\begin{aligned} ff_x &= L(Q_x)/L(P_x) = \sin\varphi \\ ff_y &= L(Q_y)/L(P_y) = \sqrt{(\cos\alpha)^2 + (\sin\alpha \cdot \cos\varphi)^2} \\ ff_z &= L(Q_z)/L(P_z) = \sqrt{(\sin\alpha)^2 + (\cos\alpha \cdot \cos\varphi)^2} \end{aligned} \quad (9.16)$$

For projection along Z- and X-axes, angle α remains the same, but angle φ is complementary to angle φ in first case. Hence, substituting φ with 90-φ Equation (9.16) becomes identical to Equation (9.12).

Example 9.5

Find the foreshortening factors for the plane whose normal vector is $N = 3i + 4j + 12k$

Normal vector $N = 3i + 4j + 12k$

Here, $a = 3$, $b = 4$, $c = 12$

Also $d = \sqrt{b^2 + c^2} = 12.65$, $e = \sqrt{a^2 + b^2 + c^2} = 13$

Thus $\cos\alpha = c/d = 12/12.65 = 0.95$, $\sin\alpha = b/d = 4/12.65 = 0.32$

$\cos\varphi - a/e - 3/13 - 0.23$, $\sin\varphi = d/e = 12.65/13 = 0.97$

From Equation (9.16):

$$ff_x = \sin\varphi = 0.97$$

$$ff_y = \sqrt{(\cos\alpha)^2 + (\sin\alpha \cdot \cos\varphi)^2} = 0.95$$

$$ff_z = \sqrt{(\sin\alpha)^2 + (\cos\alpha \cdot \cos\varphi)^2} = 0.38$$

MATLAB Code 9.5

```
clear all; clc; format compact;

a = 3; b = 4; c = 12;
d = sqrt(b^2 + c^2);
e = sqrt(a^2 + b^2 + c^2);
cosA = c/d; sinA = b/d;
cosB = a/e; sinB = d/e;
R1 = [1, 0, 0, 0 ; 0, cosA, -sinA, 0 ; 0, sinA, cosA, 0 ; 0, 0, 0, 1];
R2 = [cosB, 0, sinB, 0 ; 0, 1, 0, 0 ; -sinB, 0, cosB, 0 ; 0, 0, 0, 1];
P_YZ = [0, 0, 0, 0 ; 0, 1, 0, 0 ; 0, 0, 1, 0 ; 0, 0, 0, 1];
M = P_YZ * R2 * R1;
ffx = (sinB)
ffy = (cosA^2 + cosB^2*sinA^2)^(1/2)
ffz = (cosA^2*cosB^2 + sinA^2)^(1/2)
```

9.7 ISOMETRIC, DIMETRIC, AND TRIMETRIC

Suppose the projection is along Z-axis on XY-plane from a plane with normal $N = ai + bj + ck$. Foreshortening factor along each axis from Equation (9.16):

$$ff_x = \cos\varphi$$

$$ff_y = \sqrt{(\cos\alpha)^2 + (\sin\alpha \cdot \sin\varphi)^2}$$

$$ff_z = \sqrt{(\sin\alpha)^2 + (\cos\alpha \cdot \sin\varphi)^2}$$

If all three factors are unequal, it is referred to as trimetric projection.

This occurs when $|a|, |b|, |c|$ are all different (Marsh, 2005).

If two of the factors are equal to each other then it is referred to as dimetric projection.

This is satisfied if: $\alpha, \varphi = \pm 45°, \pm 90°, \pm 135°$

For example, putting $\alpha = 45°$, $\varphi = 90°$, we get $ffx = 0$, $ffy = 1$, $ffz = 1$.

This occurs when one of the conditions $|a| = |b|$, $|b| = |c|$ or $|c| = |a|$ is true (Marsh, 2005).

If all three factors are equal to each other then it is referred to as isometric projection. This is satisfied if: $\alpha = \pm 45°$, $\varphi = \pm 35.264°$.

For example putting $\alpha = 45°$, $\varphi = -35.265°$, we get $ffx = ffy = ffz = 0.8165$.

This occurs when $|a| = |b| = |c|$ is true (Marsh, 2005).

If the projection is along X-axis on YZ-plane then for isometric projection: $\alpha = \pm 45°$, $\varphi = 90° \pm 35.264°$.

Example 9.6

Show that axonometric projections of a unit cube at the origin, on the planes $x+y+z=5$, $x+y+2z=5$, and $x+3y+2z=5$ are, respectively, isometric, dimetric, and trimetric.

(a)

$$\text{Original coordinate matrix: } C=\begin{bmatrix} 0 & 1 & 1 & 0 & 0 & 1 & 1 & 0 \\ 0 & 0 & 1 & 1 & 0 & 0 & 1 & 1 \\ 0 & 0 & 0 & 0 & 1 & 1 & 1 & 1 \\ 1 & 1 & 1 & 1 & 1 & 1 & 1 & 1 \end{bmatrix}$$

$$P: x+y+z=5$$

$$a=1, b=1, c=1$$

$$N=[1, 1, 1, -5]$$

$$V=[1, 1, 1, 0]$$

From Equation (9.4)

Projection matrix:

$$M=\left[V^T\cdot N-\left(VN^T\right)\cdot I\right]=\begin{bmatrix} 1 \\ 1 \\ 1 \\ 0 \end{bmatrix}\begin{bmatrix} 1 & 1 & 1 & -5 \end{bmatrix}$$

$$-\begin{bmatrix} 1 & 1 & 1 & 0 \end{bmatrix}\begin{bmatrix} 1 \\ 1 \\ 1 \\ -5 \end{bmatrix}\begin{bmatrix} 1 & 0 & 0 & 0 \\ 0 & 1 & 0 & 0 \\ 0 & 0 & 1 & 0 \\ 0 & 0 & 0 & 1 \end{bmatrix}$$

$$\text{Simplifying, } M=\begin{bmatrix} -2 & 1 & 1 & -5 \\ 1 & -2 & 1 & -5 \\ 1 & 1 & -2 & -5 \\ 0 & 0 & 0 & -1 \end{bmatrix}$$

New coordinate matrix:

$$D_h=M\cdot C=\begin{bmatrix} -5 & -7 & -6 & -4 & -4 & -6 & -5 & -3 \\ -5 & -4 & -6 & -7 & -4 & -3 & -5 & -6 \\ -5 & -4 & -3 & -4 & -7 & -6 & -5 & -6 \\ -3 & -3 & -3 & -3 & -3 & -3 & -3 & -3 \end{bmatrix}\text{(homogeneous coordinates)}$$

$$\text{New coordinate matrix: } D=\begin{bmatrix} 1.67 & 2.33 & 2 & 1.33 & 1.33 & 2 & 1.67 & 1 \\ 1.67 & 1.33 & 2 & 2.33 & 1.33 & 1 & 1.67 & 2 \\ 1.67 & 1.33 & 1 & 1.33 & 2.33 & 2 & 1.67 & 2 \\ 1 & 1 & 1 & 1 & 1 & 1 & 1 & 1 \end{bmatrix}$$

(Cartesian coordinates)

$$d=\sqrt{b^2+c^2}=1.41,\ e=\sqrt{a^2+b^2+c^2}=1.73$$

$$\cos\alpha=\frac{c}{d}=0.71,\ \sin\alpha=\frac{b}{d}=0.71$$

$$\cos\varphi=\frac{d}{e}=0.82,\ \sin\varphi=\frac{a}{e}=0.58$$

$$ff_x=\cos\varphi=0.82$$

$$ff_y=\sqrt{(\cos\alpha)^2+(\sin\alpha\cdot\sin\varphi)^2}=0.82$$

$$ff_z=\sqrt{(\sin\alpha)^2+(\cos\alpha\cdot\sin\varphi)^2}=0.82$$

Since all three foreshortening factors are equal, the projection is isometric (Figure 9.8a).

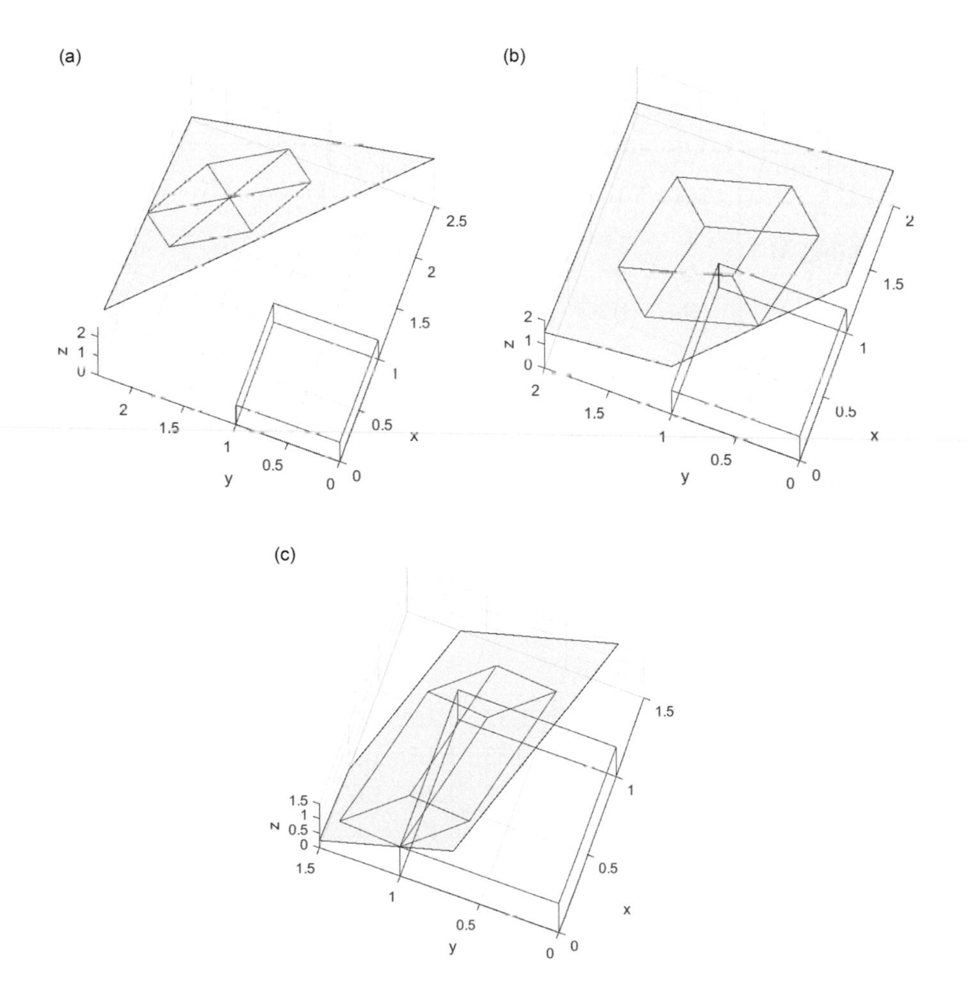

FIGURE 9.8 Plots for Example 9.6 (a) isometric projection (b) dimetric projection (c) trimetric projection.

(b)

$$P: x+y+2z=5$$

$$a=1, b=1, c=2$$

$$N=[1, 1, 2, -5]$$

$$V=[1, 1, 2, 0]$$

From Equation (9.4)

Projection matrix:

$$M=\left[V^T \cdot N-\left(VN^T\right)\cdot I\right]=\begin{bmatrix} 1 \\ 1 \\ 2 \\ 0 \end{bmatrix}\begin{bmatrix} 1 & 1 & 2 & -5 \end{bmatrix}$$

$$-\begin{bmatrix} 1 & 1 & 2 & 0 \end{bmatrix}\begin{bmatrix} 1 \\ 1 \\ 2 \\ -5 \end{bmatrix}\begin{bmatrix} 1 & 0 & 0 & 0 \\ 0 & 1 & 0 & 0 \\ 0 & 0 & 1 & 0 \\ 0 & 0 & 0 & 1 \end{bmatrix}$$

$$\text{Simplifying, } M=\begin{bmatrix} -5 & 1 & 2 & -5 \\ 1 & -5 & 2 & -5 \\ 2 & 2 & -2 & -10 \\ 0 & 0 & 0 & -6 \end{bmatrix}$$

New coordinate matrix:

$$D_h = M \cdot C=\begin{bmatrix} -5 & -10 & -9 & -4 & -3 & -8 & -7 & -2 \\ -5 & -4 & -9 & -10 & -3 & -2 & -7 & -8 \\ -10 & -8 & -6 & -8 & -12 & -10 & -8 & -10 \\ -6 & -6 & -6 & -6 & -6 & -6 & -6 & -6 \end{bmatrix}\text{(homogeneous coordinates)}$$

New coordinate matrix:

$$D=\begin{bmatrix} 0.83 & 1.67 & 1.5 & 0.67 & 0.5 & 1.33 & 1.16 & 0.33 \\ 0.83 & 0.67 & 1.5 & 1.67 & 0.5 & 0.33 & 1.16 & 1.33 \\ 1.67 & 1.33 & 1 & 1.33 & 2 & 1.67 & 1.33 & 1.67 \\ 1 & 1 & 1 & 1 & 1 & 1 & 1 & 1 \end{bmatrix}\text{(Cartesian coordinates)}$$

$$d=\sqrt{b^2+c^2}=2.24, e=\sqrt{a^2+b^2+c^2}=2.45$$

$$\cos\alpha=\frac{c}{d}=0.89, \sin\alpha=\frac{b}{d}=0.45$$

$$\cos\varphi = \frac{d}{e} = 0.91, \sin\varphi = \frac{a}{e} = 0.41$$

$$ff_x = \cos\varphi = 0.91$$

$$ff_y = \sqrt{(\cos\alpha)^2 + (\sin\alpha \cdot \sin\varphi)^2} = 0.91$$

$$ff_z = \sqrt{(\sin\alpha)^2 + (\cos\alpha \cdot \sin\varphi)^2} = 0.58$$

Since two out of three foreshortening factors are equal, the projection is dimetric. (Figure 9.8b).

(c)

$$P: x + 3y + 2z = 5$$

$$a = 1, b = 3, c = 2$$

$$N = [1, 3, 2, -5]$$

$$V = [1, 3, 2, 0]$$

From Equation (9.4)

Projection matrix:

$$M = \left[V^T \cdot N - \left(VN^T\right) \cdot I\right] = \begin{bmatrix} 1 \\ 3 \\ 2 \\ 0 \end{bmatrix} \begin{bmatrix} 1 & 3 & 2 & -5 \end{bmatrix} -$$

$$\begin{bmatrix} 1 & 3 & 2 & 0 \end{bmatrix} \begin{bmatrix} 1 \\ 3 \\ 2 \\ -5 \end{bmatrix} \begin{bmatrix} 1 & 0 & 0 & 0 \\ 0 & 1 & 0 & 0 \\ 0 & 0 & 1 & 0 \\ 0 & 0 & 0 & 1 \end{bmatrix}$$

Simplifying, $M = \begin{bmatrix} -13 & 3 & 2 & -5 \\ 3 & -5 & 6 & -15 \\ 2 & 6 & -10 & -10 \\ 0 & 0 & 0 & -14 \end{bmatrix}$

New coordinate matrix: $D_h = M \cdot C$

$$= \begin{bmatrix} -5 & -18 & -15 & -2 & -3 & -16 & -13 & 0 \\ -15 & -12 & -17 & -20 & -9 & -6 & -11 & -14 \\ -10 & -8 & -2 & -4 & -20 & -18 & -12 & -14 \\ -14 & -14 & -14 & -14 & -14 & -14 & -14 & -14 \end{bmatrix}$$ (homogeneous

coordinates)

New coordinate matrix:

$$D = \begin{bmatrix} 0.35 & 1.28 & 1.07 & 0.14 & 0.21 & 1.14 & 0.93 & 0 \\ 1.07 & 0.85 & 1.21 & 1.42 & 0.64 & 0.43 & 0.79 & 1 \\ 0.71 & 0.57 & 0.14 & 0.28 & 1.43 & 1.29 & 0.86 & 1 \\ 1 & 1 & 1 & 1 & 1 & 1 & 1 & 1 \end{bmatrix} \text{(Cartesian coordinates)}$$

$$d = \sqrt{b^2 + c^2} = 3.60, e = \sqrt{a^2 + b^2 + c^2} = 3.74$$

$$\cos\alpha = \frac{c}{d} = 0.55, \sin\alpha = \frac{b}{d} = 0.83$$

$$\cos\varphi = \frac{d}{e} = 0.96, \sin\varphi = \frac{a}{e} = 0.27$$

$$ff_x = \cos\varphi = 0.96$$

$$ff_y = \sqrt{(\cos\alpha)^2 + (\sin\alpha \cdot \sin\varphi)^2} = 0.59$$

$$ff_z = \sqrt{(\sin\alpha)^2 + (\cos\alpha \cdot \sin\varphi)^2} = 0.84$$

Since, none of the three foreshortening factors are equal to each other, the projection is trimetric (Figure 9.8c).

MATLAB Code 9.6

```
clear all; clc; format compact;
p1 = [0,0,0];
p2 = [1,0,0];
p3 = [1,1,0];
p4 = [0,1,0];
p5 = [0,0,1];
p6 = [1,0,1];
p7 = [1,1,1];
p8 = [0,1,1];

% isometric projection
fprintf('Isometric projection : \n');

C = [p1' p2' p3' p4' p5' p6' p7' p8' ;
    1 1 1 1 1 1 1 1 ];
a = 1; b = 1; c = 1;
N = [a, b, c, -5];
V = [a, b, c, 0];
M = V'*N - V*N'*eye(4);
Dh = M*C;
```

```
D = Dh/Dh(4);
d = sqrt(b^2 + c^2);
e = sqrt(a^2 + b^2 + c^2);
cosA = c/d; sinA = b/d;
cosB = d/e; sinB = a/e;
ffx = (cosB)
ffy = (cosA^2 + sinA^2*sinB^2)^(1/2)
ffz = (sinA^2 + cosA^2*sinB^2)^(1/2)
figure
syms x y z;
f = a*x + b*y + c*z - 5;
C = [p1' p2' p3' p4' p1' p5' p6' p7' p8' p5' p8' p4' p3' p7' p6' p2' ;
    1 1 1 1 1 1 1 1 1 1 1 1 1 1 1 1];
Dh = M*C; D = Dh/Dh(4);
plot3(C(1,:), C(2,:), C(3,:), 'b'); hold on; grid;
plot3(D(1,:), D(2,:), D(3,:), 'r');
fimplicit3(f, 'MeshDensity', 2, 'FaceColor', 'y', 'FaceAlpha',0.3);
xlabel('x'); ylabel('y'); zlabel('z');
axis equal; title('isometric projection');
view(-70, 80);

% dimetric projection
fprintf('Dimetric projection : \n');

C = [p1' p2' p3' p4' p5' p6' p7' p8' ;
    1 1 1 1 1 1 1 1 ];
a = 1; b = 1; c = 2;
N = [a, b, c, -5];
V = [a, b, c, 0];
M = V'*N - V*N'*eye(4);
Dh = M*C;
D = Dh/Dh(4);
d = sqrt(b^2 + c^2);
e = sqrt(a^2 + b^2 + c^2);
d = sqrt(b^2 + c^2);
e = sqrt(a^2 + b^2 + c^2);
cosA = c/d; sinA = b/d;
cosB = d/e; sinB = a/e;
ffx = (cosB)
ffy = (cosA^2 + sinA^2*sinB^2)^(1/2)
ffz = (sinA^2 + cosA^2*sinB^2)^(1/2)
figure
syms x y z;
f = a*x + b*y + c*z - 5;
C = [p1' p2' p3' p4' p1' p5' p6' p7' p8' p5' p8' p4' p3' p7' p6' p2' ;
    1 1 1 1 1 1 1 1 1 1 1 1 1 1 1 1];
Dh = M*C; D = Dh/Dh(4);
```

```
plot3(C(1,:), C(2,:), C(3,:), 'b'); hold on; grid;
plot3(D(1,:), D(2,:), D(3,:), 'r');
fimplicit3(f, 'MeshDensity', 2, 'FaceColor', 'y', 'FaceAlpha',0.3);
xlabel('x'); ylabel('y'); zlabel('z');
axis equal; title('dimetric projection');
view(-70, 80);

%trimetric projection
fprintf('Trimetric projection : \n');

C = [p1' p2' p3' p4' p5' p6' p7' p8' ;
    1 1 1 1 1 1 1 1 ];
a = 1; b = 3; c = 2;
N = [a, b, c, -5];
V = [a, b, c, 0];
M = V'*N - V*N'*eye(4);
Dh = M*C;
D = Dh/Dh(4);
d = sqrt(b^2 + c^2);
e = sqrt(a^2 + b^2 + c^2);
d = sqrt(b^2 + c^2);
e = sqrt(a^2 + b^2 + c^2);
cosA = c/d; sinA = b/d;
cosB = d/e; sinB = a/e;
ffx = (cosB)
ffy = (cosA^2 + sinA^2*sinB^2)^(1/2)
ffz = (sinA^2 + cosA^2*sinB^2)^(1/2)
figure
syms x y z;
f = a*x + b*y + c*z - 5;
C = [p1' p2' p3' p4' p1' p5' p6' p7' p8' p5' p8' p4' p3' p7' p6' p2' ;
    1 1 1 1 1 1 1 1 1 1 1 1 1 1 1 1];
Dh = M*C; D = Dh/Dh(4);
plot3(C(1,:), C(2,:), C(3,:), 'b'); hold on; grid;
plot3(D(1,:), D(2,:), D(3,:), 'r');
fimplicit3(f, 'MeshDensity', 2, 'FaceColor', 'y', 'FaceAlpha',0.3);
xlabel('x'); ylabel('y'); zlabel('z');
axis equal; title('trimetric projection');
view(-70, 80);
hold off;
```

9.8 OBLIQUE PROJECTION

When the direction of parallel projection on the viewplane is oblique i.e. not perpendicular, then it is called an oblique projection. The foreshortening factor of line segments parallel to the viewplane is 1. When the viewing direction makes an angle of 45° with the viewplane then the projection obtained is called cavalier projection. If $V = [v_1, v_2, v_3]$ be the viewpoint vector, then a cavalier projection satisfies the condition: $v_3^2 = v_1^2 + v_2^2$. Here, the image of the projected object looks much thicker than in reality as the foreshortening factor for faces perpendicular to the viewing plane is 1. To reduce the thickness, a projection type called cabinet projection can be used, which reduces this factor to 0.5. This is possible when the viewpoint vector makes an angle of 63.4° with the viewplane. If $V = [v_1, v_2, v_3]$ be the viewpoint vector, then a cabinet projection satisfies the condition: $v_3^2 = 4(v_1^2 + v_2^2)$. (Marsh, 2005)

Example 9.7

Show that the oblique viewing directions (3, 4, 5) and (3, 4, 10) produce a cavalier and cabinet projection of a unit cube on the $z = 0$ plane, respectively. Verify by computing the foreshortening factor of a line perpendicular to the viewplane in each case.

(a)

The projections are on the $z = 0$ plane, hence $N = [0, 0, 1, 0]$

For a parallel projection, viewpoint vector $V = [3, 4, 5, 0]$

Since, it satisfies the identity $v_3^2 = v_1^2 + v_2^2$ the projection produced would be a cavalier projection.

From Equation (9.4)

Projection matrix:

$$M = \left[V^T \cdot N - \left(VN^T\right) \cdot I\right] = \begin{bmatrix} 3 \\ 4 \\ 5 \\ 0 \end{bmatrix} \begin{bmatrix} 0 & 0 & 1 & 0 \end{bmatrix}$$

$$- \begin{bmatrix} 3 & 4 & 5 & 0 \end{bmatrix} \begin{bmatrix} 0 \\ 0 \\ 1 \\ 0 \end{bmatrix} \begin{bmatrix} 1 & 0 & 0 & 0 \\ 0 & 1 & 0 & 0 \\ 0 & 0 & 1 & 0 \\ 0 & 0 & 0 & 1 \end{bmatrix}$$

$$\text{Simplifying, } M = \begin{bmatrix} -5 & 0 & 3 & 0 \\ 0 & -5 & 4 & 0 \\ 0 & 0 & 0 & 0 \\ 0 & 0 & 0 & -5 \end{bmatrix}$$

Original coordinate matrix of a unit cube:

$$C = \begin{bmatrix} 0 & 1 & 1 & 0 & 0 & 1 & 1 & 0 \\ 0 & 0 & 1 & 1 & 0 & 0 & 1 & 1 \\ 0 & 0 & 0 & 0 & 1 & 1 & 1 & 1 \\ 1 & 1 & 1 & 1 & 1 & 1 & 1 & 1 \end{bmatrix}$$

New coordinate matrix:

$$D_h = M \cdot C = \begin{bmatrix} 0 & -5 & -5 & 0 & 3 & -2 & -2 & 3 \\ 0 & 0 & -5 & -5 & 4 & -4 & -1 & -1 \\ 0 & 0 & 0 & 0 & 0 & 0 & 0 & 0 \\ -5 & -5 & -5 & -5 & -5 & -5 & -5 & -5 \end{bmatrix}$$

(homogeneous coordinates)

$$\text{New coordinate matrix: } D = \begin{bmatrix} 0 & 1 & 1 & 0 & -0.6 & 0.4 & 0.4 & -0.6 \\ 0 & 0 & 1 & 1 & -0.8 & -0.8 & 0.2 & 0.2 \\ 0 & 0 & 0 & 0 & 0 & 0 & 0 & 0 \\ 1 & 1 & 1 & 1 & 1 & 1 & 1 & 1 \end{bmatrix}$$

(Cartesian coordinates)

Consider a line segment joining the origin to the point (0, 0, 1), which is perpendicular to the viewplane and has length 1. The projected coordinates of the point is obtained by multiplying M with the point and equals (−0.6, −0.8, 0), which indicates that after projection the line segment still has length 1. Hence, the foreshortening factor is 1, which is as expected for a cavalier projection. (Figure 9.9a).

(b)

For a parallel projection, viewpoint vector $V = [3, 4, 10, 0]$

Since, it satisfies the identity $v_3^2 = 4(v_1^2 + v_2^2)$ the projection produced would be a cabinet projection.

From Equation (9.4)

$$\text{Projection matrix: } M = \left[V^T \cdot N - (VN^T) \cdot I\right] = \begin{bmatrix} 3 \\ 4 \\ 10 \\ 0 \end{bmatrix} \begin{bmatrix} 0 & 0 & 1 & 0 \end{bmatrix}$$

$$- \begin{bmatrix} 3 & 4 & 10 & 0 \end{bmatrix} \begin{bmatrix} 0 \\ 0 \\ 1 \\ 0 \end{bmatrix} \begin{bmatrix} 1 & 0 & 0 & 0 \\ 0 & 1 & 0 & 0 \\ 0 & 0 & 1 & 0 \\ 0 & 0 & 0 & 1 \end{bmatrix}$$

$$\text{Simplifying, } M = \begin{bmatrix} -10 & 0 & 3 & 0 \\ 0 & -10 & 4 & 0 \\ 0 & 0 & 0 & 0 \\ 0 & 0 & 0 & -10 \end{bmatrix}$$

(a)

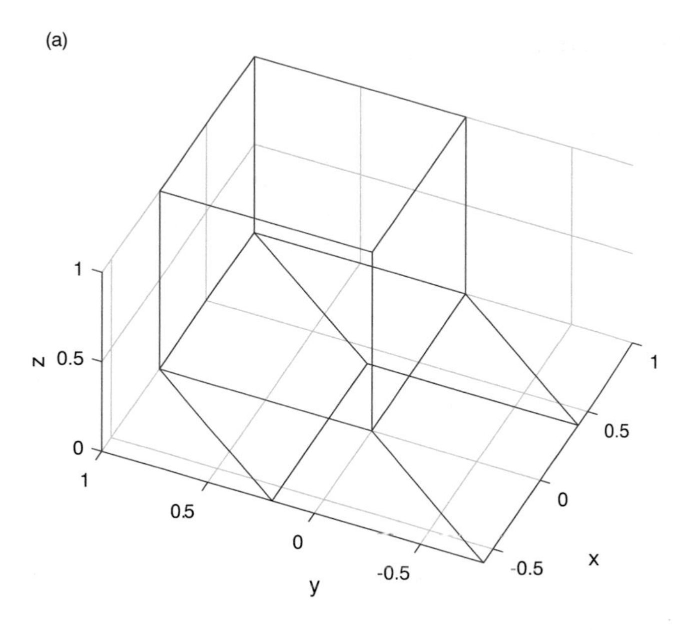

(b)

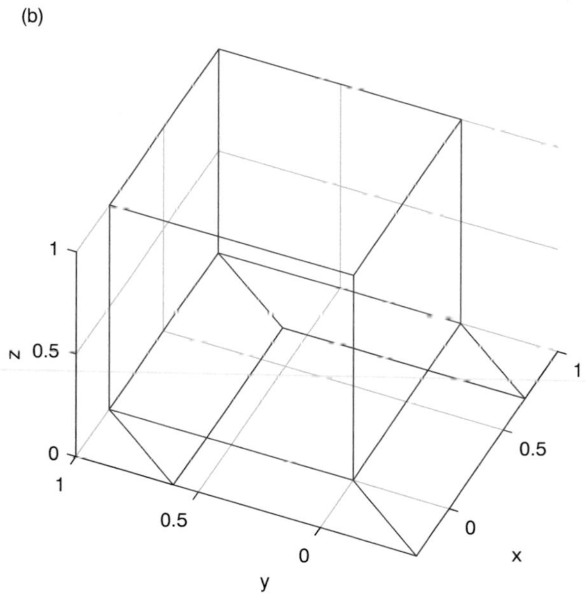

FIGURE 9.9 Plots for Example 8.9 (a) cavalier projection (b) cabinet projection.

Original coordinate matrix of a unit cube:

$$C = \begin{bmatrix} 0 & 1 & 1 & 0 & 0 & 1 & 1 & 0 \\ 0 & 0 & 1 & 1 & 0 & 0 & 1 & 1 \\ 0 & 0 & 0 & 0 & 1 & 1 & 1 & 1 \\ 1 & 1 & 1 & 1 & 1 & 1 & 1 & 1 \end{bmatrix}$$

New coordinate matrix:

$$D_h = M \cdot C = \begin{bmatrix} 0 & -10 & -10 & 0 & 3 & -7 & -7 & 3 \\ 0 & 0 & -10 & -10 & 4 & 4 & -6 & -6 \\ 0 & 0 & 0 & 0 & 0 & 0 & 0 & 0 \\ -10 & -10 & -10 & -10 & -10 & -10 & -10 & -10 \end{bmatrix}$$

(homogeneous coordinates)

$$\text{New coordinate matrix: } D = \begin{bmatrix} 0 & 1 & 1 & 0 & -0.3 & 0.7 & 0.7 & -0.3 \\ 0 & 0 & 1 & 1 & -0.4 & -0.4 & 0.6 & 0.6 \\ 0 & 0 & 0 & 0 & 0 & 0 & 0 & 0 \\ 1 & 1 & 1 & 1 & 1 & 1 & 1 & 1 \end{bmatrix}$$

(Cartesian coordinates)

Consider a line segment joining the origin to the point (0, 0, 1), which is perpendicular to the viewplane and has length 1. The projected coordinates of the point is obtained by multiplying M with the point and equals (−0.3, −0.4, 0), which indicates that after projection the line segment still has length 0.5. Hence, the foreshortening factor is 0.5, which is as expected for a cabinet projection (Figure 9.9b).

MATLAB Code 9.7

```
clear all; clc; format compact;
p1 = [0,0,0];
p2 = [1,0,0];
p3 = [1,1,0];
p4 = [0,1,0];
p5 = [0,0,1];
p6 = [1,0,1];
p7 = [1,1,1];
p8 = [0,1,1];
C = [p1' p2' p3' p4' p5' p6' p7' p8' ;
    1 1 1 1 1 1 1 1 ];

% cavalier projection
fprintf('Cavalier projection : \n')
N = [0, 0, 1, 0];
V = [3, 4, 5, 0];
M = V'*N - V*N'*eye(4);
Dh = M*C;
D = Dh/Dh(4);
figure
C = [p1' p2' p3' p4' p1' p5' p6' p7' p8' p5' p8' p4' p3' p7' p6' p2' ;
    1 1 1 1 1 1 1 1 1 1 1 1 1 1 1 1];
Dh = M*C; D = Dh/Dh(4);
plot3(C(1,:), C(2,:), C(3,:), 'b'); hold on; grid;
```

```
plot3(D(1,:), D(2,:), D(3,:), 'r');
xlabel('x'); ylabel('y'); zlabel('z');
axis equal;
view(-66, 40);
title('cavalier projection');
% verification
V1 = [0 ; 0 ; 1 ; 1];
V1p = M*V1;
V1pc=V1p/(V1p(4));
V1pcl = sqrt(V1pc(1)^2+V1pc(2)^2+V1pc(3)^2);
ff = V1pcl

%cabinet projection
fprintf('Cabinet projection : \n')
N = [0, 0, 1, 0];
V = [3, 4, 10, 0];
M = V'*N - V*N'*eye(4);
C = [p1' p2' p3' p4' p5' p6' p7' p8' ;
    1 1 1 1 1 1 1 1 ];
Dh = M*C;
D = Dh/Dh(4);
figure
C = [p1' p2' p3' p4' p1' p5' p6' p7' p8' p5' p8' p4' p3' p7' p6' p2' ;
    1 1 1 1 1 1 1 1 1 1 1 1 1 1 1 1];
Dh = M*C; D = Dh/Dh(4);
plot3(C(1,:), C(2,:), C(3,:), 'b'); hold on; grid;
plot3(D(1,:), D(2,:), D(3,:), 'r');
xlabel('x'); ylabel('y'); zlabel('z')
axis equal;
view(-66, 40);
title('cabinet projection');
% verification
V1 = [0 ; 0 ; 1 ; 1];
V1p = M*V1; V1pc = V1p/(V1p(4));
V1pcl = sqrt(V1pc(1)^2+V1pc(2)^2+V1pc(3)^2);
ff = V1pcl
hold off;
```

9.9 PERSPECTIVE PROJECTION

Under perspective projection images of parallel lines in space appear to converge to a point called the PRP or center of projection (COP). Let the point be $A(x, y, z)$ whose projection $P(x_p, y_p, 0)$ is required on a view plane coinciding with the XY-plane. Let $R(0, 0, r)$ be the PRP along Z-axis and let O be the origin. It is required to obtain the values of x_p and y_p in terms of x, y, z, and r.

Two cases are possible as follows (1) case-1: where A and R are on opposite side of the view plane and (2) case-2: where A and R are on the same side (Foley et al., 1995).

Case-1 : Opposite side of view-plane (see Figure 9.10).

From similar triangles RCB and ROQ, CB/OQ = CR/OR
Here, CB = x, OQ = x_p, CR = $r + z$, OR = r
Substituting : $x/x_p = (r+z)/r$ i.e. $x_p = \{r/(r+z)\} \cdot x$

From similar triangles ARB and PRQ, AB/PQ = BR/QR
Here, AB = y, PQ = y_p, BR = $\sqrt{(r+z)^2 + x^2}$, QR = $\sqrt{r^2 + {x_p}^2}$
Substituting : $y/y_p = \sqrt{(r+z)^2 + x^2} \Big/ \sqrt{r^2 + {x_p}^2}$ i.e. $y_p = \{r/(r+z)\} \cdot y$

NOTE:
Since C is on the negative side of the Z-axis, we substitute z with $-z$ to obtain the absolute length of CR.
Thus we get : $x_p = \{r/(r-z)\} \cdot x$ and $y_p = \{r/(r-z)\} \cdot y$

Case-2 : Same side of view-plane (see Fig. 9.11).

From similar triangles OQR and CBR, OQ/CB = OR/CR
Here, OQ = x_p, CB = x, OR = r, CR = $r - z$
Substituting : $x_p/x = r/(r-z)$ i.e. $x_p = \{r/(r-z)\} \cdot x$

From similar triangles PQR and ABR, PQ/AB = QR/BR
Here, PQ = y_p, AB = y, QR = $\sqrt{r^2 + {x_p}^2}$, BR = $\sqrt{(r-z)^2 + x^2}$

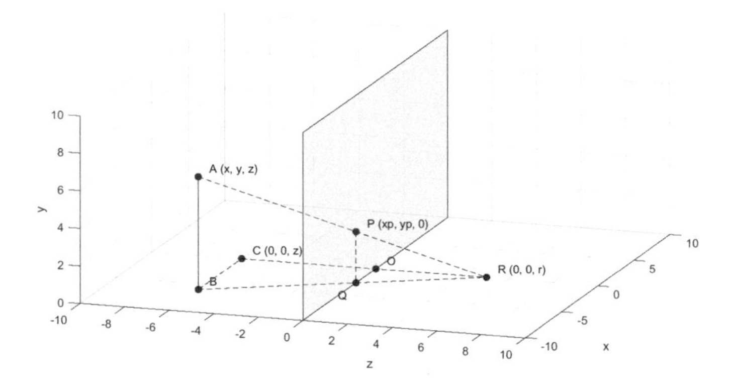

FIGURE 9.10 Perspective projection: case-1.

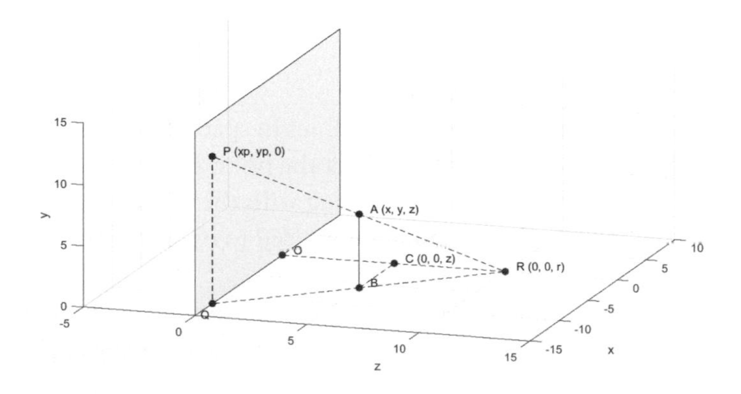

FIGURE 9.11 Perspective projection: case-2.

Substituting : $y_p/y = \sqrt{r^2 + {x_p}^2} \Big/ \sqrt{(r-z)^2 + x^2}$ i.e. $y_p = \{r/(r-z)\} \cdot y$

In each case, $x_p = \left\{\dfrac{r}{r-z}\right\} \cdot x = x \Big/ \left(1 - \dfrac{z}{r}\right),\ y_p = \left\{\dfrac{r}{r-z}\right\} \cdot y = y \Big/ \left(1 - \dfrac{z}{r}\right),\ z_p = 0$

This can be represented in terms of homogeneous coordinates $P = M_{xy} \cdot A$ as follows:

$$\begin{bmatrix} x_p \\ y_p \\ z_p \\ 1 \end{bmatrix} = \begin{bmatrix} x \\ y \\ 0 \\ 1 - \dfrac{z}{r} \end{bmatrix} = \begin{bmatrix} 1 & 0 & 0 & 0 \\ 0 & 1 & 0 & 0 \\ 0 & 0 & 0 & 0 \\ 0 & 0 & -1/r & 1 \end{bmatrix} \begin{bmatrix} x \\ y \\ z \\ 1 \end{bmatrix} \tag{9.17}$$

Here, M_{xy} is the required perspective projection matrix on the XY-plane.

Example 9.8

A cube with center at origin and vertices at (–1, 1, 1), (1, 1, 1), (1, –1, 1), (–1, –1, 1), (–1, 1, –1), (1, 1, –1), (1, –1, –1), and (–1, –1, –1) is projected on the plane $z = 3$ using perspective projection with the reference point at $z = 5$. Find its new vertices.

Here, $r = 5$

Translate view plane to origin: $T(0, 0, -3)$

Apply perspective projection on XY-plane:

$$M_{xy} = \begin{bmatrix} 1 & 0 & 0 & 0 \\ 0 & 1 & 0 & 0 \\ 0 & 0 & 0 & 0 \\ 0 & 0 & -1/r & 1 \end{bmatrix} = \begin{bmatrix} 1 & 0 & 0 & 0 \\ 0 & 1 & 0 & 0 \\ 0 & 0 & 0 & 0 \\ 0 & 0 & -1/5 & 1 \end{bmatrix}$$

Apply reverse translation: $T(0, 0, 3)$

Composite transformation: $T(0, 0, 3) \cdot M_{xy} \cdot T(0, 0, -3)$

$$\text{Original coordinate matrix: } C = \begin{bmatrix} -1 & 1 & 1 & -1 & -1 & 1 & 1 & -1 \\ 1 & 1 & -1 & -1 & 1 & 1 & -1 & -1 \\ 1 & 1 & 1 & 1 & -1 & -1 & -1 & -1 \\ 1 & 1 & 1 & 1 & 1 & 1 & 1 & 1 \end{bmatrix}$$

New coordinate matrix (homogeneous coordinates):

$$D_h = T(0, 0, 3) \cdot M_{xy} \cdot T(0, 0, -3) \cdot C = \begin{bmatrix} -1 & 1 & 1 & -1 & -1 & 1 & 1 & -1 \\ 1 & 1 & -1 & -1 & 1 & 1 & -1 & -1 \\ 4.2 & 4.2 & 4.2 & 4.2 & 5.4 & 5.4 & 5.4 & 5.4 \\ 1.4 & 1.4 & 1.4 & 1.4 & 1.8 & 1.8 & 1.8 & 1.8 \end{bmatrix}$$

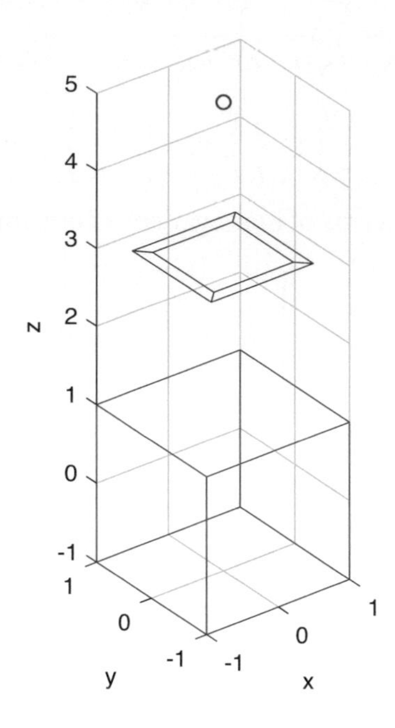

FIGURE 9.12 Plot for Example 9.8.

New coordinate matrix (Cartesian coordinates):

$$D = \begin{bmatrix} -0.71 & 0.71 & 0.71 & -0.71 & -0.55 & 0.55 & 0.55 & -0.55 \\ 0.71 & 0.71 & -0.71 & -0.71 & 0.55 & 0.55 & -0.55 & -0.55 \\ 3 & 3 & 3 & 3 & 3 & 3 & 3 & 3 \\ 1 & 1 & 1 & 1 & 1 & 1 & 1 & 1 \end{bmatrix}$$

New coordinates: (−0.71, 0.71, 3), (0.71, 0.71, 3), (0.71, −0.71, 3), (−0.71, −0.71, 3), (−0.56, 0.56, 3), (0.56, 0.56, 3), (0.56, −0.56, 3), and (−0.56, −0.56, 3) (Figure 9.12).

MATLAB Code 9.8

```
clear all; clc;
k = 3; r = 5;

p1 = [-1,1,1];
p2 = [1,1,1];
p3 = [1,-1,1];
p4 = [-1,-1,1];
p5 = [-1,1,-1];
p6 = [1,1,-1];
p7 = [1,-1,-1];
p8 = [-1,-1,-1];
C = [p1' p2' p3' p4' p5' p6' p7' p8' ;
    1 1 1 1 1 1 1 1 ];
```

```
tx = 0; ty = 0; tz = -k;
T=[1 0 0 tx ; 0 1 0 ty ; 0 0 1 tz ; 0 0 0 1];
M = [1 0 0 0 ; 0 1 0 0 ; 0 0 0 0 ; 0 0 -1/r 1];
Tr = inv(T);

Dh = Tr*M*T*C;
[rows, cols] = size(Dh);

for i=1:cols
    D(:,i) = Dh(:,i)/Dh(4,i);
end

fprintf('New vertices : \n')
for i=1:8
    fprintf('(%.2f, %.2f, %.2f) \n',D(1,i), D(2,i), D(3,i));
end

%Plotting
C = [p1' p2' p3' p4' p1' p5' p6' p7' p8' p5' p8' p4' p3' p7' p6' p2' ;
    1 1 1 1 1 1 1 1 1 1 1 1 1 1 1 1];
Dh = Tr*M*T*C;
[rows, cols] = size(Dh);

for i=1:cols
    D(:,i) = Dh(:,i)/Dh(4,i);
end

plot3(C(1,:), C(2,:), C(3,:), 'b'); hold on;
plot3(D(1,:), D(2,:), D(3,:), 'r');
plot3(0, 0, 5, 'ro'); grid;
xlabel('x'); ylabel('y'); zlabel('z');
axis equal; hold off;
```

NOTE

`size`: returns the number of rows and columns of an array

9.10 CHAPTER SUMMARY

The following points summarize the topics discussed in this chapter:

- Projection is a mapping of a higher-dimensional object to a lower-dimensional view.
- Projection can be of two types: parallel and perspective.
- In parallel projection, projection lines are parallel to each other.

- In perspective projection, projection lines appear to converge to a point called reference point.
- Parallel projection can be of two types: orthographic and oblique.
- Parallel orthographic projection can be of two types: multi-view and axonometric.
- Multi-view projections can generate three view types: top, front, and right.
- Axonometric projections can generate three view types: isometric, dimetric, and trimetric.
- Oblique projections can generate two types of views: cavalier and cabinet.
- Foreshortening factors are ratios of projected lengths to their original lengths.

9.11 REVIEW QUESTIONS

1. What is 2D and 3D projection?
2. What is the difference between parallel and perspective projection?
3. What is the difference between parallel orthographic and parallel oblique projection?
4. What is the difference between multi-view and axonometric projection?
5. What is meant by foreshortening factor? On what parameters does it depend?
6. What are the differences between isometric, dimetric, and trimetric projection views?
7. What are the differences between cavalier and cabinet projection views?
8. Explain the difference between viewline, viewpoint, viewplane, and PRP.
9. Discuss the two variations of perspective projection.
10. Explain how foreshortening factors can be used to differentiate between isometric, dimetric, trimetric, cavalier, and cabinet projections.

9.12 PRACTICE PROBLEMS

1. Consider a vector $A = 3i + 4j + 12k$. Obtain the ratios by which its projected lengths have been reduced with respect to its original length, after orthographic projection on the three principal planes.
2. A point $P(2, 1, 5)$ is projected using axonometric projection onto the plane $2x + 3y + 4z = 24$. Find its projected coordinates.
3. Determine the perspective projection of the triangle (3, 4), (5, 5), and (4, –1) onto the line $5x + y - 6 = 0$ from the viewpoint (11, 2).
4. A cube with center at origin and vertices at (–1, 1, 1), (1, 1, 1), (1, –1, 1), (–1, –1, 1), (–1, 1, –1), (1, 1, –1), (1, –1, –1), and (–1, –1, –1) is projected on a plane passing through

the origin perpendicular to the line joining points (2, 1, –2) and (3, 3, 2). Find its new vertices.

5. A vector $P = 3i + 4j + 12k$ is projected on XY-plane using perspective projection. Find projection coordinates if COP is at distance 2 from origin behind the view plane on the Z-axis. Also find foreshortening factors.

6. Given the following points: $A(3, 3, 4)$, $B(6, 10, 10)$, $C(7, 9, 12)$, $D(4, 15, 15)$, and $E(5, 6, 8)$, find out for perspective projection, which points lie on the same projection line joining A with COP at (1, 0, 0).

7. Derive the projection matrix in each case (1) perspective projection onto the viewplane $5x - 3y + 2z - 4 = 0$ from the viewpoint (2, –1, 1) and (2) parallel oblique projection onto the viewplane $2y + 3z + 4 = 0$ in the direction of the vector (1, –2, 3).

8. The point (3, –2, 4) is projected onto the plane $4x + 7y - z = 10$ such that the projection lines are along the direction of the vector $5i + j - 6k$. Find the projected coordinates.

9. Consider projection of a cube on the XY-plane using (1) $\alpha = 45°, \varphi = -35.264°$, (2) $\alpha = 45°, \varphi = -135°$, and (3) $\alpha = 20°, \varphi = -70°$. Compute projection matrix and foreshortening factors in each case.

10. Obtain a parallel projection transformation matrix onto the X–Y plane if the direction of the projection lines are along the vector $ai + bj + ck$.

Appendix I: MATLAB® Function Summary

1. `%`: signifies a comment line
2. `...`: continues the current command or function call onto the next line
3. `acos, acosd`: calculates inverse cosine in radians and degrees
4. `affine2d`: generates a 2D affine transformation of an image
5. `alpha`: sets transparency values
6. `asin, asind`: calculates inverse sine in radians and degrees
7. `axis`: controls appearance of axes, specifies ordered range of values to display
8. `clc`: clears workspace of previous text
9. `clear`: clears memory of all stored variables
10. `colorbar`: creates a color bar by appending colors in the colormap
11. `colormap`: specifies a color scheme using predefined color look-up tables
12. `cos, cosd`: calculates cosine of an angle in radians and degrees
13. `cross`: calculates cross product of vectors
14. `deg2rad`: converts degree to radian values
15. `det`: calculates determinant of a matrix
16. `diff`: calculates derivatives and partial derivatives
17. `disp`: displays the symbolic expressions without additional line gaps
18. `dot`: calculates dot product of vectors
19. `eval`: evaluates an expression
20. `eye`: generates an identity matrix of specified size
21. `ezcontour, ezcontourf`: generates a contour plot with optional filling

22. `ezmesh`: creates a mesh for the function $z = f(x, y)$
23. `ezplot, ezplot3`: plots symbolic variables directly in 2D and 3D environments
24. `ezsurf, ezsurfc`: combines a surface plot with a contour plot for the function $z = f(x, y)$
25. `figure`: generates a new window to display images and plots
26. `fill`: fills a polygon with color
27. `fimplicit, fimplicit3`: generates a 2D and 3D plot of an implicit function
28. `fliplr`: flip array in left–right direction
29. `for`: initiates a for loop for printing out all the vertices
30. `fplot, fplot3`: plot functions in 2D and 3D environments
31. `fprintf`: prints out strings and values using formatting options
32. `grid`: turns on display of grid lines in a plot
33. `hold`: holds the current graph state so that subsequent commands can add to the same graph
34. `imshow`: displays an image in a figure window
35. `imwarp, warp`: applies geometric transformation to an image for mapping it to a surface
36. `int`: integrate symbolic expression
37. `interp1`: performs 1-D interpolation
38. `inv`: computes inverse of a matrix
39. `legend`: designates different colors or line types in a graph using textual strings
40. `lightangle`: specifies lighting parameters on a surface
41. `line`: Draws a line from one point to another
42. `linspace`: creates 100 linearly spaced values between the two endpoints specified
43. `mesh`: generates a 3D mesh for plotting a function
44. `norm`: calculates the magnitude or Euclidean length of a vector
45. `patch`: generates filled polygons
46. `pchip`: performs piecewise cubic Hermite interpolation
47. `plot, plot3`: creates 2D and 3D graphical plots from a set of values
48. `polyder`: differentiates a polynomial

49. polyfit: generates a polynomial to fit a given data
50. polyval: evaluate a polynomial at a specified value
51. projective2d: generates a 2D projective transformation of an image
52. quiver, quiver3: depicts 2D and 3D vectors as arrows with direction and magnitude
53. rad2deg: converts radian to degree values
54. roots: finds roots of polynomial equation
55. scatter: type of plot where the data is represented by colored circles
56. sign: returns sign of the argument +1, 0, or −1
57. simplify: simplifies an equation by resolving all intersections and nestings
58. sin, sind: calculates sine of an angle in radians and degrees
59. size: returns the number of rows and columns of an array
60. solve: generates solution of equations
61. spline: performs cubic spline interpolation
62. subplot: displays multiple plots within a single figure window
63. subs: substitutes symbolic variable with a matrix of values
64. surf: generates a 3D surface for plotting a function
65. syms: declares the arguments following as symbolic variables
66. text: inserts text strings at specific locations within a graphical plot
67. title: displays a title on top of the graph
68. view: specifies the horizontal and vertical angles for viewing a 3D scene
69. vpa: displays symbolic values as variable precision floating point values
70. xlabel, ylabel, zlabel: puts text labels along the corresponding primary axes
71. zeros: generates a matrix filled with zeros

Appendix II: Answers to Practice Problems

Problem	Answers
1.1	$y=-3x-12$
1.2	$-0.405x^2+1.27x$
1.3	$x=11.67t^2-9.67t,\ y=1.67t^2-5.67t+2$
1.4	$x=1.5t^2+2.5t-2,\ y=-7.5t^2+5.5t+1$
1.5	$y=-0.157x^3+1.57x^2-3.24x+1.83$
1.6	$x=61.38t^3-125.3t^2+61.92t+3,\ y=-142.3t^3+221t^2-80.68t+2$
1.7	$y_A=0.0513x^3+0.377x^2+0.92x-0.405,$
	$y_B=-0.0414x^3+0.0986x^2+0.642x-0.498,$
	$y_C=0.381x^3-6.23x^2+32.3x-53.2$
1.8	$y=k-0.5x-1.5kx^2$
1.9	$k=0.5$
1.10	$x_A=t+1,\ y_A=4.07t^3-10.1t^2+5t-2,$
	$x_B=t+2,\ y_B=-0.2t^3+2.13t^2-2.93t-3,$
	$x_C=t+3,\ y_C=-3.27t^3+1.53t^2+0.733t-4$
2.1	$a=-3b/4$
2.2	(i) $x=-5.3t^3+8t^2+0.3t+1,\ y=-3.9t^3+5.5t^2+0.4t+1$
	(ii) $x=0.3t^3+2.7t+1,\ y=-3.1t^3+1.5t^2+3.6t+1$
2.3	$B=\begin{bmatrix} 1 & 0 & a & -3 \\ -a & 2b & -2 & 0 \\ b & -c & 3 & -a \\ -2c & a & 0 & -c \end{bmatrix}$
2.4	$(-23.62,-112.92),(3,-4),(1,0),(73.34,120.35)$
	$x=70t^3-90t^2+17t+3,\ y=166t^3-233t^2+79t-4$
2.5	$x=-2t^2-4t+2,\ y=6t^2-4t+2$
2.6	$x=t^3-3t^2-3t+3,\ y=10t^3-30t^2+21t-4$
2.9	$(1,4),(2,6.5),(6,15)$
2.10	$(4,-45),(3,-4),(-2,-3),(-9,-22)$

(Continued)

Problem	Answers
3.1	$x(t)=\begin{cases} t & (0\le t<1) \\ -2t+3 & (1\le t<2) \\ 2t-5 & (2\le t<3) \\ -t+4 & (3\le t<4) \end{cases}$
	$y(t)=\begin{cases} 0 & (0\le t<1) \\ t-1 & (1\le t<2) \\ -2t+5 & (2\le t<3) \\ t-4 & (3\le t<4) \end{cases}$
3.2	$B_{0,2}=\begin{cases} t & (0\le t<1) \\ 2-t & (1\le t<2) \end{cases}$
	$B_{1,2}=\begin{cases} t-1 & (1\le t<2) \\ 3-t & (2\le t<3) \end{cases}$
	$B_{2,2}=\begin{cases} t-2 & (2\le t<3) \\ 4-t & (3\le t<4) \end{cases}$
	$B_{3,2}=\begin{cases} t-3 & (3\le t<4) \\ 5-t & (4\le t<5) \end{cases}$
3.3	$B_{0,3}=\begin{cases} \left(\frac{1}{2}\right)t^2 & (0\le t<1) \\ -t^2+3t-\frac{3}{2} & (1\le t<2) \\ \left(\frac{1}{2}\right)(t-3)^2 & (2\le t<3) \end{cases}$
	$B_{1,3}=\begin{cases} \left(\frac{1}{2}\right)(t-1)^2 & (1\le t<2) \\ -(t-1)^2+3(t-1)-\frac{3}{2} & (2\le t<3) \\ \left(\frac{1}{2}\right)(t-4)^2 & (3\le t<4) \end{cases}$
	$B_{2,3}=\begin{cases} \left(\frac{1}{2}\right)(t-2)^2 & (2\le t<3) \\ -(t-2)^2+3(t-2)-\frac{3}{2} & (3\le t<4) \\ \left(\frac{1}{2}\right)(t-5)^2 & (4\le t<5) \end{cases}$

(Continued)

Problem	Answers

$$B_{3,3} = \begin{cases} \left(\frac{1}{2}\right)(t-3)^2 & (3 \le t < 4) \\ -(t-3)^2 + 3(t-3) - \frac{3}{2} & (4 \le t < 5) \\ \left(\frac{1}{2}\right)(t-6)^2 & (5 \le t < 6) \end{cases}$$

$$B_{4,3} = \begin{cases} \left(\frac{1}{2}\right)(t-4)^2 & (4 \le t < 5) \\ -(t-4)^2 + 3(t-4) - \frac{3}{2} & (5 \le t < 6) \\ \left(\frac{1}{2}\right)(t-7)^2 & (6 \le t < 7) \end{cases}$$

3.4 $10t^2$

$-13.3t^2 + 9.33t - 0.933$

3.5 $P_0 \cdot (5 - 10t) + P_1 \cdot (10t - 4)$

3.6 $P_0 \cdot t/4$

$P_0 \cdot (5 - t) + P_1 \cdot (t - 4)$

3.7

$$x(t) = \begin{cases} t^2 & (0 \le t < 1) \\ 2t - 1 & (1 \le t < 2) \\ -0.5t^2 + 4t - 3 & (2 \le t < 3) \\ 0.5t^2 - 2t + 6 & (3 \le t < 4) \\ -4.5t^2 + 38t - 74 & (4 \le t < 5) \\ 3.5(t-6)^2 & (5 \le t < 6) \end{cases} \qquad y(t) = \begin{cases} 2.5t^2 & (0 \le t < 1) \\ -5.5t^2 + 16t - 8 & (1 \le t < 2) \\ 7.5t^2 - 36t + 44 & (2 \le t < 3) \\ -11t^2 + 75t - 122.5 & (3 \le t < 4) \\ 9t^2 - 85t + 197.5 & (4 \le t < 5) \\ -2.5(t-6)^2 & (5 \le t < 6) \end{cases}$$

3.8

$$x(t) = \begin{cases} 0.1t^2 & (0 \le t < 1) \\ 0.1t^2 & (1 \le t < 2) \\ -0.083t^2 + 1.83t - 4.58 & (2 \le t < 3) \\ 0.15t^2 - 1.9t + 10.4 & (3 \le t < 4) \\ -0.392t^2 + 7.85t - 33.5 & (4 \le t < 5) \\ 0.583(t-15)^2 & (5 \le t < 6) \end{cases} \qquad y(t) = \begin{cases} 0.25t^2 & (0 \le t < 1) \\ -2.5t^2 + 22t - 44 & (1 \le t < 2) \\ 1.25t^2 - 15.5t + 49.8 & (2 \le t < 3) \\ -4.85t^2 + 82.1t - 341 & (3 \le t < 4) \\ 0.858t^2 - 20.6t + 122 & (4 \le t < 5) \\ -0.417(t-15)^2 & (5 \le t < 6) \end{cases}$$

3.9

$$x(t) = \begin{cases} 0.333t^3 & (0 \le t < 1) \\ -0.333t^3 + 2t^2 - 2t + 0.667 & (1 \le t < 2) \end{cases}$$

$$y(t) = \begin{cases} 0 & (0 \le t < 1) \\ 0.166(t-1)^3 & (1 \le t < 2) \end{cases}$$

3.10

$$B_{0,4} = \begin{cases} 0.1t^3 & (0 \le t < 1) \\ -0.2t^3 + 0.9t^2 - 0.9t + 0.3 & (1 \le t < 2) \\ 0.05t^3 - 0.6t^2 + 2.1t - 1.7 & (2 \le t < 3) \\ -0.05(t-6)^3 & (3 \le t < 4) \end{cases}$$

(Continued)

Problem	Answers
	$B_{1,4}=\begin{cases} 0.05(t-1)^3 & (1\le t<2) \\ -0.0472t^3+0.433t^2-1.02t+0.728 & (2\le t<3) \\ 0.147t^3-2.48t^2+13.6t-23.6 & (3\le t<4) \\ -0.0278(t-8)^3 & (4\le t<5) \end{cases}$
4.1	$M_1=R(45)\cdot F_x\cdot R(-45)=\begin{bmatrix} 0 & 1 & 0 \\ 1 & 0 & 0 \\ 0 & 0 & 1 \end{bmatrix}$ $M_2=R(90)\cdot F_x=\begin{bmatrix} 0 & 1 & 0 \\ 1 & 0 & 0 \\ 0 & 0 & 1 \end{bmatrix}$
4.2	$\begin{bmatrix} 0.5 & 1 & 0 \\ 0 & 0.5 & 2 \\ 0 & 0 & 1 \end{bmatrix}$
4.3	$(4.80,-0.60),(7.00,-1.00),(6.80,0.40)$
4.4	$\begin{bmatrix} 0.98 & 0 & 0.1 \\ 0 & 0.98 & 0.1 \\ 0 & 0 & 1 \end{bmatrix}$
4.5	$\begin{bmatrix} 0 & 0 & 5 \\ 0 & 0 & 4 \\ 0 & 0 & 1 \end{bmatrix}$
4.6	$M_s=\begin{bmatrix} 1 & 0 & 1 \\ 0 & 1 & 1 \\ 0 & 0 & 1 \end{bmatrix}, M_t=\begin{bmatrix} 0.5 & 0 & 1.5 \\ 0 & 1 & 1 \\ 0 & 0 & 1 \end{bmatrix}$
4.7	$y=-\frac{x}{m}-\frac{c}{m}$
4.8	$\begin{bmatrix} a & b & 0 \\ ab & a+b^2 & 0 \\ 0 & 0 & 1 \end{bmatrix}$
4.9	$\begin{bmatrix} 0.2 & 0.1 & -0.3 \\ -0.2 & 0.4 & -0.2 \\ 0 & 0 & 1 \end{bmatrix}$
4.10	(a) (7.00, –4.00), (–2.00, 2.00), (–10.00, –3.00), (–1.00, –9.00) (b) (7.00, –4.00), (–0.50, 0.50), (–1.67, –0.50), (–0.33, –3.00)
5.1	12.407

(Continued)

Problem	Answers
5.2	0.88
5.3	$\frac{1}{3}$
5.4	18
5.5	$(1.15,-2.05),(-1.15,2.05),(0,0)$
5.6	$i+2j,-2i+j,\ y=2x-1$
5.7	$\frac{3i-j}{\sqrt{10}},\frac{i+3j}{\sqrt{10}}$
5.8	$x+2y-3=0, i+2j$
5.9	$i+j,-i+j$
5.10	$0.416,(0.48,0.43)$
6.1	u is the unit vector along the direction of p
6.2	$(4-4t)i+(3t)j$
6.3	$s=-3, t=-2$
6.4	$\frac{x-2}{3}=\frac{-y+3}{1}=\frac{z+4}{2}$
6.5	$\frac{x}{4}=\frac{-y+3}{3}, z=0$
6.6	L is perpendicular to P
6.7	$(0,5,2)$
6.8	L is parallel to P
6.9	$r=(0,-3,8)+t\cdot(20,21,-68)$
6.10	$\begin{bmatrix} -0.62 & 0.78 & 0 \\ -0.78 & -0.62 & 0 \\ 0 & 0 & 1 \end{bmatrix}, \begin{bmatrix} -0.78 & 0.62 & 0 \\ -0.62 & -0.78 & 0 \\ 0 & 0 & 1 \end{bmatrix}$
7.1	$(-0.32,3.45,0),(0.89,2.73,1.93)$
7.2	$(-0.37k,-k,1.37k)$
7.3	$\begin{bmatrix} 0.65 & 0 & 0.76 & -0.41k \\ 0 & 1 & 0 & 0 \\ -0.76 & 0 & 0.65 & 1.11k \\ 0 & 0 & 0 & 1 \end{bmatrix}$
7.4	$(4.28,0.67,2.38)$
7.5	$\begin{bmatrix} 1/3 & -2/3 & -2/3 & 0 \\ -2/3 & 1/3 & -2/3 & 0 \\ -2/3 & -2/3 & 1/3 & 0 \\ 0 & 0 & 0 & 1 \end{bmatrix}$

(Continued)

Problem	Answers
7.6	$\begin{bmatrix} 1/9 & 4/9 & -8/9 & 8/9 \\ 4/9 & 7/9 & 4/9 & -4/9 \\ -8/9 & 4/9 & 1/9 & 8/9 \\ 0 & 0 & 0 & 1 \end{bmatrix}$
7.7	(2, 7, 1), (2, 7, 0), (3, 10, 0), (3, 10, 1), (0, 0, 1), (0, 0, 0), (1, 3, 0), (1, 3, 1)
7.8	$\begin{bmatrix} 0.7454 & -0.2981 & -0.5963 & 0 \\ 0 & 0.8944 & -0.4472 & 0 \\ 0.6667 & 0.3333 & 0.6667 & 0 \\ 0 & 0 & 0 & 1 \end{bmatrix}$ $\begin{bmatrix} 0.7454 & 0.2981 & -0.5963 & 0 \\ 0 & 0.8944 & 0.4472 & 0 \\ 0.6667 & -0.3333 & 0.6667 & 0 \\ 0 & 0 & 0 & 1 \end{bmatrix}$ $\begin{bmatrix} 0.7454 & 0.2981 & 0.5963 & 0 \\ 0 & -0.8944 & 0.4472 & 0 \\ 0.6667 & -0.3333 & -0.6667 & 0 \\ 0 & 0 & 0 & 1 \end{bmatrix}$
7.9	$\begin{bmatrix} 0.8165 & 0 & 0.5774 & 0 \\ -0.4082 & 0.7071 & 0.5774 & 0 \\ -0.4082 & -0.7071 & 0.5774 & 0 \\ 0 & 0 & 0 & 1 \end{bmatrix}$
8.1	$\left(-s^2-1, 0, t-1\right)$
8.2	$5u^2+2v^2 \geq 10$
8.3	$\left(u, v, \frac{u^2}{a}+\frac{v^2}{b}\right)$
8.4	$z=0.6x+0.8y$
8.5	$-4i-12j+2k$
8.6	$\frac{153\pi}{5}$
8.7	$\frac{512\pi}{21}$
8.8	$\left(-0.0221 \quad 0.4204 \quad 0.9071\right)$, (0.892, 0.45, −0.0369), 41.094°
8.9	$\left(-0.2673 \quad -0.8018 \quad -0.5345\right)$, 70.89°
8.10	33.3%

(Continued)

Problem	Answers
9.1	0.3846, 0.9515, 0.9730
9.2	(1.7931,0.6897,4.5862)
9.3	(0.2632,4.6842), (−0.3333,7.6667), (1.6053,−2.0263)
9.4	(−1.24,0.52,0.05), (0.67,0.33,−0.33), (0.86,−1.29,0.43), (−1.05,−1.10,0.81), (−0.86,1.29,−0.43), (1.05,1.10,−0.81), (1.24,−0.52,−0.05), (−0.67,−0.33,0.33)
9.5	(0.4286,0.5714,0), 0.1429, 0.1429, 0
9.6	C, E
9.7	$\begin{bmatrix} -1 & -6 & 4 & -8 \\ -5 & -8 & -2 & 4 \\ 5 & -3 & -9 & -4 \\ 5 & -3 & 2 & -15 \end{bmatrix}, \begin{bmatrix} -5 & 2 & 3 & 4 \\ 0 & -9 & -6 & -8 \\ 0 & 6 & 4 & 12 \\ 0 & 0 & 0 & -5 \end{bmatrix}$
9.8	(5.4242, −1.5152, 1.0909)
9.9	$\begin{bmatrix} 0.8165 & 0.4082 & 0.4082 & 0 \\ 0 & 0.7071 & -0.7071 & 0 \\ 0 & 0 & 4 & 0 \\ 0 & 0 & 0 & 1 \end{bmatrix}$, 0.8165, 0.8165, 0.8165 $\begin{bmatrix} -0.7071 & 0.5 & 0.5 & 0 \\ 0 & 0.7071 & -0.7071 & 0 \\ 0 & 0 & 4 & 0 \\ 0 & 0 & 0 & 1 \end{bmatrix}$, −0.7071, 0.8660, 0.8660 $\begin{bmatrix} 0.3420 & 0.3214 & 0.8830 & 0 \\ 0 & 0.9397 & -0.3420 & 0 \\ 0 & 0 & 4 & 0 \\ 0 & 0 & 0 & 1 \end{bmatrix}$, −0.3420, 0.9931, 0.9469
9.10	$\begin{bmatrix} -c & 0 & a & 0 \\ 0 & -c & b & 0 \\ 0 & 0 & 0 & 0 \\ 0 & 0 & 0 & -c \end{bmatrix}$

References

Chakraborty, Suman. 2010. *Fundamentals of Computer Graphics*. Everest Publishing House, Maharastra, India. (https://www.bookganga.com/eBooks/Books/Details/4992321229101922852)

Foley, James D., Andries van Dam, Steven K. Feiner, et al. 1995. *Computer Graphics—Principles and Practice*. Addison-Wesley Professional, Boston, MA. (https://www.amazon.com/Computer-Graphics-Principles-Practice-2nd/dp/0201848406/)

Hearn, Donald and M. Pauline Baker. 1996. *Computer Graphics, C Version (2nd edition)*. Prentice Hall, Upper Saddle River, NJ. (https://www.amazon.com/Computer-Graphics-C-Version-2nd-dp-0135309247/dp/0135309247/)

Marchand, Patrick. 2002. *Graphics and GUIs with MATLAB*. CRC Press, Boca Raton, FL (https://www.amazon.com/Graphics-GUIs-MATLAB/dp/1584883200)

Marsh, Duncan. 2005. *Applied Geometry for Computer Graphics and CAD*. Springer, London, UK. (https://www.amazon.com/Geometry-Computer-Graphics-Springer-Paperback/dp/B00F2RJUBC/)

Mathews, John H. and Curtis D. Fink. 2004. *Numerical Methods Using MATLAB*. Prentice Hall, New Delhi, India. (https://www.amazon.com/Numerical-Methods-Using-Matlab-4th/dp/0130652482/)

Olive, Jenny. 2003. *Maths: A Student's Survival Guide*. Cambridge University Press, Cambridge, UK. (https://www.amazon.com/Maths-Students-Survival-Self-Help-Engineering/dp/0521017076/)

O'Rourke, Michael. 2003. *Principles of Three Dimensional Computer Animation*. W. W. Norton & Company, New York. (https://www.amazon.com/Principles-Three-Dimensional-Computer-Animation-Third/dp/0393730832/)

Rovenski, Vladimir. 2010. *Modeling of Curves and Surfaces with MATLAB*. Springer, Heidelberg, Germany. (https://www.amazon.com/Modeling-Surfaces-Undergraduate-Mathematics-Technology-dp-0387712771/dp/0387712771/)

Shirley, Peter. 2002. *Fundamentals of Computer Graphics*. AK Peters/CRC Press, Natick, MA/Boca Raton, FL. (https://www.amazon.com/Fundamentals-Computer-Graphics-Peter-Shirley/dp/1568811241/)

Index